Springer Tracts in Modern Physics
Volume 142

Springer-Verlag Berlin Heidelberg GmbH

Springer Tracts in Modern Physics

Covering reviews with emphasis on the fields of Elementary Particle Physics, Solid-State Physics, Complex Systems, and Fundamental Astrophysics

Manuscripts for publication should be addressed to the editor mainly responsible for the field concerned:

Gerhard Höhler
Institut für Theoretische Teilchenphysik
Universität Karlsruhe
Postfach 6980
D-76128 Karlsruhe
Germany
Fax: +49 (7 21) 37 07 26
Phone: +49 (7 21) 6 08 33 75
Email: gerhard.hoehler@physik.uni-karlsruhe.de

Johann Kühn
Institut für Theoretische Teilchenphysik
Universität Karlsruhe
Postfach 6980
D-76128 Karlsruhe
Germany
Fax: +49 (7 21) 37 07 26
Phone: +49 (7 21) 6 08 33 72
Email: johann.kuehn@physik.uni-karlsruhe.de

Thomas Müller
IEKP
Fakultät für Physik
Universität Karlsruhe
Postfach 6980
D-76128 Karlsruhe
Germany
Fax:+49 (7 21) 6 07 26 21
Phone: +49 (7 21) 6 08 35 24
Email: mullerth@vxcern.cern.ch

Roberto Peccei
Department of Physics
University of California, Los Angeles
405 Hilgard Avenue
Los Angeles, California 90024-1547
USA
Fax: +1 310 825 9368
Phone: +1 310 825 1042
Email: robertop@college.ucla.edu

Frank Steiner
Abteilung für Theoretische Physik
Universität Ulm
Albert-Einstein-Allee 11
D-89069 Ulm
Germany
Fax: +49 (7 31) 5 02 29 24
Phone: +49 (7 31) 5 02 29 10
Email: steiner@physik.uni-ulm.de

Joachim Trümper
Max-Planck-Institut
für Extraterrestrische Physik
Postfach 1603
D-85740 Garching
Germany
Fax: +49 (89) 32 99 35 69
Phone: +49 (89) 32 99 35 59
Email: jtrumper@mpe-garching.mpg.de

Peter Wölfle
Institut für Theorie
der Kondensierten Materie
Universität Karlsruhe
Postfach 69 80
D-76128 Karlsruhe
Germany
Fax: +49 (7 21) 69 81 50
Phone: +49 (7 21) 6 08 35 90/33 67
Email: woelfle@tkm.physik.uni-karlsruhe.de

Tobias Ruf

Phonon Raman Scattering in Semiconductors, Quantum Wells and Superlattices

Basic Results and Applications

With 143 Figures

Springer

Dr. Tobias Ruf
MPI für Festkörperforschung
Heisenbergstr. 1
D-70569 Stuttgart
Email: ruf@cardix.mpi-stuttgart.mpg.de

Cataloging-in-Publication Data applied for
Die Deutsche Bibliothek – CIP-EInheitsaufnahme
Ruf, Tobias: Phonon Raman scattering in semiconductors, quantum wells and superlattices: basic results and applications / Tobias Ruf.

(Springer tracts in modern physics; Vol. 142)
ISBN 978-3-662-14765-8 ISBN 978-3-540-69420-5 (eBook)
DOI 10.1007/978-3-540-69420-5

Physics and Astronomy Classification Scheme (PACS):
78.66–w, 63.22+m, 73.20.Dx, 78.30.–j, 78.20.Ls, 07.79.Fc, 78.20.–e, 78.30.Fs,
81.05.–t, 78.35.+c, 78.55.–m, 71.70.Di, 71.38.+i

ISBN 978-3-662-14765-8

Originally published by Springer-Verlag Berlin Heidelberg New York in 1998
Softcover reprint of the hardcover 1st edition 1998

Typesetting: Camera-ready copy by the author using a Springer TEX macro-package
Cover design: *design & production* GmbH, Heidelberg
SPIN: 10569098 56/3144-5 4 3 2 1 0 – Printed on acid-free paper

Preface

This book presents recent results of basic research in the field of Raman scattering by optic and acoustic phonons in semiconductors, quantum wells and superlattices. It also describes various new applications in analytical materials research which have emerged alongside the scientific progress. Trends in Raman techniques and instrumentation and their implications for future developments are illustrated.

We discuss Raman scattering in ultra-short-period superlattices, where interface effects become important. New possibilities for research and applications in isotopically pure elemental and compound semiconductors are outlined. Isotopic superlattices are presented as model systems for phonon confinement. The vibrational properties of superlattices grown along high-index crystallographic directions are highlighted. Resonant Raman scattering, partly in connection with high magnetic fields, is of key importance to understand the continuous emission effect of acoustic phonons in quantum wells and superlattices. We discuss its relation to coherent and incoherent Raman scattering processes and the role of crystal-momentum conservation in the presence of interface disorder. Various examples illustrate the use of resonant magneto-Raman spectroscopy with optic and acoustic phonons for the investigation of electronic structure, electron–phonon interaction, phonon and electron confinement, and zone-folding effects in bulk and low-dimensional semiconductors. The related phenomenon of resonant magneto-luminescence is also discussed.

Applications include the characterization of heterostructure interface quality, the determination of homogeneous and inhomogeneous broadenings of electronic states, the analysis of defects and doping, the measurement of stress-profiles in silicon-based devices, investigations on damage mechanisms in quantum-well lasers, and the quality-control of protective coatings made from diamond-like carbon. Various examples show the use of Raman scattering as an analytical tool in the development of ZnSe- and GaN-based materials for optoelectronics. It can be applied even as an in-situ technique during semiconductor epitaxial growth.

Nano-Raman spectroscopy and light scattering in the optical near-field regime make it possible to combine the high chemical sensitivity of the method with a spatial resolution below the diffraction limit. We discuss a new generation of compact imaging and high-throughput Raman instruments, based on advanced optical elements, such as holographic notch- or acousto-optic bandpass-filters, and outline their potential for applications and basic research.

I am grateful to M. Cardona for his generous support and valuable contributions to the work described in this book. In many discussions he shared his enthusiasm as well as his rich treasure of experience and deep insight into physical phenomena. It is a pleasure to thank many colleagues and collaborators with whom I had the chance to pursue various projects over the years, in particular V. F. Sapega, J. Spitzer, V. I. Belitsky, R. T. Phillips, C. Trallero-Giner, F. Iikawa, S. I. Gubarev, A. Cros, G. Goldoni, A. Fainstein, A. Göbel, D. Wolverson, Z. V. Popović, P. Y. Yu, K. Ploog, G. Abstreiter, K. Eberl, F. Briones, A. Cantarero, H.-R. Trebin, and U. Rössler. Without their contributions this work would not have been possible. I would like to thank H. Hirt, M. Siemers, the late P. Wurster, and P. Hießl for first-class technical assistance with the experiments. Thanks are due to G. Schneider, A. Dworschak, T. Strach, and T. Strohm, for expert and patient help with computer and TEXnical problems, to S. Birtel for taking care of many secretarial tasks, and to Springer-Verlag for editorial support. I would like to express my gratitude to my wife Annette and our daughters Hannah and Debora for their patience and understanding during the writing of this book.

Stuttgart, July 1997 *Tobias Ruf*

Contents

1. Introduction

In the following we outline the history and the development of inelastic light scattering and present an overview of the topics discussed in this book.

1.1 General Features of Inelastic Light Scattering

The basic phenomena of waves and vibrations as well as light scattering were investigated as early as the first quarter of the 20th century by C. V. Raman in Calcutta [1.1]. As one of the scientific pioneers of India he began to perform such studies as a leisure-time activity in the laboratories of the Indian Association for the Cultivation of Science and later continued his work as a professor at the university there. Initially, his investigations concentrated on optical polarization and anisotropy effects which arise from the interaction of light with molecules. Dedicated work on a broad range of substances ranging from gases and liquids to amorphous solids and crystals as well as the idea to search for an optical analog of the Compton effect led him to the discovery of inelastic light scattering, for which he was awarded the Nobel prize in physics in 1930. The changes of photon frequencies due to the emission or absorption of elementary excitations, observed by Raman, were considered to be among the most convincing arguments for the quantum theory of light at the time [1.1]. A characteristic feature of Raman's approach to physics is the use of simple methods to obtain answers on specific, important questions which he was able to address by the means at his disposal, a careful performance and analysis of experiments as well as great enthusiasm for the projects he had embarked on [1.1]. This approach has not lost its actuality despite the specialization of scientific disciplines, which is nowadays commonplace. Raman's life and scientific achievements have been summarized in various biographies [1.2, 1.3] and in a collection of his publications [1.4, 1.5]. The lecture of these original articles allows one a glimpse into a fascinating epoch of modern physics. Figure 1.1 shows Raman with a model of the experimental setup used for the discovery of the Raman effect [1.3]. At that time sunlight or mercury vapor lamps were used to excite spectra. Radiation dispersed in a prism monochromator was detected on photographic plates. This required long exposure times, and careful reference measurements were necessary in order to suppress stray light.

Fig. 1.1. Sir C. V. Raman with an apparatus used for inelastic light scattering experiments (from [1.3])

Nowadays inelastic light scattering is a unique method in solid-state physics for the investigation of elementary excitations and electronic structure. A plethora of developments has contributed to this advancement: ion lasers with a multitude of discrete lines as well as tunable dye or solid-state lasers are readily available as monochromatic and very intense light sources over a broad spectral range, for both continuous wave and pulsed excitation. Modern Raman spectrometers with superior stray-light rejection properties allow one to observe weak scattering signals even for small frequency shifts from the excitation line. In addition to the conventional single-channel detection scheme which has advanced due to the development of long focal-length double monochromators and improvements in low-noise and broad spectral-range photodetectors, multi-channel systems with parallel acquisition of spectra by charge-coupled device (CCD) or photo-diode array detectors have become particularly important for the observation of weak signals [1.6, 1.7]. These improvements have made more precise measurements possible, and the range of applications for Raman spectroscopy has become broader.

Among the most important topics in semiconductor physics which are being investigated by inelastic light scattering are lattice vibrations and low-energy electronic excitations. An understanding of these phenomena in elemental and compound semiconductors as well as in artificial crystals, which are made out of these constituents by molecular-beam epitaxy (MBE) and other growth methods, is quite important for both basic science and applications [1.8]. The controlled preparation of superlattices (SL) and quantum well (QW) structures with a present accuracy of the order of atomic monolayers for the separation of different materials allows one to deliberately control and modulate physical parameters such as electronic band gaps, effective masses, dielectric or magnetic properties. This offers the possibility for systematic investigations of physical effects which arise from changes in dimensionality on one hand, while, on the other hand, the properties of devices can be optimized. To mention but one example, the recent advances in the development of the quantum cascade laser for the efficient generation of coherent radiation in the infrared were only possible due to the precise control and manipulation of such subtle effects as the tunneling rates between QW states and the relaxation of carriers due to electron–phonon interaction, issues which have emerged from investigations of basic scientific character not too long ago [1.9].

The knowledge of lattice-dynamical properties contributes significantly to our understanding of the structure of solids. Phonon frequencies, scattering intensities and selection rules which can be determined, for example, by Raman spectroscopy, lead to conclusions concerning microscopic parameters such as bonding and structure as well as deviations from the ideal crystalline lattice. These properties also allow one to investigate the influence of external parameters and therefore to perform thorough tests of theoretical models by which they can eventually be further improved.

Lattice vibrations cause a periodic modulation of the electronic structure of a solid. Raman signals can thus be interpreted as being due to changes of the dielectric function under the normal coordinate of an acoustic or optic phonon [1.10, 1.11]. It is for this reason that Raman scattering has been interpreted as a form of modulation spectroscopy [1.12], and it is thus sensitive also to effects of the electronic structure. In semiconductors the scattering intensity is enhanced near critical points of the band structure. Resonant Raman scattering exploits this effect for a precise determination of electronic resonances and to amplify otherwise weak signals in strongly absorbing materials [1.11, 1.13]. Resonances of the Raman signal are also influenced by the broadening of electronic states. This allows one to investigate carrier lifetime and relaxation phenomena.

Apart from lattice vibrations, inelastic light scattering is also possible from the excitations of an electron gas. This *electronic* Raman effect yields information on charge- and spin-density fluctuations in doped semiconductors, QWs or carrier accumulation layers at heterostructure interfaces [1.14, 1.15]. Under some circumstances the scattering signal is proportional to the imaginary part of the dielectric function which acts as the interface between microscopic theoretical models and macroscopic quantities [1.14].

1.2 Overview of Topics

In this book we discuss different aspects of Raman spectroscopy in semiconductors, quantum wells, and superlattices.

Chapter 2 gives an introduction to the basic features of inelastic light scattering by phonons in SLs using three specific examples [1.16–1.18]. In the experiments discussed in this context, contrary to the following chapters, crystal-momentum conserving scattering processes, allowed by the selection rules, are considered.

In Sect. 2.1 we review the general properties of Raman scattering and discuss the electronic and lattice-dynamical properties of SLs and QWs.

Section 2.2 presents results on confined longitudinal-optic phonons in GaAs/AlAs SLs with ultra-short periods [1.16]. We discuss the concept of backfolding these modes onto the bulk dispersion relations and make comparisons with theoretical predictions. Discrepancies between previous measurements and calculations are resolved by a combination of better sample quality and the consideration of nonideal interfaces in theoretical models.

While both, lattice-dynamical and electronic properties are being modulated in SLs made from different materials, isotopic SLs allow one to investigate confined phonons in crystals which have an essentially bulk-like electronic structure. In Sect. 2.3 we present measurements on optic phonons in isotopic ^{70}Ge/^{74}Ge and ^{69}GaAs/^{71}GaAs SLs [1.17]. The results confirm calculations of phonon frequencies according to the planar bond-charge model and scattering intensities using bond polarizabilities. We identify mixtures of modes with vibrational amplitudes confined mostly in different material layers and effects of isotopic disorder at the interfaces.

Section 2.4 presents results on acoustic phonons in GaAs/AlAs SLs grown along high-index crystallographic directions such as, e.g., [311] [1.18]. For this orientation the degeneracy of pure transverse and mixed transverse-longitudinal acoustic dispersion branches is lifted and all three kinds of modes can be observed. The backfolding of the bulk phonon dispersion in a SL allows one to test the predictions of the elastic continuum model for mode frequencies and scattering intensities.

Chapter 3 is dedicated to the continuous emission by acoustic phonons [1.19–1.27]. This effect which appears as a characteristic background signal in SLs and multiple quantum well (MQW) structures in the range of small Raman shifts is due to disorder-induced crystal-momentum nonconserving scattering of acoustic modes. We demonstrate that the basic understanding of this phenomenon leads to new possibilities for the determination of material parameters. Superlattices and MQWs made from GaAs/AlAs and GaAs/Al$_x$Ga$_{1-x}$As are used as model systems in this context.

In Sect. 3.1 we summarize various observations of the continuous emission under different circumstances [1.19, 1.23, 1.24]. The effect occurs in QWs and SLs in resonance with electronic transitions and can thus be influenced, e.g., by external magnetic [1.19, 1.26] or electric fields [1.27]. As a function of the excitation energy one observes a characteristic behavior of crystal-momentum conserving Raman scattering by folded acoustic phonons which is different from that found for crystal-momentum nonconserving continuous emission.

In order to explain the origin of the continuous emission we introduce in Sect. 3.2 a model for Raman scattering of acoustic phonons in a single QW and show that in this case crystal-momentum conservation is relaxed and phonons from the whole Brillouin zone can participate in scattering processes [1.19]. This model leads to predictions concerning the spectral shape of the continuous emission which are experimentally verified [1.19, 1.20]. We also illustrate the connection between the isolated QW, for which the model has been developed, and the MQW structures and SLs investigated experimentally. Deviations of these structures from the ideal case due to nonperfect interfaces are of central importance.

In Sect. 3.3 we discuss characteristic intensity anomalies of the continuous emission which appear at energies corresponding to band gaps of the acoustic phonon dispersion [1.19,1.21–1.24]. They can be explained by the dependence of the Raman intensity on crystal momentum. These features not only appear for band gaps at the edge or in the center of the Brillouin zone but also at internal gaps where resonant coupling of modes with mixed longitudinal and transverse character may occur. We discuss possible scattering mechanisms for these effects.

In Sect. 3.4 we show that the continuous emission spectra and the intensity anomalies depend on scattering processes in which electronic intermediate states are scattered elastically in the QW plane [1.21–1.23]. The wavevectors involved depend on the lateral extension of potential fluctuations which may be caused, e. g., by interface roughness and layer thickness fluctuations. We determine the lateral extension of such defects from continuous emission spectra and compare them to results obtained by other techniques.

The relationship between crystal-momentum-conserving Raman scattering by folded acoustic phonons and the crystal-momentum nonconserving continuous emission is discussed in more detail in Sect. 3.5 [1.23]. The scattering intensities of these two processes around resonances depend on homogeneous and inhomogeneous broadenings of electronic intermediate states which can thus be determined from Raman spectra in a simple way. This procedure is illustrated by two examples: In indirect-gap SLs with short periods the continuous emission under resonant excitation is so strong that the sharp lines due to folded phonons cannot be observed anymore. For known inhomogeneous broadening the homogeneous linewidth in this case can be obtained from fits of a theoretical model to the spectra. In direct-gap MQWs we investigate the resonance behavior of both effects at interband transitions between Landau levels. An external magnetic field allows us to vary the resonance conditions in a continuous way and makes systematic investigations possible from which both, homogeneous and inhomogeneous broadenings can be determined.

Temperature-dependent effects are discussed in Sect. 3.6 [1.23, 1.25]. Changes of the homogeneous linewidth with temperature influence the continuous emission as well as the crystal-momentum conserving scattering by acoustic phonons. From this we determine coupling constants of electronic states with optic and acoustic phonons and analyze the dependence of the electron–phonon interaction on QW thickness. We also discuss the limits of the method [1.25].

Section 3.7 describes magneto-Raman experiments in QWs in which we use acoustic modes in the continuous emission spectrum instead of optic phonons [1.19, 1.26]. In this case the Raman shift is smaller and to some extent tunable. This has the advantage that individual interband transitions between electron and hole Landau levels can be investigated separately. It is therefore not necessary, as in the case of optic phonons, to treat a multitude of transitions at the same time in order to model the observed intensity profiles. Additionally, strong signals are obtained since, due to the small phonon frequencies involved, the Raman processes considered are close to double resonance. We analyze fan plots of resonance energies vs. magnetic field and determine the anisotropy of the electron mass due to band nonparabolicity.

Chapter 4 presents an overview of resonant magneto-Raman scattering by optic phonons in bulk semiconductors and QWs. The basic principles for the resonant enhancement of the scattering intensity in a magnetic field are laid out [1.28–1.39]. Different examples illustrate how this technique allows one to investigate effects due to the electronic structure and the electron–phonon interaction.

In Sect. 4.1 we give an introduction to magneto-Raman scattering [1.28–1.35]. We explain the resonant enhancement of the Raman intensity at magneto-optical interband transitions between Landau levels and basics of the semiconductor electronic structure in a magnetic field. The various possible Raman processes and their selection rules are discussed.

Section 4.2 is devoted to examples from bulk semiconductors. We discuss fan plots of incoming, outgoing and double resonances in GaAs as well as the nonparabolicity of its conduction band [1.28–1.35]. The complicated results are explained in terms of simple models for the electronic structure. Resonant magneto-polarons in InP, observed by magneto-Raman scattering, are also mentioned [1.36, 1.37]. Special attention is paid to the Landau-index dependence of the level splitting and the coupling of neighboring levels. The exchange interaction in semimagnetic $Cd_{0.95}Mn_{0.05}Te$ leads to large splittings of the exciton states in a magnetic field. We demonstrate that this effect can also be used to achieve doubly resonant magneto-Raman scattering [1.38].

Section 4.3 is dedicated to magneto-Raman scattering by optic phonons in MQWs [1.39]. The experimental fan plots measured for different well widths are fitted to a theoretical model. This allows us to assign the resonances and to determine material properties such as anisotropy effects of the conduction electron effective mass or the hole dispersion perpendicular to the growth direction. We also calculate magneto-Raman intensity profiles vs. magnetic field and discuss the limits for the theoretical description of the observations.

Chapter 5 is about resonant magneto-luminescence in GaAs and in GaAs/$Al_xGa_{1-x}As$ QWs [1.40, 1.41, 1.43, 1.44]. For optical excitation close to the band edge one observes emission from interband transitions between electron and hole Landau levels which is partially nonthermal. A connection to Raman scattering, the main topic of this book, arises from the fact that this effect is only important for excitation of spectra with excess energies of electrons or holes up to that of a longitudinal-optic (LO) phonon. Carriers excited this way therefore cannot relax and thermalize quickly via emission of optical phonons. Scattering mechanisms and effects of electron–phonon interaction as encountered in Raman scattering, notably by acoustic phonons, become important.

In Sect. 5.1 we discuss results of resonant magneto-luminescence in bulk GaAs. A fine structure of Landau levels is observed and associated to the effects of valence band mixing [1.40, 1.41]. For excitation very close to a Landau level transition we observe quite strong resonances of the magneto-luminescence. In high-quality samples this allows us to perform measurements in the quasiclassical limit of large Landau quantum numbers and small magnetic fields [1.42, 1.43]. Nonparabolicities, effective masses and g factors can be determined this way.

In Sect. 5.2 we extend these investigations to QWs [1.44]. For excitation close to a Landau level transition we find strong emission from lower-lying transitions with smaller Landau indices. At the same time we observe photoexcited electronic Raman scattering. To explain these effects we discuss mechanisms for scattering of carriers between Landau levels with the aid of acoustic phonons and interface roughness.

In Chapter 6 we illustrate the widespread use and versatility of phonon Raman scattering in applied semiconductor research by recent examples from a variety of areas and topics. We also discuss trends and developments in methods and instrumentation for Raman spectroscopy. This is the basis for an outlook on future developments.

Section 6.1 describes various applications of Raman scattering as an analytical tool for the characterization of bulk and low-dimensional semiconductors. From optic- and acoustic-phonon spectra one obtains detailed information about the interface quality of superlattices and quantum wells. The sensitivity to interface phenomena can be enhanced by exploiting resonance effects of the Raman efficiency. Isotopic superlattices allow one to study self-diffusion via changes in the phonon spectrum. Phonon-frequency shifts due to stress in microelectronics devices can be mapped by Raman scattering. The technique yields valuable information for the epitaxy of new optoelectronic materials and can be used to determine defect and doping concentrations. It is employed for the quality control of hard-disk coatings, in the investigation of damage mechanisms in quantum well lasers and for many other purposes.

Another large area where Raman scattering is applied intensively due to its chemical selectivity is the in-situ monitoring of epitaxial growth.

In Sect. 6.2 we describe recent and future developments of Raman spectroscopy. New techniques like near-field optics and other methods using small apertures allow one to obtain spectral information with subwavelength spatial resolution. We discuss the possibilities and potential improvements for Raman scattering on the nanoscale and summarize recent encouraging results in this direction. The future development of Raman spectroscopy at high resolution as well as for many analytical purposes is closely connected to progress in instrumentation. High-throughput spectrometers can be built with the newly available holographic and acousto-optic filters which even allow one to do true imaging. These compact and portable Raman set-ups with no moving parts open the door for many new applications in research as well as in industry.

2. Raman Scattering in Semiconductor Superlattices

In this chapter we discuss the basic features of inelastic light scattering by phonons in SLs. To be specific, we use as examples the optical phonons in ultra-short period SLs [1.16] and isotope SLs [1.17] as well as acoustic vibratons in SLs grown along high-index crystallographic directions [1.18]. In this context, contrary to the following chapters, we consider crystal-momentum-conserving scattering processes, allowed by the selection rules.

2.1 Basic Properties

2.1.1 Raman Scattering

Raman spectroscopy allows one to observe elementary excitations of solids by the frequency shift between incident and scattered photons on a sample. From the viewpoint of quantum mechanics the first step in a Raman process is the absorption of a photon $\hbar\omega_l$. This leads to an electronic intermediate state which may interact with elementary excitations such as, e. g., phonons with quantum energy $\hbar\Omega$, via several mechanisms, and elementary excitations are created or annihilated. Finally, the scattered intermediate state recombines by emitting a photon with a different frequency $\hbar\omega_s$. All possible Raman processes arise from temporal permutations of these three steps. For one-phonon scattering which can be described by third-order perturbation theory one therefore obtains six terms [1.10]. Energy is conserved in the whole process requiring that

$$\hbar\omega_l = \hbar\omega_s \pm \hbar\Omega \ . \tag{2.1}$$

The emission/absorption of elementary excitations is called Stokes or anti-Stokes scattering. For each intermediate step crystal-momentum has to be conserved which is expressed by

$$\boldsymbol{k}_l = \boldsymbol{k}_s \pm \boldsymbol{q} \ . \tag{2.2}$$

In this equation $\boldsymbol{k}_l$ and $\boldsymbol{k}_s$ are photon wavevectors, and $\boldsymbol{q}$ is the crystal-momentum vector of the elementary excitation. Since $|\boldsymbol{k}_l|$ and $|\boldsymbol{k}_s|$ in the visible range of the spectrum are small compared to reciprocal lattice vectors, only elementary excitations with $|\boldsymbol{q}| \simeq 0$ participate in Raman processes, i. e.,

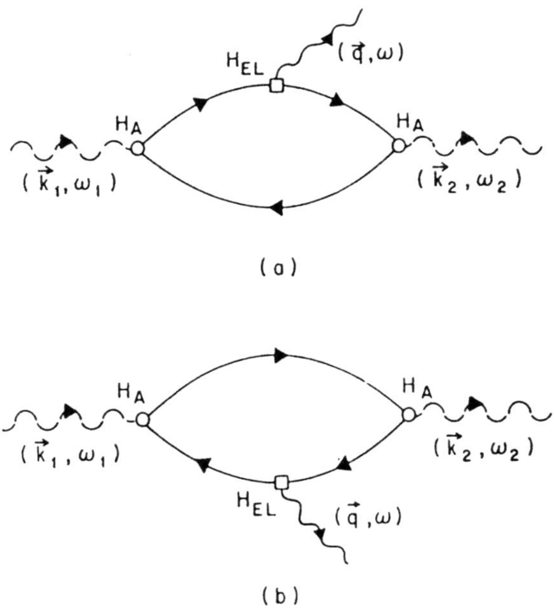

Fig. 2.1. First-order Stokes-Raman processes. The vertices H_A label steps mediated by the electron-photon, H_{EL} those due to the electron–phonon interaction for **(a)** electron and **(b)** hole scattering (from [1.10])

only states close to the Brillouin zone center can be observed. For forward scattering $|\boldsymbol{q}| = |\boldsymbol{k}_l - \boldsymbol{k}_s| \simeq 0$ holds rather well. In this case the transfer of crystal-momentum onto the elementary excitation is so small that polariton effects can be observed [1.11]. However, the Raman spectra of absorbing materials are mostly measured in backscattering geometry where the crystal-momentum transfer is $|\boldsymbol{q}| \simeq 2|\boldsymbol{k}_l| = 4\pi n(\lambda_l)/\lambda_l$ and depends on the refractive index $n(\lambda_l)$ and the laser wavelength λ_l. The value of $|\boldsymbol{q}|$ is still small but the fact that it is nonzero is essential for some scattering mechanisms. Under special circumstances it is even possible to investigate dispersion effects of the elementary excitations [1.11, 1.13, 1.15].

In semiconductors the absorption of photons by excitation across the band gap leads to intermediate states in the form of electron–hole pairs which are scattered and recombine. Due to the electron–photon and electron–phonon interactions the Raman intensity and its resonance behavior depends on the underlying electronic structure. For a theoretical description of these effects in the vicinity of critical points it is usually sufficient to consider the term which leads to the strongest resonances. An example is given in Fig. 2.1 for scattering of electrons (a) and holes (b), respectively. The Raman intensity of the processes in Fig. 2.1 is proportional to

$$I \sim \left| \sum_{\mu_i, \mu_j} \frac{\langle F|H_A|\mu_j\rangle\langle\mu_j|H_{EL}|\mu_i\rangle\langle\mu_i|H_A|I\rangle}{(\hbar\omega_s - E_{\mu_j} + i\Gamma_{\mu_j})(\hbar\omega_l - E_{\mu_i} + i\Gamma_{\mu_i})} \right|^2 . \tag{2.3}$$

The initial and final state of the system are given by $|I\rangle$ and $|F\rangle$. The summation extends over all possible intermediate states $|\mu\rangle$ with energies E_μ and broadenings Γ_μ. Resonances of the Raman intensity occur whenever the energy denominators in (2.3) become small for a suitable choice of $\hbar\omega_l$ oder $\hbar\omega_s$. In semiconductors this is the case for excitation at band gaps and other critical points of the electronic structure. Resonances of $\hbar\omega_l$ with E_μ are called *incoming*; those of $\hbar\omega_s$ with E_μ are called *outgoing*. One distinguishes between processes in which only electronic states of one electron and one hole band participate, the so-called two-band terms, and processes which involve three different bands and are thus called three-band terms [1.10]. In this case *double resonance* may occur if the separation between two bands just matches the energy of an elementary excitation.

Symmetry properties of the interactions involved in a Raman process lead to selection rules which are conveniently summarized in the form of Raman tensors [1.11]. Contraction of these second-rank tensors with the electric-field polarization vectors of incident and scattered photons yields the scattering intensity for a given configuration. In this picture nonvanishing tensor elements, the Raman polarizabilities, are the interface between a macroscopic and a microscopic description of the effect. According to group theory, a Raman process is allowed in principle if the participating phonon transforms according to an irreducible representation of the direct product of the polarization vectors under the point group of the crystal [1.11]. For optic phonons in polar semiconductors two mechanisms of electron–phonon coupling are important. The short-range *deformation-potential* coupling mediates scattering processes via the periodic modulation of the electronic structure caused by the relative sublattice displacement of the ions under the phonon [1.11]. With the Raman polarizability d the tensor for backscattering of LO phonons from an [001]-oriented surface of a tetrahedral semiconductor is [1.11, 1.13]

$$\mathbf{R_D} = \begin{pmatrix} 0 & d & 0 \\ d & 0 & 0 \\ 0 & 0 & 0 \end{pmatrix} . \tag{2.4}$$

The long-range *Fröhlich interaction* has its origin in the macroscopic electric field which in a polar semiconductor is connected with a longitudinal lattice vibration [1.11]. Backscattering of LO phonons from an [001]-oriented surface of, e. g., GaAs, is described by the following Raman tensor:

$$\mathbf{R_F} = \begin{pmatrix} a_F & 0 & 0 \\ 0 & a_F & 0 \\ 0 & 0 & a_F \end{pmatrix} . \tag{2.5}$$

The Raman polarizability is a_F. For polarization vectors $\hat{e_l}$ and $\hat{e_s}$ the Raman scattering intensity is proportional to

$$I \sim |\hat{e_s}^* \mathbf{R}\, \hat{e_l}|^2 . \tag{2.6}$$

2.1.2 Electronic Structure

In superlattices and quantum wells made from semiconductors one finds significant changes in the electronic structure and the lattice-dynamical properties as compared to the bulk. Varying band gaps of the different constituents cause periodic potentials across samples [2.1–2.3]. The simplest case is the Kronig–Penney square-well potential for electrons and holes [2.1–2.4]. For sufficiently high or thick barriers the electronic states are confined in isolated potential wells. This case is illustrated schematically in Fig. 2.2 [1.8]. As shown in Fig. 2.2 (a), the wave functions in the well plane can be regarded approximately as standing waves which, however, penetrate somewhat into the adjacent barrier material. In the limit of infinitely high barriers the energies E of these states are given by

$$E = \frac{\hbar^2}{2m^*}\left(n\frac{\pi}{d}\right)^2 , \tag{2.7}$$

where d is the layer thickness and m^* is the effective mass of the carriers. The index $n = 1, 2, 3, \ldots$ labels the number of half wavelengths in the wave function of a confined state; $n(\pi/d)$ can be regarded as a wavevector. For even (odd) n the wave functions have odd (even) parity with respect to the center of the well. While the continuous bulk dispersion is split into discrete levels along the direction of the potential modulation, the movement of carriers in the well plane is free. Figure 2.2 (b) shows a parabolic dispersion of subbands for wavevectors in the QW plane. The electronic density of states is given in Fig. 2.2 (c). One finds the step functions typical for a two-dimensional system.

To illustrate how these changes in the electronic structure influence the Raman intensity, Fig. 2.3 shows the dependence of the LO phonon signal on

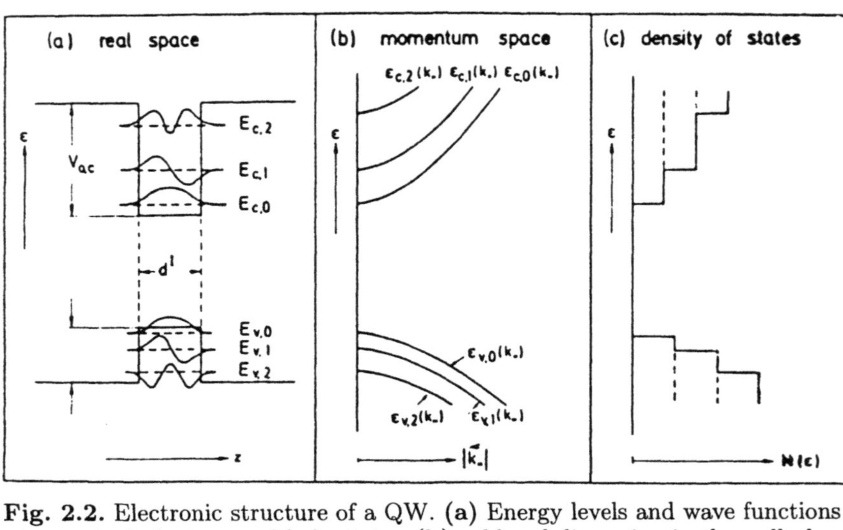

Fig. 2.2. Electronic structure of a QW. **(a)** Energy levels and wave functions of quantized electron and hole states, **(b)** subband dispersion in the well plane, **(c)** density of states of the two-dimensional system (from [1.8])

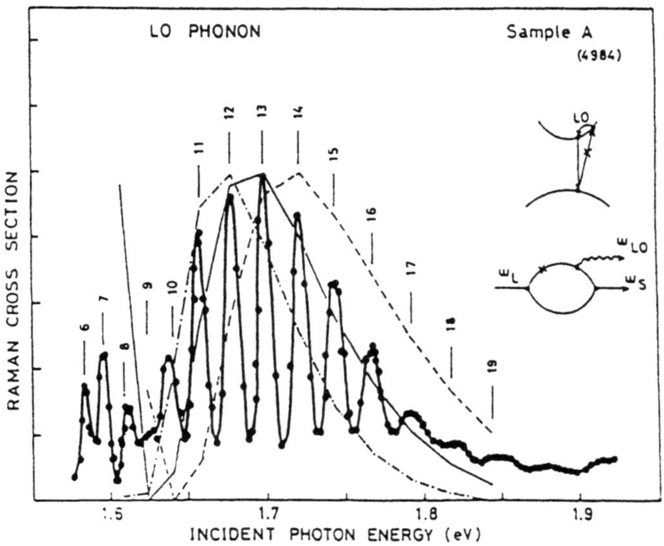

Fig. 2.3. Resonance profile of the LO phonon Raman intensity in 745 Å GaAs QWs for excitation at different subbands (from [2.5])

the excitation energy for GaAs QWs with a width of 745 Å [2.5]. One observes a series of oscillations with maximum intensity for excitation into an electronic subband. The resonances can be attributed to electron states with indices between $n = 6$ and 19. Deviations of the expected quadratic behavior of the resonance energies on n for higher levels allow one to determine the band nonparabolicity. The envelope of the intensity maxima arises from an asymmetric p-type modulation doping of the sample and ensuing modifications of the hole wave functions. The thin lines represent calculations of this envelope.

For finite barriers the wave functions confined to individual wells penetrate into the neighboring material. Figure 2.4 summarizes this case [1.8]. In analogy to the superposition of electronic states and the formation of bands in connection with the transition from single atoms to a bulk crystal, the minibands of a SL, shown in Fig. 2.4 (a), form when the overlap between single QW functions increases [2.1–2.3]. Layer thickness control with an accuracy of atomic monolayers achieved by modern epitaxy gives the unique possibility of tailoring the electronic structure in order to achieve new physical effects and to optimize device properties [1.8]. While only certain discrete values of the wavevector, determined by the confinement in the layer, are important for single QWs, the SL states, as shown in Fig. 2.4 (b), have a dispersion along the growth direction. Instead of the bulk lattice constant a_0 the SL period $d = d_A + d_B$, i. e., the sum of the layer thicknesses for the two constituents A and B, appears as a new length scale. The reduced SL mini-Brillouin zone is limited by $\pm\pi/d$. The SL dispersion can be regarded as a backfolding of the bulk dispersion with multiples of these new reciprocal lattice vectors. This

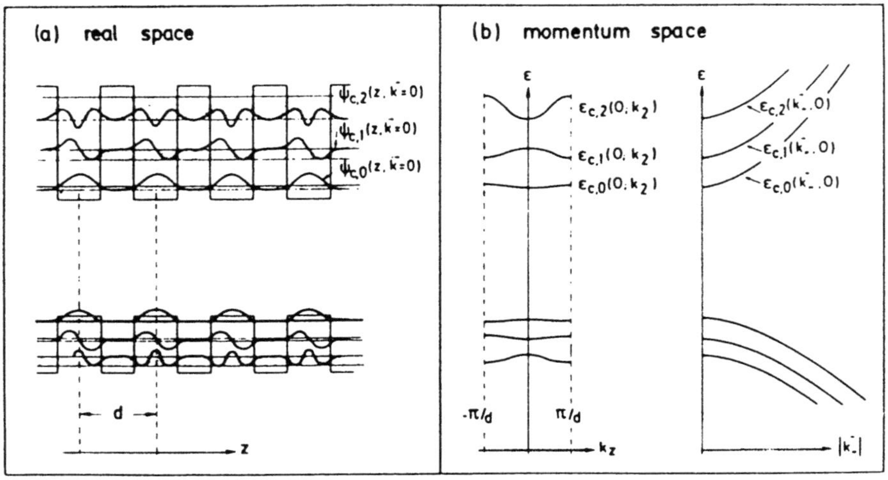

Fig. 2.4. Electronic structure of a SL: **(a)** Envelope of the miniband wave functions at the Brillouin zone center, **(b)** left: miniband dispersion for electrons and holes; right: dispersion along the well planes (from [1.8])

leads to a modification of the dispersion, and gaps appear at those points of the mini-Brillouin zone which differ by reciprocal SL vectors.

2.1.3 Superlattice Phonons

In order to understand the lattice-dynamical properties of SLs and the modes which can be observed in these systems by Raman spectroscopy two aspects are of importance. In analogy to the potential modulation and the resulting quantization of electronic states or the formation of minibands for sufficiently strong coupling, the question arises, whether lattice modes of one material can also exist in the neighboring layers or not [1.13, 2.6–2.9]. For vanishing dispersion overlap standing waves are formed in the individual layers. Modes propagating in both materials can be described by an average dispersion which reflects their properties and the relative layer thicknesses. Furthermore the approximation of negligible crystal-momentum transfer even for backscattering which works well for bulk crystals may no longer hold in all cases for SLs with a much larger period and a significantly reduced mini-Brillouin zone. On the other hand, dispersion effects can be investigated in SLs by varying the laser excitation energy [1.13, 2.6–2.9]. For the common system $(GaAs)_n/(AlAs)_m$, consisting of n monolayers (ml) GaAs and m monolayers AlAs, the first case occurs for the optical branches while the latter one is realized for the acoustic modes. Figure 2.5 shows a comparison of the dispersions for longitudinal and transverse phonons in a $(GaAs)_5/(AlAs)_4$ SL with those of the bulk materials [2.6]. The calculations were performed using a linear-chain model. The solid lines present the bulk dispersion of GaAs, the dashed ones that of AlAs. Due to the larger period of 9 ml, the SL Brillouin

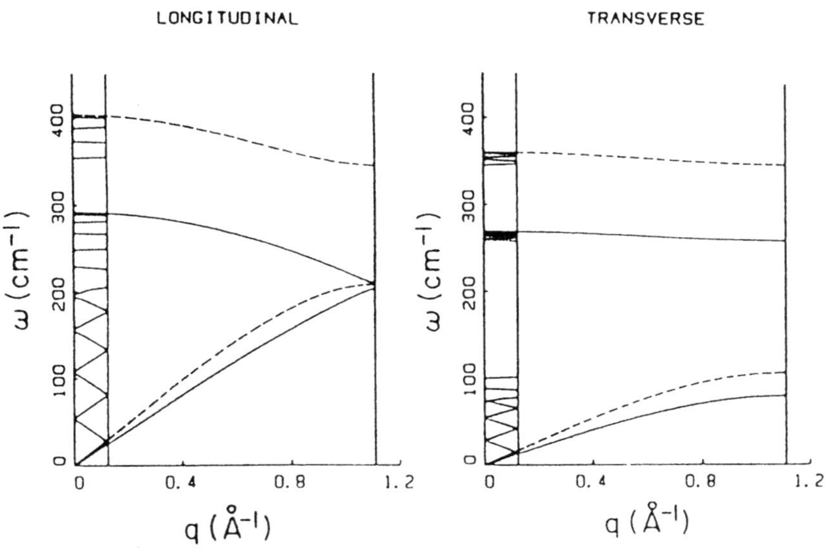

Fig. 2.5. Dispersions of the longitudinal and transverse phonons in a $(GaAs)_5/(AlAs)_4$ SL compared to those of the bulk semiconductors. *Solid lines*: GaAs, *dashed lines*: AlAs (from [2.6])

zone is reduced to 1/9 of its size in the bulk. The optical branches do not overlap. One obtains confined modes which are almost dispersionless. According to the number of monolayers there are four branches for AlAs and five for GaAs. The frequencies of acoustic phonons overlap in a wide range. Their SL dispersion can therefore be described in a first approximation by an average bulk dispersion "backfolded" by reciprocal SL vectors. Contrary to the optic vibrations, the acoustic modes of all nine branches cause atomic displacements in both materials.

Optic Phonons. In analogy to the electronic QW states, the confined phonons in a SL correspond to standing vibrations with a certain number of half wavelengths in a well [1.13, 2.7–2.10]. Figure 2.6 shows the calculated displacements (u_z) for the three highest-frequency GaAs-like LO modes in a [001]-oriented $(GaAs)_{20}/(AlAs)_{20}$ MQW structure at the center of the Brillouin zone [2.11]. The modes labeled LO_m stand for confined phonons with m half waves. The dashed vertical lines give the positions of the As interface atoms. Note their nonvanishing displacements and the penetration of the vibrations into the neighboring material. These calculations were performed using a microscopic model with interatomic force constants derived from ab-initio pseudopotential calculations [2.11]. The LO_m modes of a QW with n monolayers (lattice constant a_0) can be associated with an effective wavevector q_m which allows one to map their frequencies onto the bulk dispersion [2.10, 2.12, 2.13]:

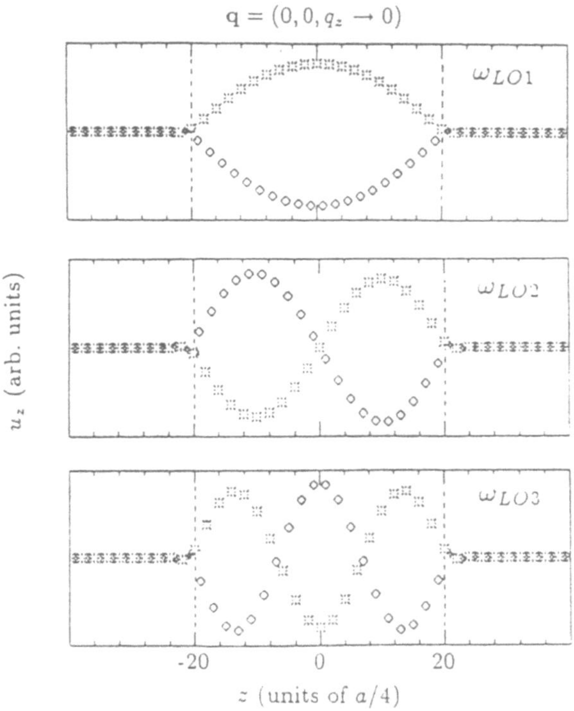

Fig. 2.6. Atomic displacements of the confined GaAs-like phonons in a $(GaAs)_{20}/(AlAs)_{20}$ MQW The *diamonds*, *asterisks*, and *dashed squares* label the Ga, Al, and As layers, respectively (from [2.11])

$$q_m = \frac{m\pi}{(n+\delta)a_0}, \quad m = 1, 2, ..., n \ . \tag{2.8}$$

The correction δ accounts for the penetration of the vibration into the neighboring layers. For [001]-oriented SLs one finds, as further discussed in Sect. 2.2, to a good approximation $\delta \simeq 1$. Figure 2.7 shows optic-phonon Raman spectra of a [001]-oriented $(GaAs)_{10}/(AlAs)_{10}$ SL for resonant and non-resonant excitation and different polarizations of incident and scattered photons [2.14]. In addition to the confined phonons LO_m one observes transverse-optic modes (TO) and interface vibrations (IF). GaAs/AlAs SLs with the growth direction [001] belong to the point group D_{2d} and thus have reduced symmetry as compared to the bulk (T_d). For odd or even indices m the LO_m modes have B_2 or A_1 symmetry, and their atomic displacements have odd (even) parity with respect to the center of the well [1.13]. Their Raman tensors have the same form as in (2.4) (B_2) and (2.5) (A_1). Therefore scattering from modes with odd index n is observed in crossed polarizations ($z(x,y)\bar{z}$), mediated by deformation-potential interaction, and scattering by modes with even index occurs for parallel polarizations ($z(x,x)\bar{z}$) via Fröhlich coupling. In the notations used we give from left to right the direction of propagation of the incident photons, their polarization, the polarization of the detected photons, and their propagation direction. For excitation away from resonance these selection rules are obeyed well, as can be seen in Fig. 2.7. Resonance

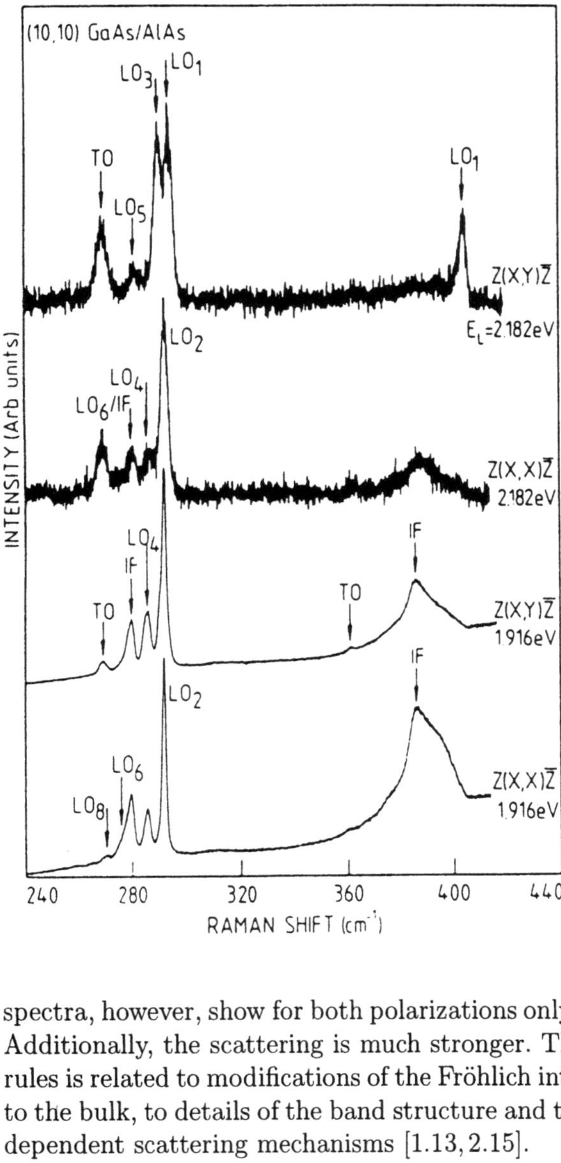

Fig. 2.7. Raman spectra of optic phonons in a $(GaAs)_{10}/(AlAs)_{10}$ SL for parallel and crossed polarizations of incident and scattered photons as well as for resonant (1.916 eV) and nonresonant (2.182 eV) excitation (from [2.14])

spectra, however, show for both polarizations only LO modes with even index. Additionally, the scattering is much stronger. This violation of the selection rules is related to modifications of the Fröhlich interaction in SLs as compared to the bulk, to details of the band structure and to further crystal-momentum dependent scattering mechanisms [1.13, 2.15].

Acoustic Phonons. The acoustic branches of GaAs and AlAs overlap to a considerable degree. As has been mentioned already above, the SL dispersion of such modes (see, e.g., Fig. 2.5) corresponds in first approximation to an average bulk dispersion "backfolded" by reciprocal SL vectors [2.6]. The long-wavelength character of acoustic phonons in comparison to the lattice constant allows one, in a linear approximation, to describe their dispersion in the framework of a continuum model with spatially varying densities $\rho(z)$ and elastic constants $C(z)$ as material parameters [1.13, 2.6, 2.8, 2.9]. Solving the wave equation for the atomic displacement $u(z)$ along the growth direction

$$\frac{\partial}{\partial t}\left[\rho(z)\frac{\partial u}{\partial t}\right] = \frac{\partial}{\partial z}\left[C(z)\frac{\partial u}{\partial z}\right] \tag{2.9}$$

for piecewise constant material parameters in the layers A and B with the boundary conditions

$$u_A(z_i) = u_B(z_i), \quad C_A\frac{\partial u_A}{\partial z}\Big|_{z_i} = C_B\frac{\partial u_B}{\partial z}\Big|_{z_i} \tag{2.10}$$

at the interfaces z_i and periodic boundary conditions for the whole structure, one obtains the following implicit expression for the dispersion $\omega(q)$ which was first derived by Rytov [2.16] in a geophysical context describing the propagation of elastic waves in layered media [1.13, 2.6]:

$$\cos(qd) = \cos\left[\omega\left(\frac{d_A}{v_A} + \frac{d_B}{v_B}\right)\right] - \frac{\epsilon^2}{2}\sin\left(\omega\frac{d_A}{v_A}\right)\sin\left(\omega\frac{d_B}{v_B}\right) . \tag{2.11}$$

For SL wavevectors $-\pi/d \leq q \leq \pi/d$ it depends mostly on the traversal time $t = d_A/v_A + d_B/v_B$ of an acoustic wave through one period. A small correction, characterized by

$$\epsilon = \frac{\rho_B v_B - \rho_A v_A}{\sqrt{\rho_B v_B \rho_A v_A}} , \tag{2.12}$$

removes the degeneracy of the folded dispersion branches at the boundary and the center of the mini-Brillouin zone. One obtains band gaps of the acoustic phonons whose size depends on the difference between the acoustic impedances $Z = \rho v$ in the materials A and B. The sound velocities $v = \sqrt{C/\rho}$ for longitudinal and transverse modes as well as for different propagation directions are determined by different linear combinations of elastic constants.

Figure 2.8 shows a Raman spectrum for a GaAs/$Al_{0.3}Ga_{0.7}As$ SL in the acoustic-phonon range [2.6]. The layer thicknesses of GaAs and $Al_{0.3}Ga_{0.7}As$ in this structure are 42 and 8 Å, respectively. In the following we indicate the thickness of SL layers by the notation (42/8) Å. For parallel polarizations one observes three sharp doublets of lines which correspond to modes of the folded longitudinal-acoustic (LA) phonon dispersion. The inset shows the dispersion calculated according to (2.11). Measured Raman shifts are indicated as crosses at a crystal-momentum transfer of $q = 4\pi n(\lambda_l)/\lambda_l$ which corresponds to the excitation wavelength of $\lambda_l = 5145$ Å. For the calculation of q one has to take into account the refractive index $n(\lambda_l)$ of the SL. The measurements were performed at a temperature of $T = 300$ K. For crossed polarizations no folded-phonon doublets are observed in this case. The arrows in Fig. 2.8 label calculated Raman shifts using structural data from X-ray diffraction measurements. They agree well with the experiment.

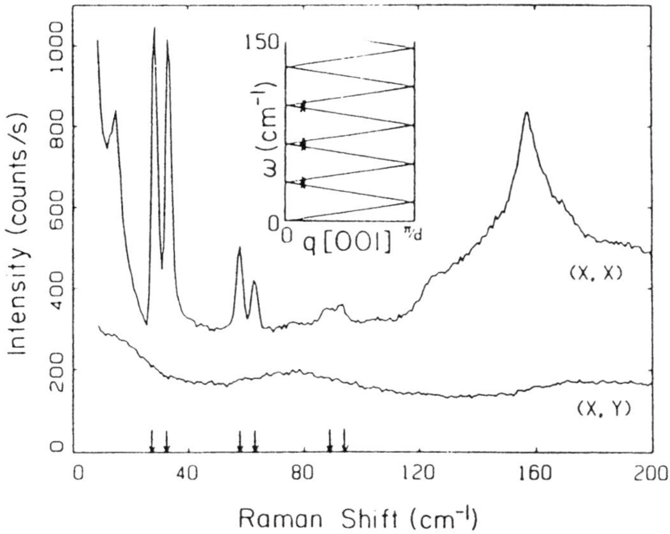

Fig. 2.8. Raman spectra of a (42/8) Å GaAs/$Al_{0.3}Ga_{0.7}As$ SL. The *vertical arrows* indicate the folded-phonon doublets expected from the dispersion in the inset (from [2.6])

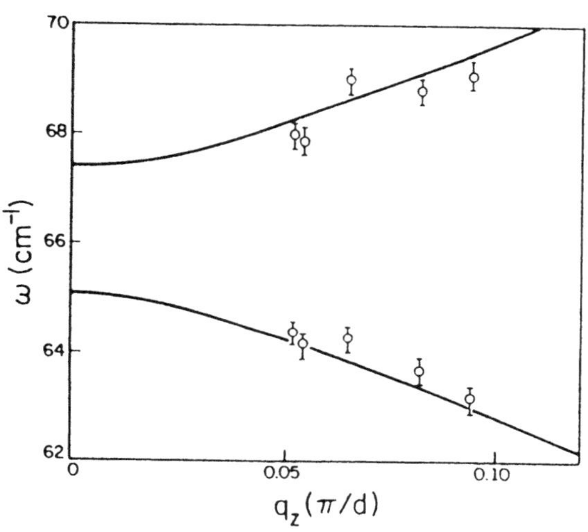

Fig. 2.9. Folded phonons in a $(GaAs)_5/(AlAs)_4$ SL (*open circles*) compared to the calculated dispersion (*solid lines*) (from [2.6])

Changing the excitation wavelength allows one to vary the crystal-momentum transfer q onto the phonons. Figure 2.9 shows folded-phonon frequencies in the vicinity of the first band gap at the zone center for a $(GaAs)_5/(AlAs)_4$ SL. Excitation wavelengths between 4579 and 6764 Å were used in these measurements. The solid lines were calculated in a linear-chain model [2.6]. In SLs with larger periods the range of wavevectors accessible may extend across the mini-Brillouin zone edge which makes the detailed measurement of phonon band gaps possible [2.17].

Interface Phonons. A special feature of long-wavelength optic phonons in polar semiconductors is their associated electrostatic field [1.13, 2.7–2.9, 2.18]. In SLs this causes a coupling of modes in neighboring QWs and thus a dispersion. For infrared-active confined LO_m modes with an odd index this effect can be treated, in analogy to the case of single layers [2.19, 2.20] or single or double heterostructures [2.21] made from polar constituents, in the framework of the dielectric continuum model by considering a periodic variation of the dielectric function in the SL [2.22]. In this model one solves the macroscopic Maxwell equations

$$\nabla \times \boldsymbol{E} = 0, \quad \nabla \cdot \boldsymbol{D} = 0 \tag{2.13}$$

for the electric field $\boldsymbol{E}$ and the dielectric displacement $\boldsymbol{D} = \epsilon(\omega)\boldsymbol{E}$ with a spatially varying dielectric function

$$\epsilon(\omega) = \epsilon(\infty)\,\frac{\omega^2 - \omega_{LO}^2}{\omega^2 - \omega_{TO}^2} \,. \tag{2.14}$$

In the bulk (2.13) are simultaneously fulfilled for $\epsilon = 0$ ($\boldsymbol{D} = 0$) as well as for $\epsilon = \infty$ ($\boldsymbol{E} = 0$), and one obtains longitudinal- (LO) and transverse-optic (TO) phonons, respectively [2.9]. At the interfaces of a SL the electrostatic boundary conditions (see below) require the continuity of the tangential component of $\boldsymbol{E}$ and the normal component of $\boldsymbol{D}$. Together with a Bloch-condition for the SL this leads to the dispersion relation for the so-called interface modes [1.13, 2.7–2.9, 2.22]

$$\cos(q_z d) = \cosh(q_\| d_A)\cosh(q_\| d_B) + \kappa \sinh(q_\| d_A)\sinh(q_\| d_B) \tag{2.15}$$

with crystal-momentum components $\boldsymbol{q} = (q_\|, q_z)$ in the well plane and perpendicular to it. The parameter κ is given by

$$\kappa = \frac{1}{2}\left(\frac{\epsilon_B}{\epsilon_A} + \frac{\epsilon_A}{\epsilon_B}\right) \,. \tag{2.16}$$

For each $\boldsymbol{q}$ one obtains four solutions which originate in the LO and TO vibrations of the polar constituent materials. In the center of the SL mini-Brillouin zone they correspond to GaAs- and AlAs-like LO_1 and TO_1 phonons. The frequencies of the interface modes for $\boldsymbol{q} \to 0$ depend on the angle $\Theta = \arctan(q_\|/q_z)$ between the two crystal-momentum components. Changes in the charges induced by the vibrations influence the LO–TO splitting which is strongly reduced for increasing Θ. This leads to a strong dispersion of the interface modes which is shown in Fig. 2.10 for different relative layer thicknesses [2.9].

As an ansatz to solve the Maxwell equations (2.13) one can also choose an electric field

$$\boldsymbol{E} = -\nabla\phi + \frac{1}{\epsilon}\nabla \times \boldsymbol{A} \tag{2.17}$$

with the skalar potential ϕ and the vector potential $\boldsymbol{A}$ [1.13, 2.9]. This leads to the Poisson equation

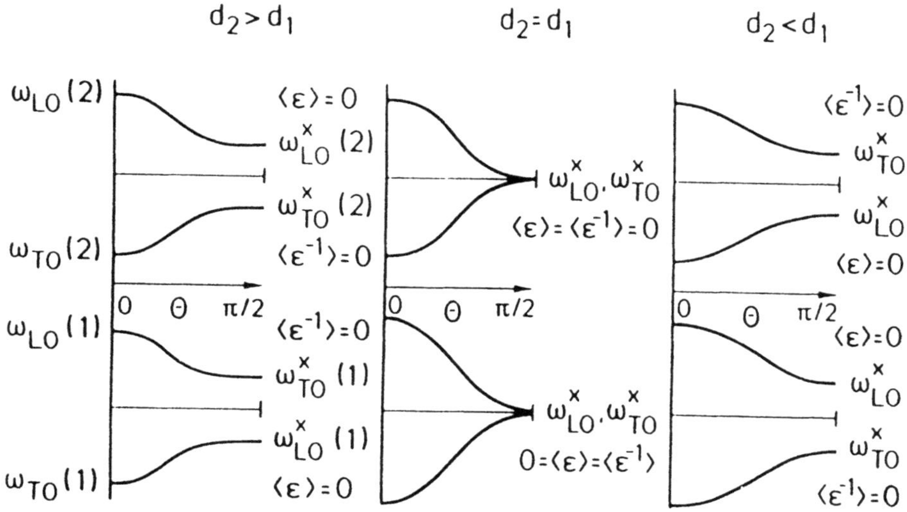

Fig. 2.10. Angle dependence of the frequencies of long-wavelength interface modes ($q \to 0$) in SLs made from two polar semiconductors. The dispersions are given for three cases with different relative layer thicknesses (from [2.9])

$$\epsilon\nabla^2\phi = 0 \tag{2.18}$$

for each layer. Solutions of (2.18) have the following general form [1.13, 2.9]

$$\begin{aligned} \phi_1(x,z) &= \phi_0\, e^{iq_\parallel x}\cos(q_z z) \\ \phi_2(x,z) &= \phi_0\, e^{iq_\parallel x}\sin(q_z z)\ , \end{aligned} \tag{2.19}$$

where ϕ_1 (ϕ_2) is symmetric (antisymmetric) with respect to the well center ($z = 0$) of material A with thickness d_A. For confined LO phonons in A ($\epsilon_A = 0$) the electric field in material B has to vanish. Thus $\boldsymbol{E}_B = 0$ and $\boldsymbol{D}_B = 0$. At the interfaces ($z = \pm d_A/2$) the continuity of the tangential component of the electric field E_x leads to the conditions

$$\begin{aligned} &i\, q_\parallel\, \phi_0\, e^{iq_\parallel x}\cos(q_z d_A/2) \equiv 0 \qquad (\phi_1) \\ &i\, q_\parallel\, \phi_0\, e^{iq_\parallel x}\sin(q_z d_A/2) \equiv 0 \qquad (\phi_2)\ , \end{aligned} \tag{2.20}$$

while the continuity of the normal component of the dielectric displacement D_z requires

$$\begin{aligned} &\epsilon_A\, q_z\, \phi_0\, e^{iq_\parallel x}\sin(q_z d_A/2) \equiv 0 \qquad (\phi_1) \\ &\epsilon_A\, q_z\, \phi_0\, e^{iq_\parallel x}\cos(q_z d_A/2) \equiv 0 \qquad (\phi_2)\ . \end{aligned} \tag{2.21}$$

These *electrostatic boundary conditions* are fulfilled for $\epsilon_A = 0$ and $q_\parallel = 0$ [1.13, 2.9].

However, the mechanical displacement $\boldsymbol{u}$ of a mode at an interface also needs to be continuous [1.13,2.9]. The displacement field $\boldsymbol{u}$ leads to a polarization $\boldsymbol{P}$, and for $\epsilon_A = 0$ ($\boldsymbol{D} = 0$) one has $\boldsymbol{u} \sim \boldsymbol{P} \sim \boldsymbol{E} \sim -\nabla\phi$. The continuity condition for u_x is equivalent to that for E_x (see (2.20)). For $u_z \sim E_z$ one obtains

$$q_z\,\phi_0\,e^{iq_\parallel x}\sin(q_z d_A/2) \equiv 0 \qquad (\phi_1)$$
$$q_z\,\phi_0\,e^{iq_\parallel x}\cos(q_z d_A/2) \equiv 0 \qquad (\phi_2)\,, \tag{2.22}$$

the so-called *mechanical boundary conditions.* For $q_\parallel = 0$ there exist, as mentioned above, solutions corresponding to confined phonons LO_m which can be associated with effective crystal-momentum vectors

$$q_z^m = m\pi/d_A, \qquad m = 2,4,6... \qquad (\phi_1)$$
$$q_z^m = m\pi/d_A, \qquad m = 1,3,5... \qquad (\phi_2)\,. \tag{2.23}$$

However, for $q_z \neq 0$ and $q_\parallel \neq 0$ the mechanical boundary conditions are incompatible with the electrostatic ones for E_x (see (2.20)) since the sine- and cosine-terms are interchanged. Thus one set of boundary conditions requires zeroes of the potential at the interface whereas the other one is fulfilled only for maxima. In the literature, confined phonons with a potential which, in fulfilment of the electrostatic boundary conditions, vanishes at the interface are called "slab modes" [2.19,2.20]. However, the atomic displacement patterns u_z of these vibrations have extrema at the interface and are thus discontinuous. The so-called "guided modes" which fulfill the mechanical boundary conditions have nodes at the interface and extrema in the potential [2.10,2.23]. This apparent discrepancy of the results obtained with the two kinds of boundary conditions for confined phonons and their relation to interface modes has recently been clarified by a more careful analysis of macroscopic models [2.24,2.25], by comparisons of macroscopic models with microscopic calculations [2.26], and by lattice-dynamical studies for SLs [2.11,2.27,2.28].

Results obtained by the different models are shown in Fig. 2.11 [2.11]. In microscopic calculations it is found that both, the electrostatic and the mechanical boundary conditions are fulfilled. This result is also obtained in a continuum model if the dispersion effects and the coupling of the atomic displacements to the electrostatic field are carefully considered which allows one to describe confined and interface phonons in a unified way [2.24,2.25]. As mentioned above, one finds that confined LO_1 and TO_1 modes for each material at $\boldsymbol{q} = 0$ change character and become interface modes when the crystal-momentum component $q_\parallel$ is different from zero. This dependence of the interface phonons on $q_\parallel$ which is obtained in microscopic calculations is shown in Fig. 2.12 for the AlAs-like modes of a $(\mathrm{GaAs})_{20}/(\mathrm{AlAs})_{60}$ SL [2.11].For $q_\parallel = q_x \to 0$, IF1 goes over into the AlAs LO_1 mode. As a consequence of the simultaneous fulfilment of the electrostatic and mechanical boundary conditions the displacement pattern of this mode is significantly deformed compared to a pure sine-like half wave and shifted towards the interfaces. For $\boldsymbol{q} = 0$, IF2 is the TO_1 phonon in AlAs and has no displacements along the growth direction. For increasing $q_x \neq 0$ the atomic displacements of these modes are restricted to the immediate vicinity of the interfaces, hence their name.

Figure 2.13 shows Raman spectra of a (100/135) Å GaAs/AlAs MQW, measured after polishing a thick sample under different angles with respect

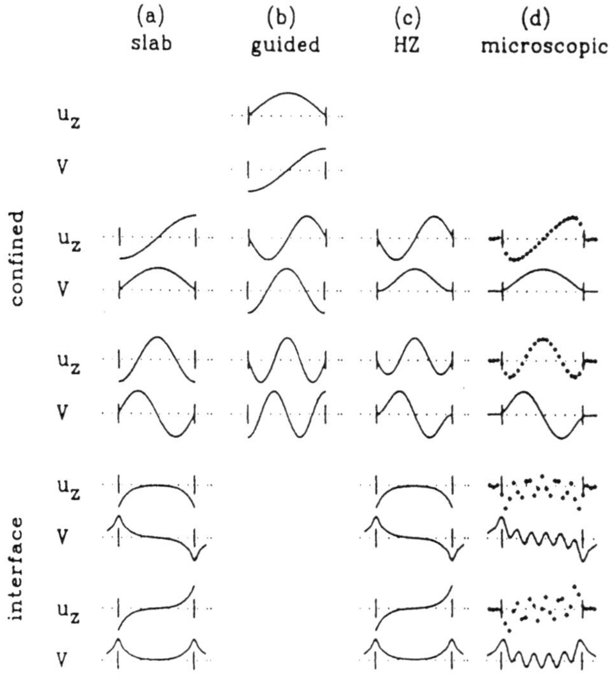

Fig. 2.11. Atomic displacements (u_z) and potentials (V) of GaAs-like optic phonons in a [001]-oriented GaAs QW with a thickness of 56 Å between AlAs barriers. The interfaces are indicated by vertical lines. From top to bottom, the results are given for the highest-energy "confined" and the two "interface" modes as obtained by different models. Modes from macroscopic calculations fulfill the electrostatic **(a)** or the mechanical boundary conditions **(b)**. In a macroscopic model **(c)** which was obtained by comparison with a microscopic theory or in ab-initio calculations **(d)** it is found that the atomic displacements as well as the potentials of confined phonons vanish at the interface. The results in **(d)** were calculated for $q_z = 0$ and $q_{\|} = 0.15\,\text{Å}^{-1}$. The points indicate the anion displacements (from [2.11])

to the growth direction [2.29]. The different scattering geometries are given schematically next to each spectrum. This approach allows one to vary the amount of crystal-momentum transfer $q_{\|}$ in the QW planes. The arrows mark interface modes which for increasing $q_{\|}$ move towards each other, in good agreement with the predictions of the dielectric continuum model. When $q_{\|}$ becomes large in symmetric SLs the two interface modes for each material tend towards the respective average of LO and TO frequencies (see also Fig. 2.10). This strong dispersion of the interface phonons leads to crossings with other LO_m modes. Vibrations with even index are not infrared-active. They have no long-range electric fields and are therefore not influenced by the interface modes. Confined phonons with odd indices, however, interact with the interface modes and internal gaps of the phonon dispersion appear. This theoretically expected behavior is shown by the dotted lines in Fig. 2.14 for the GaAs-like region of a $(GaAs)_{12}/(AlAs)_{12}$ SL [2.30]. The phonon frequen-

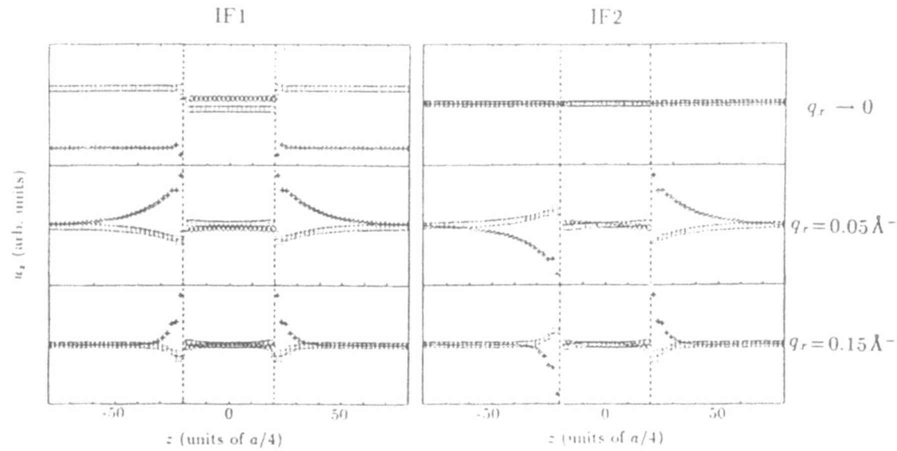

Fig. 2.12. Atomic displacements u_z of AlAs-like interface modes in a $(GaAs)_{20}/(AlAs)_{60}$ SL for different crystal-momentum vectors $\boldsymbol{q} = (q_x, 0, 0)$. The elements are labeled as in Fig. 2.7 (from [2.11])

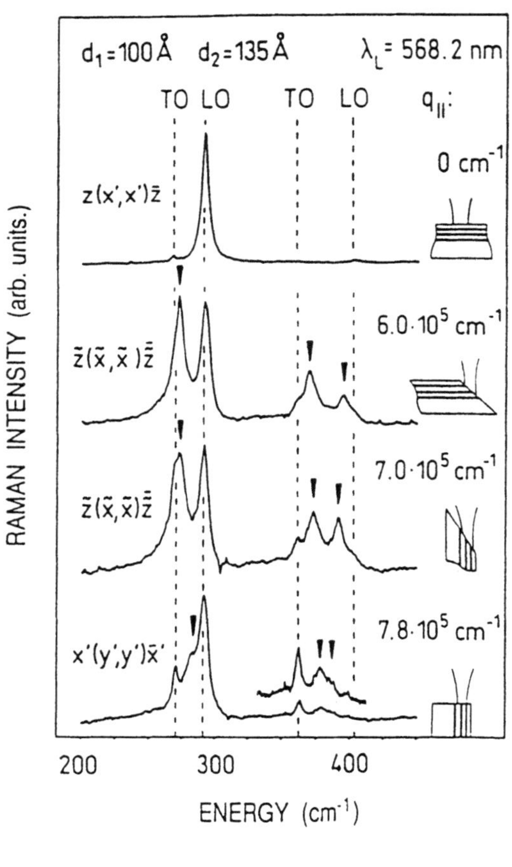

Fig. 2.13. Raman spectra of a (100/135) Å GaAs/AlAs MQW structure for different angles with respect to the growth direction. The *arrows* denote the interface phonons whose position between the LO and TO frequencies of bulk GaAs and AlAs (*dashed vertical lines*) depends on $q_{\|}$ (from [2.29])

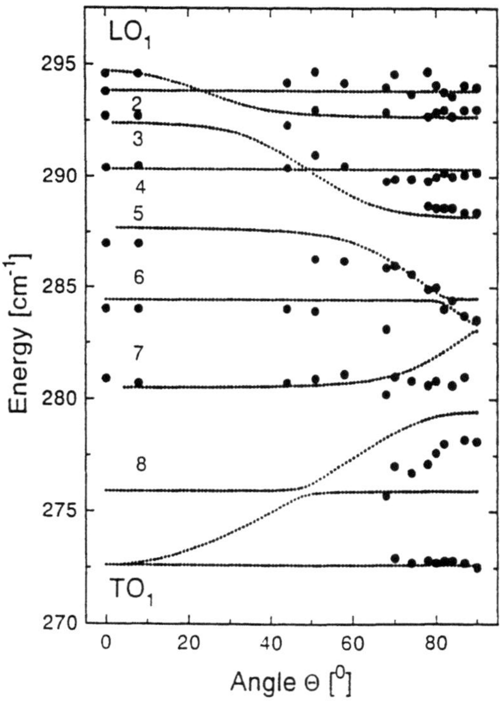

Fig. 2.14. Angle-dependent theoretical (*dotted lines*) and experimental (*filled dots*) phonon dispersions for a $(GaAs)_{12}$/$(AlAs)_{12}$ SL in the GaAs range (from [2.30])

cies are given vs. the angle of the wavevector with respect to the growth axis. Careful micro-Raman measurements on side-polished samples in analogy to those presented in Fig. 2.13 confirm these predictions (filled dots) [2.30].

Crystal-momentum nonconserving Raman processes also allow one to observe interface modes and dispersion effects even for backscattering along the growth direction, when $q_{\|}$ should ideally be zero [2.18]. This effect, due to interface disorder, may have a rather strong influence on the measured SL Raman spectra, especially for resonant excitation [2.31]. Figure 2.15 shows the calculated dispersion of the GaAs-like phonons in a (46/46) Å GaAs/AlAs MQW structure (a) and Raman spectra which were measured under different resonance conditions (b) [2.31]. The dispersion of the interface modes in Fig. 2.15 (a), as calculated with the dielectric continuum model, is shown by the dashed lines. The crystal momentum in the plane, $q_{x,y}$, is given in units of the inverse well width a. In the calculations a SL wavevector of $q_z = 0.5/(2a)$ was used. The solid lines correspond to confined LO_m modes whose indices are indicated. They were calculated in a continuum model which describes both, the confined and the interface modes [2.25]. Dispersion gaps of LO_m modes with odd index which arise from the interaction with the interface phonons are marked by thick bars. Figure 2.15 (b) shows experimental Raman spectra for different resonance conditions [2.31]. Spectrum (i) was measured far from electronic resonances using crossed polarizations. A series of confined phonons with odd indices is observed whose intensities decrease according to

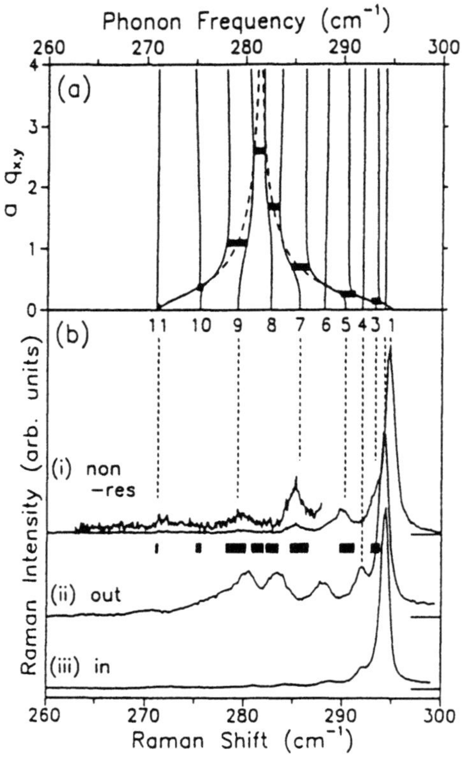

Fig. 2.15. (a) Calculated phonon dispersion and **(b)** measured Raman spectra of a (46/46) Å GaAs/AlAs MQW (from [2.31])

the expected m^{-2} behavior [1.13, 2.31]. Spectra (*ii*) and (*iii*) were measured for outgoing resonance with the first transition between quantized light-hole and electron states (e1-lh1(1s)) and in incoming resonance with the first transition between heavy-hole and electron levels (e1-hh1(1s)) for parallel polarizations [2.31]. Apparently, these spectra are significantly different from (*i*). The discrepancies, however, cannot be explained by the fact that in resonance one should observe modes with even indices and off resonance those with odd indices for these scattering geometries, as was the case in Fig. 2.7 [2.10, 2.31]. Note that characteristic intensity minima appear at frequencies which coincide with gaps of the calculated dispersion in Fig. 2.15 (a) which, for clarity, are given once more by thick bars [2.31]. In resonance crystal-momentum nonconserving scattering processes become important, and modes from a considerable part of the Brillouin zone participate. The spectra therefore represent an integrated phonon density of states which has minima at dispersion gaps arising from resonant coupling of interface modes with confined phonons. These ideas are further supported by systematic variations of the relative GaAs and AlAs layer thicknesses [2.31, 2.32].

Finally, it remains to remark on the fact that the nomenclature of layered semiconductor structures as superlattices or multiple quantum wells varies from case to case. Probably most widespread is a distinction according to criteria based on the electronic structure. Thus, noncoupling systems are called single or multiple quantum wells. Systems with overlapping wave functions of adjacent wells which lead to the formation of minibands are called superlattices. For lattice-dynamical properties this distinction no longer holds in a strong sense, since the acoustic modes generally exhibit (phonon) superlattice properties, even in systems with high or thick (electronic) barriers, while the optic modes may remain confined even in systems with rather short periods and hardly interact with each other. For interface phonons, however, this is not true. In this book we use the electronic structure criteria to denote specific systems. For convenience, however, both types of structures are called superlattices in the general context.

2.2 Optic Phonons in Ultra-Short-Period GaAs/AlAs Superlattices

2.2.1 Confined Phonons in Very Thin Layers

Many investigations of confined LO phonons in GaAs/AlAs superlattices have qualitatively confirmed the basic concept of mapping these modes onto the bulk dispersions via the effective wavevector given by (2.8) [1.16, 2.10, 2.14, 2.33–2.36]. However, systematic deviations between the "backfolded" dispersions and those of the bulk were found which are even more pronounced when compared to the highly precise calculations now available [2.37–2.41]. Figure 2.16 shows theoretical dispersions for GaAs and AlAs in comparison to experimental values from confined phonons in different GaAs/AlAs SLs. The growth direction for all SLs considered in this section is [001]. The theoretical dispersions in Fig. 2.16 are based on a dynamical matrix and interatomic force constants which were calculated by density-functional theory

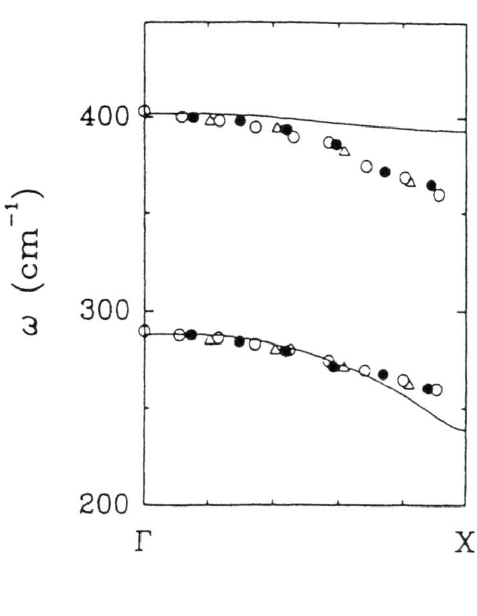

Fig. 2.16. Theoretical LO phonon dispersions for GaAs and AlAs (*solid lines*) in comparison with experimental values from $(GaAs)_n$/ $(AlAs)_n$ SLs. ○: n=8, •: n=6, △: n=4. Data at the Γ point were measured in bulk samples (from [2.38])

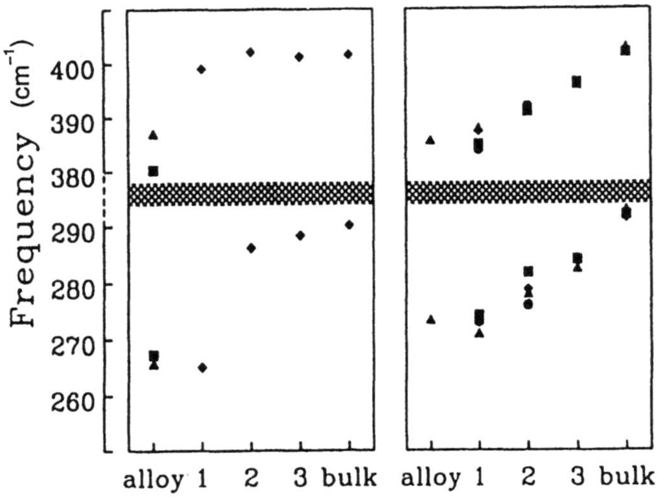

Fig. 2.17. GaAs- and AlAs-like LO_1 modes in $(GaAs)_n/(AlAs)_n$ SLs with $n = 1-3$, $Al_{0.5}Ga_{0.5}As$ ("alloy"), GaAs, and AlAs ("bulk"). Theoretical values are given in the left, measured ones in the right plot. See text for details (from [2.37])

using pseudopotentials [2.37–2.41]. Macroscopic electrostatic fields of the LO phonons were considered ab-initio. This type of calculations presently yields the most precise microscopic description of the phonon dispersions in bulk III–V and group-IV semiconductors and is used as a starting point for investigations of SLs. The experimental values in Fig. 2.16 are from [2.35], they are, however, representative also for other investigations [2.10,2.14,2.33,2.36]. For large wavevectors the backfolded experimental GaAs dispersion systematically tends towards higher frequencies than theoretically expected while the opposite behavior is found for AlAs.

Another problem in connection with the confined optic phonons in SLs concerns the behavior of the LO_1 mode for ultra-short-period systems [2.37, 2.38, 2.40]. Figure 2.17 shows a comparison between the calculated and measured LO_1 frequencies for a series of ultra-short-period GaAs/AlAs SLs as well as for $Al_{0.5}Ga_{0.5}As$, GaAs, and AlAs [2.37]. Theoretical phonon frequencies are given in the left plot. The diamonds denote results of the abovementioned theory for SLs, GaAs, and AlAs ("bulk"). The triangle and the square for $Al_{0.5}Ga_{0.5}As$ ("alloy") are results of a supercell calculation for the alloy and the chalkopyrite structure for $GaAlAs_2$, respectively [2.37]. Measured LO_1 frequencies are given in the right plot. Diamonds, squares, triangles, and circles mark the results of [2.34], [2.42], [2.33], and [2.43], respectively. The calculated AlAs LO_1 modes for all SLs show hardly any change as compared to the bulk. Their frequencies, however, are significantly larger than in $Al_{0.5}Ga_{0.5}As$. For the GaAs LO_1 modes the calculations predict a jump between $n = 1$ and $n = 2$ with the value for $n = 1$ being close to that of $Al_{0.5}Ga_{0.5}As$ and that for $n = 2$ slightly below the bulk GaAs frequency. In contrast, the experiments in the right plot of Fig. 2.17 show for both materi-

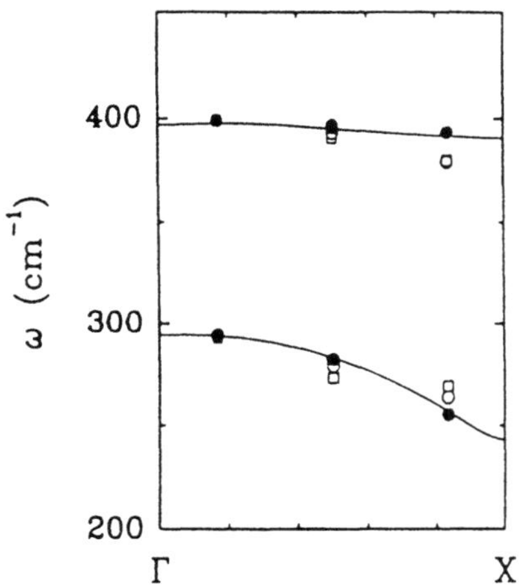

Fig. 2.18. Backfolding of Raman-active LO_1, LO_3 and LO_5 frequencies of $(GaAs)_5/(AlAs)_5$ SLs onto the bulk dispersions of GaAs and AlAs (*solid lines*). •: ideal SL, ○ (□): one (two) $Al_{0.5}Ga_{0.5}As$ alloy layers at the interfaces (from [2.38])

als a strong dependence of the LO_1 modes on the number of monolayers and a continuous change between the bulk frequencies and those in $Al_{0.5}Ga_{0.5}As$.

Both problems, that of dispersion mapping and the behavior of LO_1 for ultra-short periods, point to the existence of deviations between the experimental confined-phonon frequencies, measured by Raman spectroscopy, and the theoretical predictions which are believed to be rather precise. It has to be taken into account, however, that these calculations were performed for ideal SLs. An important hint to the influence of deviations from the ideal behavior in real samples comes from measurements on specimen with different interface quality [2.36, 2.44]. By different means, such as growth interruption or varying the substrate temperature, the interface properties of MBE-grown samples can be influenced. In [2.36] it was shown that in wide GaAs/AlAs SLs the number of LO modes observed and their frequencies depend strongly on the interface quality which can also be characterized by other techniques such as, e. g., luminescence. Especially the modes with high index are quite sensitive to interface roughness, and in the best samples the agreement between the backfolded dispersion and the theory can be improved [2.36]. In [2.44] the dependence of the confined-phonon frequencies on the substrate temperature during MBE growth was systematically investigated in ultra-short-period GaAs/AlAs SLs; the results allow one to make conclusions about concentration profiles and interface properties.

The influence of roughness can be taken into account theoretically by considering alloy layers, i. e., cation mixing of Ga and Al at the interfaces [2.38, 2.40]. Figure 2.18 shows the result of backfolding the calculated Raman active LO_1, LO_3 and LO_5 modes onto the bulk dispersions for an ideal $(GaAs)_5/(AlAs)_5$ SL (filled circles) and the layer sequences $(GaAs)_{5-n}(Al_{0.5}Ga_{0.5}As)_n/(AlAs)_{5-n}(Al_{0.5}Ga_{0.5}As)_n$ for $n = 1$ (open circles) and $n = 2$ (open squares), i. e., with one or two alloy layers at the interfaces [2.38]. In the ideal case the backfolded dispersion points agree well with those calculated for the bulk which shows that the concept of assigning an effective wavevector to the LO_m modes can be applied down to such thin layers. With increasing interface disorder systematic deviations of the backfolded dispersions compared to those of GaAs and AlAs appear. In agreement with the experimental trend of Fig. 2.16 the AlAs modes for large wavevectors tend towards lower frequencies whereas the GaAs phonons have higher energies than expected from the bulk.

The LO-phonon dispersion of AlAs is rather flat and shifts strongly towards lower frequencies when Ga is added. It does therefore not overlap with the AlAs-like LO band of an $Al_{0.5}Ga_{0.5}As$ alloy [2.38, 2.40]. Consequently the effective wavelength of the confined LO phonons is reduced when alloy layers at the interfaces are being considered. Deviations from the bulk dispersion towards lower frequencies thus occur if the wavevectors are still assigned as in an ideal system [2.38, 2.40]. For GaAs modes and adjacent $Al_{0.5}Ga_{0.5}As$ layers the dispersion overlap is not yet completely removed [2.38, 2.40]. Up to values of $q^* \simeq (3/4) \cdot (2\pi/a_0)$ the overlap of the bulk dispersions of GaAs and $Al_{0.5}Ga_{0.5}As$ vanishes and, in analogy to the case of GaAs and AlAs, one obtains a reduction of the confined-phonon frequencies compared to the GaAs dispersion. This explains the lower GaAs frequencies in Fig. 2.18 in the middle of the Brillouin zone which essentially reflect modes with a larger wavevector. For wavevectors larger than q^*, however, the dispersion branches still overlap. The effective wavelengths of these confined phonons therefore increase and their crystal momentum is reduced. Consequently their frequencies appear increased compared to those in bulk GaAs [2.38, 2.40].

2.2.2 Real Superlattices: Interface Phenomena

The problems just illustrated motivated studies on ultra-short-period GaAs/AlAs SLs which are described in the following [1.16]. The aim of these investigations was to test the theoretical predictions more precisely than before by using different samples with optimized interface properties. In addition to a comparison with the calculated LO_1 frequencies such systems also appeared attractive for determining the limit of the backfolding concept which had previously been given by $n \geq 4$ [2.38, 2.40].

A series of $(GaAs)_n/(AlAs)_n$ SLs with nominal layer thicknesses between $1 \leq n \leq 5$, an $Al_{0.5}Ga_{0.5}As$ alloy, and the bulk semiconductors GaAs and AlAs were used. Different SLs were grown by MBE with a substrate temperature of 580° C and with atomic-layer MBE (ALMBE) at lower temperatures of 380° C. We thus used modern MBE concepts in order to improve the interface properties which consist of choosing suitable growth conditions and allowing for surface diffusion [2.36, 2.44, 2.45]. The periodicity of the samples which had total thicknesses between 100 and 400 GaAs/AlAs sequences was checked by double-crystal X-ray diffraction. The samples were further characterized by luminescence, photoreflectivity and ellipsometry. For $n \geq 2$ we found good agreement with the expected dependence of the direct $\Gamma - \Gamma$ and

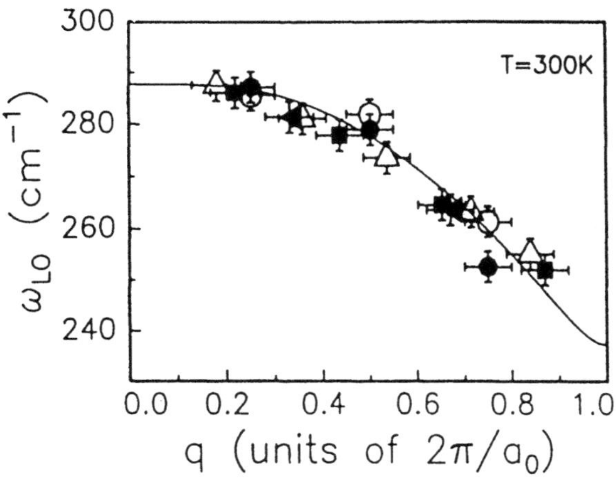

Fig. 2.19. LO-phonon dispersion in GaAs determined from the confined modes in ultra-short-period [001]-oriented $(GaAs)_n/(AlAs)_n$ SLs. ◀: n=2, (○, •): n=3, ■: n=4, △: n=5. The *solid line* is the theoretical dispersion

the indirect $\Gamma - X$ interband transitions on the layer thickness for GaAs/AlAs SLs [2.46].

In Figure 2.19 we show the backfolded LO-phonon dispersion of GaAs as determined by Raman scattering from $(GaAs)_n/(AlAs)_n$ SLs with $2 \leq n \leq 5$ [1.16] in comparison to the theory [2.38,2.39]. The effective wavevectors q were determined according to (2.8) with $\delta = 1$. In these SLs the LO frequency of the underlying GaAs substrate was observed at $(291.7 \pm 0.3)\,\mathrm{cm}^{-1}$ (300 K). This line was used for an accurate determination of the LO_m modes nearby. The theoretical dispersion, calculated at low temperatures, was shifted by $-2.3\,\mathrm{cm}^{-1}$, according to the frequency change of the GaAs LO phonon between 10 and 300 K, in order to compare it with the experimental data. The theoretical LO frequency for $q = 0$ is thus $288\,\mathrm{cm}^{-1}$, about 1% smaller than the experimental value [2.39]. A similar deviation of $2.8\,\mathrm{cm}^{-1}$ results from a comparison of the theory [2.37, 2.39] with neutron data [2.47] and can be considered as a measure of its accuracy. As can be seen in Fig. 2.19 the agreement between the theoretical dispersion along the [001]-direction and the experiments is good over the whole Brillouin zone. Note that there are no systematic deviations for large wavevectors. This result shows that confined phonons in GaAs/AlAs SLs can indeed be backfolded onto the bulk dispersion and that this concept cannot be applied only to thick layers but also to ultra-short-period systems down to a thickness of two monolayers.

In Figure 2.20 we compare the measured LO_1 frequencies of these SLs and the LO modes in $Al_{0.5}Ga_{0.5}As$, GaAs, and AlAs with the theory [1.16]. The triangles label the experimental values. The circles are calculations for ideal structures from [2.37, 2.38]. The deviations between theory and experiment for $n \geq 3$ are significantly reduced compared to Fig. 2.17, and one finds, as expected theoretically, almost no dependence of the LO_1 frequencies on

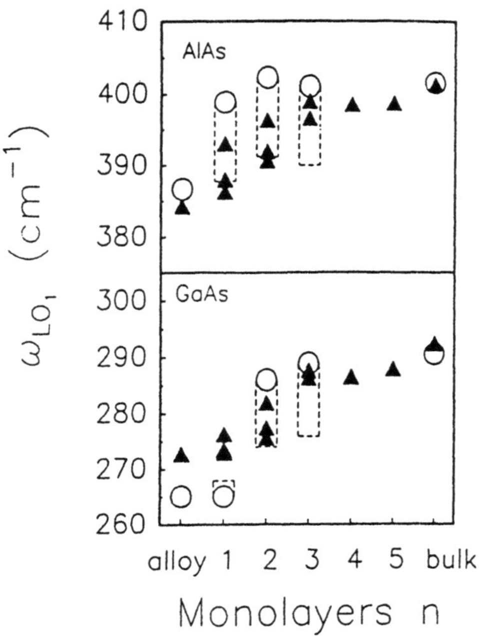

Fig. 2.20. Experimental LO_1 frequencies (*triangles*) in $(GaAs)_n/(AlAs)_n$ SLs, $Al_{0.5}Ga_{0.5}As$, GaAs, and AlAs compared to the theoretical values (*circles*: ideal SLs, *dashed bars*: $(Al_{1-x}Ga_xAs)_n/(Al_xGa_{1-x}As)_n$ SLs with $0.67 \leq x \leq 1$)

the layer thickness. For $n \leq 2$ the frequencies measured on different samples scatter considerably. We find, however, that the samples grown at a lower substrate temperature of 380 K with ALMBE, having the best structural properties according to X-ray diffraction, also show better agreement of the LO_1 frequencies with the calculations for ideal SLs. This dependence of the phonon frequencies on the growth conditions and the better agreement with the theory for lower substrate temperatures have also been observed in [2.44]. To illustrate the influence of cationic intermixing at the interfaces the calculated spread of the LO_1 frequencies in $(Al_{1-x}Ga_xAs)_n/(Al_xGa_{1-x}As)_n$ SLs for $0.67 \leq x \leq 1$ is given by the dashed bars in Fig. 2.20 [2.38]. This approximation of interface roughness by alloy SLs with changes in the composition of whole layers is certainly a strong simplification of the problem. Nevertheless, these systems show the important trends and the model should thus be applicable to ultra-short periods. The phonon frequencies measured for $n = 2$ and 3 can be explained by the range of values chosen for x. For thinner layers the influence of disorder increases as can be seen from the larger scatter of the data points for $n = 3$ than for $n = 2$. For $n = 1$ the GaAs modes fall out of this range of x and are quite close to the value measured for $Al_{0.5}Ga_{0.5}As$. which, however, also deviates from the calculations. In the AlAs range one sample shows for $n = 1$ a larger LO_1 frequency than the alloy. This is in agreement with theory which predicts a discontinuous change in the frequencies between the $(GaAs)_1/(AlAs)_1$ SL and $Al_{0.5}Ga_{0.5}As$ (see Fig. 2.17). The theoretically expected change in the frequencies of the GaAs modes between $n = 2$ and $n = 1$ is also observed if only the measurement for the best sample is considered for $n = 2$. This particular ($n = 2$)-sample was also used for backfolding the dispersion in Fig. 2.19. In contrast to earlier measurements these results thus confirm the theory in another important point.

The discrepancies between theory and experiment concerning the confined-mode backfolding onto the bulk dispersions and the LO_1 frequencies in GaAs/AlAs SLs mentioned at the beginning could thus be clarified. Progress on either side has contributed to this achievement: Samples with better interfaces made it possible to approach the ideal limit experimentally. An extension of the theory to real interfaces, on the other hand, was necessary to calculate the deviations expected in these, still nonideal, systems.

2.3 Optic Phonons in Isotopic Superlattices

2.3.1 Stable Isotopes in Semiconductor Physics

The large amounts of isotopically pure stable elements which have become available in recent years have also had an impact on semiconductor physics and made it possible to investigate effects which were previously not accessible [1.17, 2.48–2.58]. The physical properties of semiconductors with pure or deliberately mixed isotope composition have been reviewed in [2.48, 2.49, 2.59–2.61]. As can be seen from the natural stable isotope composition of several elements relevant to semiconductor physics compiled in Table 2.1 [2.62], isotope disorder is a quite common phenomenon, and most elements contain a considerable mixture of them. However, due to the lack of isotopically pure crystals, isotope effects in bulk semiconductors remained largely unexplored until recently.

Nuclear Mass Effects. Semiconductors are influenced by isotope effects in different ways [2.48, 2.49]. The phonon frequencies $\omega = \sqrt{k/M}$ and mean-square amplitudes $\langle u^2 \rangle = \hbar(2n_B(\omega) + 1)/(4\sqrt{kM})$ depend on the average atomic mass M of a crystal. In these expressions k is a force constant of the crystal lattice and $n_B(\omega)$ is the Bose–Einstein factor. Via the dependence of $\langle u^2 \rangle$ on M, isotope effects change quantities which are influenced by anharmonic effects. Among these are the lattice constant, the elastic properties as well as the phonon linewidths and lifetimes [2.49, 2.51]. The mass fluctuations around the average value M in crystals containing different isotopes also perturb the translational invariance of the lattice. This causes scattering processes which lead to shifts in the phonon frequencies and to changes in their linewidths. In theoretical treatments these effects are expressed by a complex self-energy. They can be studied in detail by comparing isotopically pure systems and crystals with controlled isotope disorder. This leads to a better understanding of semiconductors with the natural isotope abundance. The influence of mass fluctuations on the translational invariance was

Table 2.1. The natural distribution of stable isotopes in elements relevant for semiconductor physics. Small concentrations are not always indicated (from [2.62])

group	element	average mass	stable isotopes
I	Cu	63.546	^{63}Cu: 69.17%, ^{65}Cu: 30.83%
	Ag	107.8682	^{107}Ag: 51.84%, ^{109}Ag: 48.16%
II	Mg	24.3050	^{24}Mg: 78.99%, ^{25}Mg: 10.00%, ^{26}Mg: 11.01%
	Zn	65.39	^{64}Zn: 48.6%, ^{66}Zn: 27.9%, ^{67}Zn: 4.1%, ^{68}Zn: 18.8%
	Cd	112.411	^{110}Cd: 12.51%, ^{111}Cd: 12.81%, ^{112}Cd: 24.13%, ^{113}Cd: 12.22%, ^{114}Cd: 28.72%, ^{116}Cd: 7.47%
	Ba	137.327	^{134}Ba: 2.417%, ^{135}Ba: 6.592%, ^{136}Ba: 7.854%, ^{137}Ba: 11.23%, ^{138}Ba: 71.70%
III	B	10.81	^{10}B: 19.9%, ^{11}B: 80.1%
	Al	26.98154	^{27}Al: 100%
	Ga	69.723	^{69}Ga: 60.11%, ^{71}Ga: 39.89%
	In	114.82	^{113}In: 4.3%, ^{115}In: 95.7%
IV	C	12.011	^{12}C: 98.90%, ^{13}C: 1.10%
	Si	28.0855	^{28}Si: 92.23%, ^{29}Si: 4.67%, ^{30}Si: 3.10%
	Ge	72.61	^{70}Ge: 20.5%, ^{72}Ge: 27.4%, ^{73}Ge: 7.8%, ^{74}Ge: 36.5%, ^{76}Ge: 7.8%
	Sn	118.710	^{112}Sn: 0.97%, ^{114}Sn: 0.65%, ^{116}Sn: 14.53%, ^{117}Sn: 7.68%, ^{118}Sn: 24.22%, ^{119}Sn: 8.58%, ^{120}Sn: 32.59%, ^{122}Sn: 4.63%, ^{124}Sn: 5.79%
V	N	14.00674	^{14}N: 99.63%, ^{15}N: 0.37%
	P	30.973762	^{31}P: 100%
	As	74.9216	^{75}As: 100%
	Sb	121.75	^{121}Sb: 57.3%, ^{123}Sb: 42.7%
VI	O	15.9994	^{16}O: 99.76%, ^{18}O: 0.20%
	S	32.066	^{32}S: 95.02%, ^{34}S: 4.21%
	Se	78.96	^{76}Se: 9.2%, ^{78}Se: 23.7%, ^{80}Se: 49.8%, ^{82}Se: 8.8%
	Te	127.60	^{122}Te: 2.57%, ^{124}Te: 4.76%, ^{125}Te: 7.10%, ^{126}Te: 18.89%, ^{128}Te: 31.73%, ^{130}Te: 33.97%
VII	Cl	35.4527	^{35}Cl: 75.77%, ^{37}Cl: 24.23%
	Br	79.904	^{79}Br: 50.69%, ^{81}Br: 49.31%
	I	126.90447	^{127}I: 100%

also demonstrated by Raman measurements in Ge [2.53] and diamond [2.55], where density-of-states-related side peaks of the main LO-phonon line appear due to the relaxation of crystal-momentum conservation by isotopic disorder. Via the electron–phonon interaction, changes of the lattice-dynamical properties with isotope composition also affect the electronic structure of semiconductors such as their band gaps and homogeneous linewidths [2.49,2.63,2.64].

Nuclear Structure Effects. Apart from these phenomena, which depend on the nuclear mass, there are also isotope effects on the nuclear structure [2.49]. Various isotopes of an element may have rather different nuclear spins which can be used, e. g., in spin-resonance investigations. An important application of isotopically pure semiconductors which also had an impact on solid-state spectroscopy was the investigation of the double-beta decay in ^{76}Ge for which single crystals with masses of several kilograms were required [2.65]. Once the material was available, small pieces could also be used for light-scattering experiments. Another example of isotope effects nourishing the interaction between different areas of physics are linewidth measurements of nuclear transitions in ^{13}C [2.66]. The precise determination of these resonances around 3.5 MeV is important for the calibration of nuclear-structure data. However, this method can also be used, in addition to Raman spectroscopy, as a destruction-free method for determining the ^{13}C-content in artificial diamonds [2.67]. It has also been proposed that SLs of ^{73}Ge/^{70}Ge could be used to create highly monochromatic synchrotron radiation by nuclear diffraction at the 13.3 keV-nuclear resonance line of ^{73}Ge. This could be applied for Mössbauer spectroscopy [2.68]. Another nuclear degree of freedom which is presently gaining importance in semiconductor physics is the neutron capture cross-section which can be very different for various isotopes of an element. This allows one to perform neutron scattering experiments in isotopically pure crystals which are impossible in the natural ones due to the presence of strongly absorbing isotopes. The most prominent examples from this area are semiconductors containing Cd. While phonons in zincblende CdTe [2.69] had been studied earlier, neutron dispersion data for wurtzite CdS, made from ^{114}Cd to avoid the strongly absorbing ^{113}Cd, have been obtained only very recently [2.70], despite the important model character of this compound for the II-VI materials and the wurtzite structure. Unstable isotopes with decay times of several hours or days have been used to identify the chemical origin of different defect lines in the luminescence of Si or GaAs by monitoring the changes in the relative signal intensities [2.71–2.73].

Phonons. In order to illustrate isotope mass effects on the phonons Fig. 2.21 shows Raman spectra of various isotopically pure Ge crystals in comparison to the line observed in natural Ge [2.48,2.74]. The frequencies of these modes are inversely proportional to the square root of the average atomic mass. Even though Ge, as can be seen in Table 2.1, contains large fractions of various isotopes with significantly different phonon frequencies, natural Ge only exhibits one sharp line at the energy corresponding to the average mass,

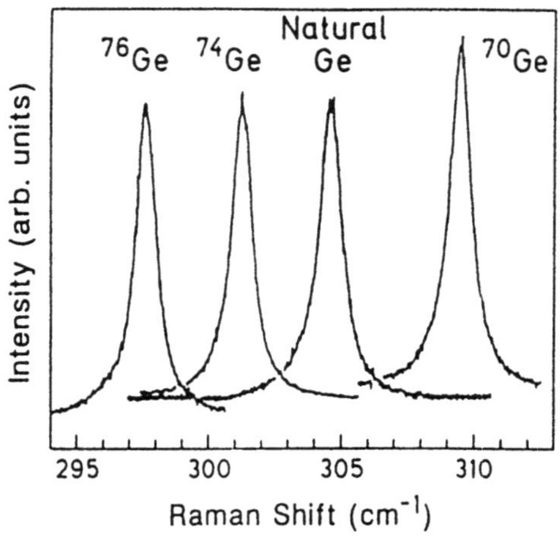

Fig. 2.21. Raman spectra of the optic phonon in isotopically pure and natural Ge (from [2.48, 2.74])

except for a small correction which is mostly due to the isotope disorder [2.48, 2.49]. This can be explained by the fact that the mass fluctuations $\Delta M/M$ in natural Ge only lead to small frequency changes $\Delta\omega = (\Delta M/2M)\omega \simeq 12\,\mathrm{cm}^{-1}$ compared to the band width of the optic-phonon dispersion which is almost ten times larger [2.48, 2.49]. Therefore Anderson localization of these phonons does not occur [2.48, 2.49].

Despite the lack of phonon localization in bulk Ge an important question in the lattice-dynamical properties of semiconductors is that of optic-phonon confinement in isotopic SLs whose layers consist of different pure isotopes [1.17, 2.54]. Such structures have been proposed earlier [2.75, 2.76] and are now available for experiments [1.17]. General arguments about confinement effects in such systems can be derived from the phonon dispersions of the isotopically pure bulk materials. Figure 2.22 shows the calculated dispersion curves of longitudinal modes in ^{70}Ge and ^{74}Ge [2.77]. For $k = 0$ the LO frequency in ^{70}Ge is about $9\,\mathrm{cm}^{-1}$ larger than in ^{74}Ge. The dispersion overlap is removed for wavevectors up to $k \simeq 0.45 \cdot 2\pi/a_0$. In this frequency range, according to the considerations of Sect. 2.1 and 2.2, one expects confined modes to appear. In isotopic SLs one should therefore find a different behavior for lattice vibrations and electrons: While both, electronic and lattice-dynamical properties, have two-dimensional character in SLs made from different materials, in these systems only the phonons change their character, the electronic structure remains almost unaffected. Isotopic SLs thus offer the unique possibility to investigate dimensionality effects on the lattice vibrations without significantly changing the electronic properties [1.17, 2.75].

The dispersions in Fig. 2.22 were calculated in a planar bond-charge model [2.78, 2.79] which is shown in Fig. 2.23. It is based on a linear force-constant model for the ionic planes. The additional bond-charge planes whose mass is assumed to vanish allow one to consider the electronic degrees of freedom which improves the agreement with the experiment and reduces the number of parameters [2.80, 2.81]. For longitudinal modes along the [001]-

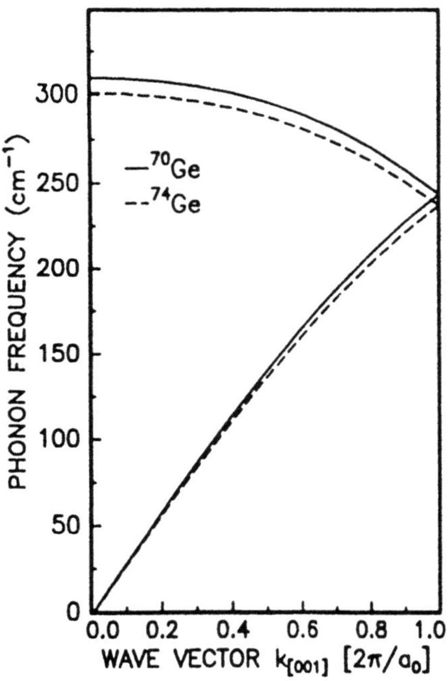

Fig. 2.22. Calculated dispersion curves of longitudinal phonons in ^{70}Ge and ^{74}Ge for the [001]-direction (from [2.77])

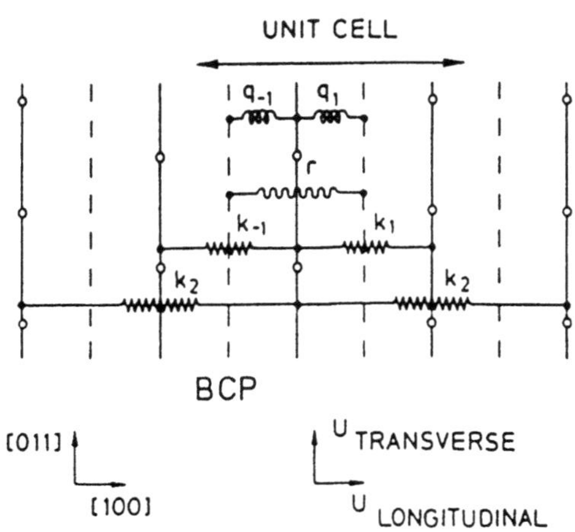

Fig. 2.23. Planar bond-charge model for the phonon dispersion in Ge along [001]. The *solid lines* represent the ion planes, the *dashed lines* stand for the bond-charge planes (BCP). See the text for a description of the couplings and force constants considered (from [2.78])

direction the model is particularly simple. One has only three force constants, $q_1{=}q_{-1}$, r, and $k_1{=}k_{-1}$ [2.78] which describe the coupling between the neighboring ion planes (k_1), between the ion and the bond-charge planes (q_1) and between the bond-charge planes (r). In [2.78] they were determined by fitting neutron data for natural Ge. Isotope effects can be treated in this model by changing the ion masses. One has to consider, however, that the fits of force constants to neutron data did not take into account the self-energy effects on the phonon frequencies due to isotopic disorder [2.82]. These contributions are responsible for an increase of the LO frequency at the Γ point in natural

Ge of about 0.6 cm^{-1} [2.82]. Together with inaccuracies of the fits the calculated LO frequencies in isotopically pure crystals are 2 cm^{-1} larger than those measured by Raman spectroscopy. This discrepancy is taken into account for comparisons with experiments by a constant shift of all frequencies. Using a larger unit cell the bond-charge model can readily be extended to isotopic SLs. Theoretical calculations of such systems for Ge were reported in [2.54].

The Raman scattering intensities of the modes determined this way can be calculated with the bond-polarizability model [1.13]. In this model parameterized expressions for the Raman polarizabilities of atomic crystal bonds and their changes with bond length for displacements of atoms parallel and perpendicular to the bond direction are used [1.13, 2.83]. The Raman tensor thus reflects the influence of the phonons on bond lengths and bond angles. The Raman intensity depends on the sum of the displacement differences between neighboring atom pairs [1.13, 2.84]. For longitudinal-optic phonons at $k = 0$ one finds that backscattering only occurs for crossed polarizations. For parallel polarizations scattering is forbidden, in accordance with the selection rules, except for a small k-dependent contribution [2.54]. The Raman intensities in isotopic SLs should be well described by this model since the electronic character of the bonds hardly varies over the whole system, in contrast to SLs from different III–V semiconductors. The scattering intensities only depend on one bond polarizability which should allow one to predict the relative intensities of different modes without additional parameters [1.13, 2.54].

2.3.2 Symmetric Superlattices

Different $^{70}Ge/^{74}Ge$ SLs, isotopically pure bulk layers of both materials, and $^{70}Ge_{0.5}{}^{74}Ge_{0.5}$ alloys were grown by MBE on natural [001]-oriented Ge substrates [1.17, 2.85]. To prepare atomically flat surfaces for SL growth 230 Å thick buffer layers of ^{70}Ge were first grown on the substrates. At a substrate temperature of 350° C the SLs were then grown with a rate of 4 Å/min up to a total thickness of about 400 monolayers (ml) (1 ml = 1.41 Å).

Figure 2.24 shows the measured and calculated Raman spectra of a series of symmetric $^{70}Ge_n/^{74}Ge_n$ SLs [1.17, 2.77]. The experimental spectra in Fig. 2.24 (a) were measured at 10 K with an excitation wavelength of 5145 Å. The bulk LO frequencies are indicated by the arrows at the top of Fig. 2.24 (a). The spectra exhibit a series of lines which depend on the SL period in a characteristic way. By comparison with the calculations they can be assigned to confined LO_m modes in ^{70}Ge and ^{74}Ge, as indicated by the labels in Fig. 2.24 (a). The theoretical spectra were calculated with the planar bond-charge model for the frequencies, and the bond-polarizability model was used for the scattering intensities. To facilitate the comparison with the experiment the lines in Fig. 2.24 (b) were broadened by Lorentzian profiles with widths of 1.2 cm^{-1} and, as outlined above, shifted by 2 cm^{-1} towards lower frequencies. One finds good agreement of these calculations with the measurements.

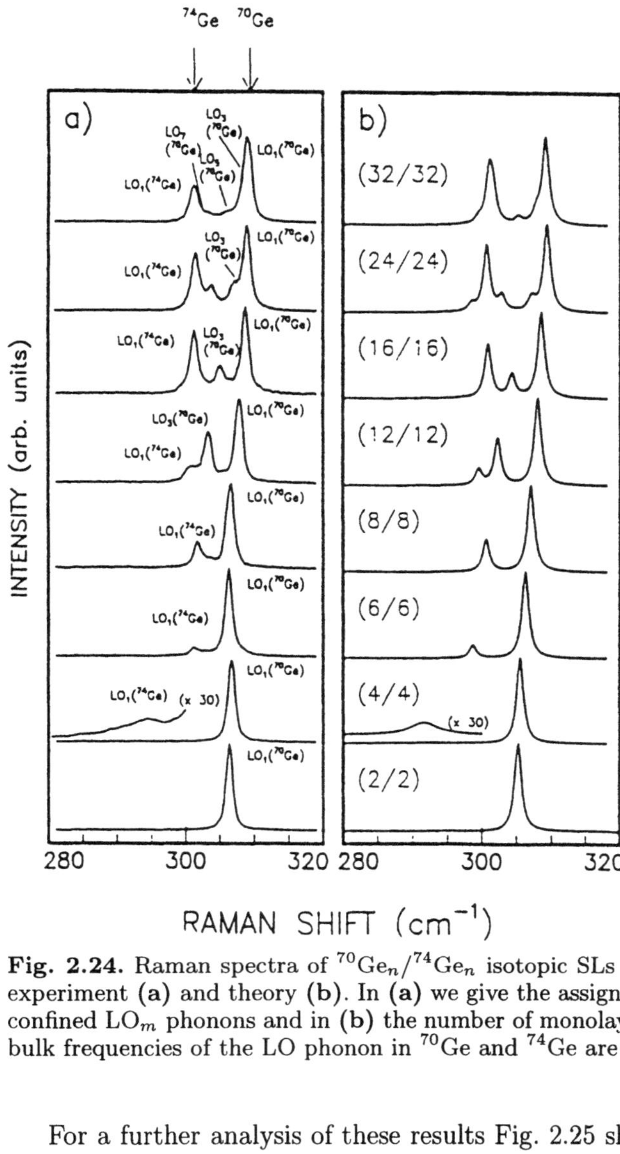

Fig. 2.24. Raman spectra of $^{70}Ge_n/^{74}Ge_n$ isotopic SLs in a comparison between experiment **(a)** and theory **(b)**. In **(a)** we give the assignment of spectral peaks to confined LO_m phonons and in **(b)** the number of monolayers of each material. The bulk frequencies of the LO phonon in ^{70}Ge and ^{74}Ge are also indicated

For a further analysis of these results Fig. 2.25 shows the calculated frequencies of the Raman-active odd-index LO_m modes in symmetric $^{70}Ge_n/$ $^{74}Ge_n$ SLs vs. layer thickness [1.17, 2.77]. Figure 2.26 shows squared phonon displacement patterns for various SLs [2.77]. For the largest periods the confinement of modes in ^{70}Ge or ^{74}Ge is clearly visible. With decreasing layer thickness the mode frequencies decrease according the bulk dispersion. Resonant couplings occur where ^{70}Ge modes overlap with ^{74}Ge vibrations. As can be seen in Fig. 2.25, characteristic splittings appear at the expected crossing points, e. g., at n=12, 20, and 30. The displacements of the modes involved are mixed at these points (see Fig. 2.26) which decreases their confinement. An example of this phenomenon are the coupled $LO_3(^{70}Ge)/LO_1(^{74}Ge)$ modes

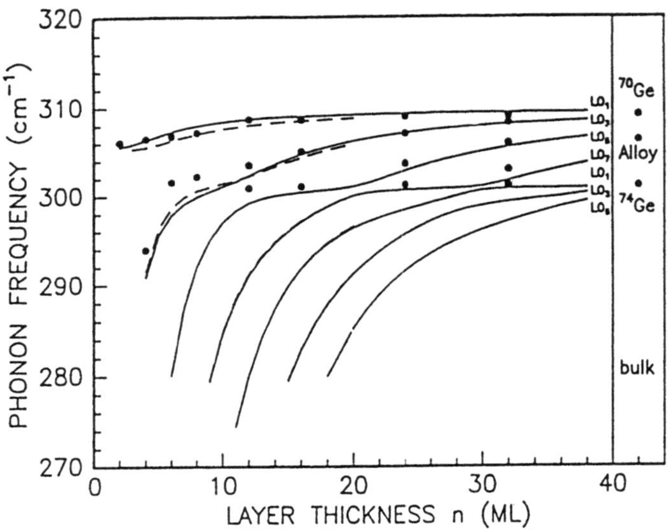

Fig. 2.25. Phonon frequencies of the Raman-active LO_m modes in symmetric $^{70}Ge_n/^{74}Ge_n$ isotopic SLs vs. layer thickness. The *solid lines* represent calculations for ideal structures. The *dashed lines* were calculated taking into account isotopic mixing effects at the interfaces. The *points* label the experimental results of Fig. 2.24 (a). The LO frequencies of different bulk crystals are indicated on the right side of the figure

for n=12. For modes with frequencies smaller than $LO_1(^{74}Ge)$ the bulk dispersions of the two materials overlap. Therefore they have displacements in both layers, and with decreasing period it becomes increasingly harder to identify their origin in confined modes of either material. From the calculated frequencies and the displacement patterns one arrives at the mode assignment of Fig. 2.24 (a) and the experimental results which are indicated by the dots in Fig. 2.25.

Further insight about the phonons in samples with different layer thickness can be gained from the scattering intensities. Figure 2.27 shows the calculated signal strengths in the $^{70}Ge_n/^{74}Ge_n$ SLs normalized to the respective $LO_1(^{70}Ge)$ phonon [1.17, 2.77]. For the largest periods the LO_m modes of ^{70}Ge show the expected intensity decrease proportional to m^{-2}. Due to mode mixing with the $LO_1(^{74}Ge)$ phonon at certain layer thicknesses, however, the intensities can significantly deviate from this behavior. In resonance the character of the mixed modes is about one half $LO_m(^{70}Ge)$ and one half $LO_1(^{74}Ge)$. This leads to an increase of the $LO_m(^{70}Ge)$ mode intensities which can then even exceed that of the ^{70}Ge phonons with lower indices even though these signals are expected to be much smaller due to their higher index m. For $n = 32$, e. g., the $LO_7(^{70}Ge)$ line has a larger intensity than the nonresonant $LO_5(^{70}Ge)$ and $LO_3(^{70}Ge)$ modes. The intensity of the $LO_1(^{74}Ge)$ mode, however, is suppressed at this layer thickness. The measured scattering

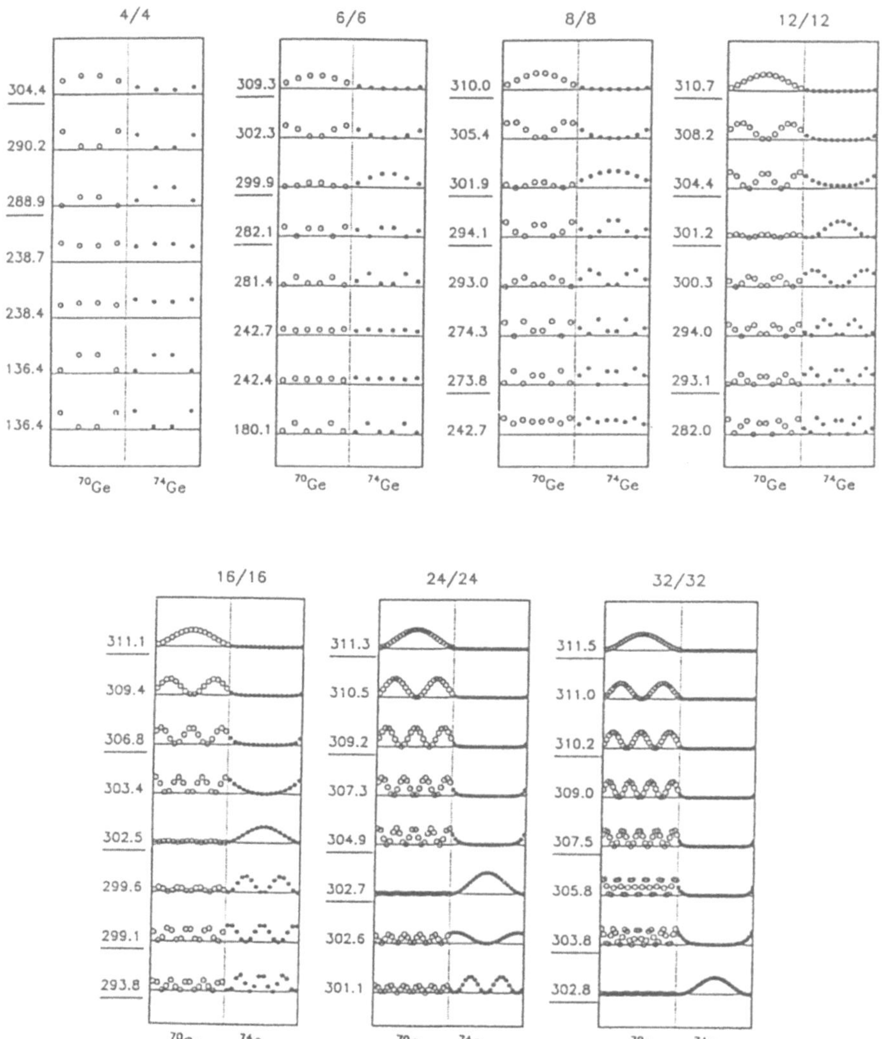

Fig. 2.26. Squared displacement patterns for phonons in $^{70}\mathrm{Ge}_n/^{74}\mathrm{Ge}_n$ isotopic SLs. The layer thicknesses are indicated above each block in units of monolayers. The mode frequencies are given on the left side of each pattern (in cm^{-1}); Raman-active modes are underlined

intensities, obtained by fits to the spectra in Fig. 2.24, are shown by the open and filled circles in Fig. 2.27.

In agreement with the experiment the calculated intensities also decrease strongly with the layer thickness for phonons whose frequency is lower than that of $\mathrm{LO}_1(^{74}\mathrm{Ge})$. This effect is related to the increasing relaxation of the phonon confinement in the range where the disperisons of ^{70}Ge and ^{74}Ge

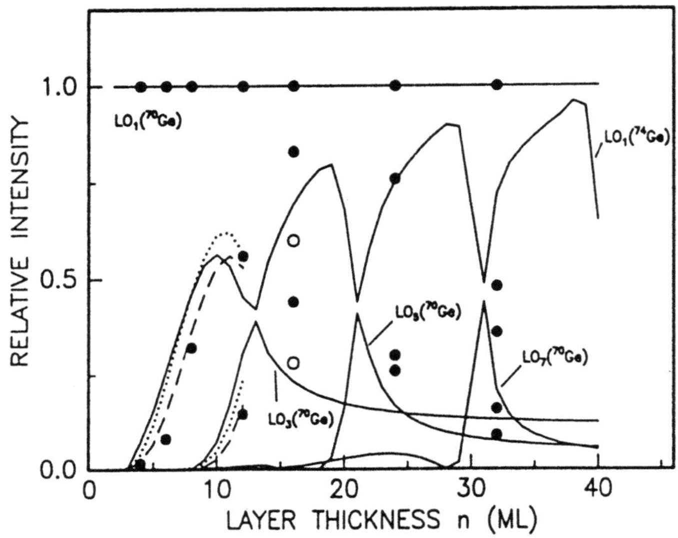

Fig. 2.27. Measured (•: λ_l = 5145 Å, o: λ_l = 6471 Å) and calculated relative Raman intensities of optic phonons in $^{70}Ge_n/^{74}Ge_n$ isotopic SLs. All intensities are normalized to the respective $LO_1(^{70}Ge)$ mode. The *solid lines* are calculations for ideal systems, the *dashed* and *dotted lines* take isotopic mixing at the interfaces into account

overlap. For this reason no modes lower than $LO_1(^{74}Ge)$ are observed in Fig. 2.24 (a). The higher-index $LO_m(^{74}Ge)$ modes ($m \geq 3$) which can be identified for larger periods in Fig. 2.25 strongly mix for smaller layer thicknesses and cannot be observed. An example of such intensity changes during the transition of a mode into the dispersion continuum is given by the behavior of the $LO_3(^{70}Ge)$ mode for the three samples with $n = 16$, 12 and 8. For $n = 16$ the $LO_3(^{70}Ge)$ frequency is larger than that of the $LO_1(^{74}Ge)$ phonon. Figure 2.24 shows two lines which are clearly separated from each other, and the intensity of $LO_1(^{74}Ge)$ is about twice as large as that of $LO_3(^{70}Ge)$. The displacements of both modes show strong confinement (see Fig. 2.26). For $n = 12$ resonance occurs. The $LO_3(^{70}Ge)$ line is now about three times as strong as $LO_1(^{74}Ge)$. The mode displacements show less confinement as compared to $n = 16$. For $n = 8$ the $LO_3(^{70}Ge)$ mode is below the $LO_1(^{74}Ge)$ frequency and thus in the continuum. Only the $LO_1(^{74}Ge)$ line is observed and, as can be seen from the displacement patterns, confined.

Apart from these resonance effects at crossing points of $LO_m(^{70}Ge)$ and $LO_1(^{74}Ge)$ modes and the reduction of the scattering intensity for phonons entering the continuum, further phenomena can be observed in Figs. 2.25 and 2.27. The $LO_1(^{70}Ge)$ and $LO_1(^{74}Ge)$ frequencies calculated for ideal SLs are located above and below the measured values for the shorter-period samples ($n \leq 8$). Also, the calculated relative intensities are larger than those observed experimentally. These systematic deviations for smaller layer thicknesses point to the influence of nonideal interfaces. To investigate their influence we have performed calculations assuming two monolayers of $^{70}Ge_{0.5}{}^{74}Ge_{0.5}$ at each interface. In a first model these isotopically mixed layers were taken into account via their contribution to the average mass of each of the two material layers. This gives the dashed lines in Figs. 2.25 and 2.27. The calculation reproduces the trends of the frequency and intensity changes. For the intensity changes we obtain good agreement with the experiment. In another calculation isotopic mixing at the interfaces was treated by diagonalization of a supercell consisting of four layers, and the mixed regions were taken into account according to their average mass. This leads to the dotted lines in Fig. 2.27. Only small intensity changes are found in comparison to the ideal SLs. The calculated frequency shifts are about half of those found in the first model. According to the first model, the measurements are compatible with isotopically mixed layers of two monolayers at each interface. Raman scattering appears to be one of the few experimental methods to investigate such effects in short-period isotopic systems. Recently, the interface properties of these samples and the self-diffusion of Ge have been studied more carefully by annealing experiments where the transition between the different confined SL phonons and only one line in the alloy was monitored by Raman spectra (see Sect. 6.1.1) [2.86]. Interferences in the neutron total reflection, which arise from different coherent scattering lengths of the isotopes, also appear as a promising technique for investigations of interface effects in these samples [2.87].

For the $^{70}Ge_{16}{}^{74}Ge_{16}$ sample one finds significant deviations of the observed scattering intensities (filled circles) from the calculated values in Fig. 2.27 which cannot be explained by interface disorder. For another excitation energy (open circles), however, good agreement with the theory is obtained. This indicates that resonance effects are important. A systematic investigation of the scattering intensities for various wavelengths shows a different resonance behavior of the $LO_1(^{70}Ge)$ and $LO_1(^{74}Ge)$ modes in the range of the E_1 and $E_1+\Delta_1$ band gaps [2.77,2.88]. It was found that the maximum of the intensity ratio $LO_1(^{74}Ge)/LO_1(^{70}Ge)$ depends linearly on the SL period [2.77,2.88]. For an excitation wavelength of $\lambda_l = 5145$ Å (filled circles in Fig. 2.27) one finds a strong resonance for the sample with 16 monolayers which affects the comparison with the calculated scattering intensities. Resonances for the other sample do not occur for this wavelength, and their intensities can thus be compared to the theory. The origin of these resonances is not yet completely understood. Supposedly, crystal-momentum nonconserving steps in Raman processes which arise from the finite penetration depth of light in the vicinity of the E_1 band gap play a certain role [2.77,2.88]. Similar effects in Ge were reported also in [2.89] and [2.90] where strongly dispersive modes in the range of acoustic [2.89] and optic phonons [2.90] were observed for excitation around the E_1 and $E_1 + \Delta_1$ band gaps.

2.3.3 Asymmetric Superlattices

For further investigations of phonon confinement effects in short-period isotopic SLs asymmetric samples of $^{70}Ge_n/^{74}Ge_2$ and $^{70}Ge_2/^{74}Ge_n$ with $n = 6$, 8 and 10 were grown. Their Raman spectra are shown in Fig. 2.28 [2.77].

A comparison of the measured phonon frequencies with calculations in the planar bond-charge model is given in Fig. 2.29 [2.77]. The squared displacement patterns of the modes are compiled in Fig. 2.30 [2.77].

The Raman spectra of the samples with two monolayers ^{74}Ge show a confined $LO_1(^{70}Ge)$ mode whose frequency decreases for smaller layer thicknesses of ^{70}Ge (see the upper three spectra in Fig. 2.28). According to the displacement patterns in Fig. 2.30 (a) ^{74}Ge remains a good phonon barrier for this vibration even for thin layers. The measured frequency decrease of $LO_1(^{70}Ge)$ can thus be explained by the increase of the effective wavevector for smaller layer thicknesses and backfolding onto lower-energy points of the bulk dispersion, in good agreement with the theoretically expected behavior (see Fig. 2.29 (a)). In these samples no confined ^{74}Ge modes appear. The spectra for $n = 10$ and 8, however, show weak lines which can be attributed to $LO_3(^{70}Ge)$ vibrations.

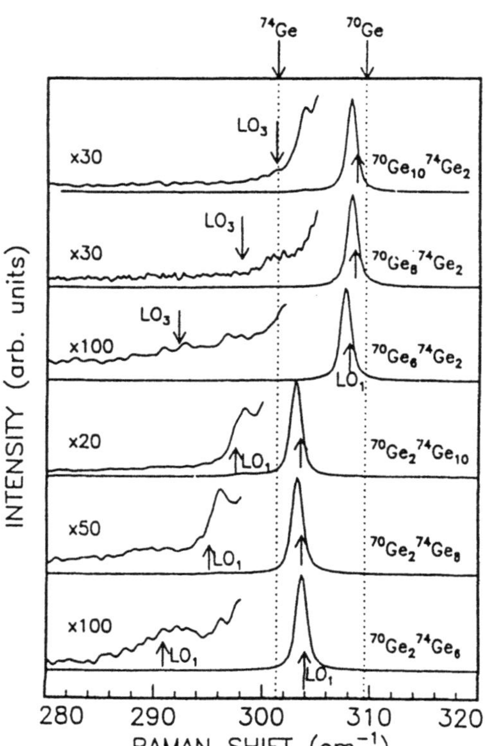

Fig. 2.28. Raman spectra of asymmetric $^{70}Ge/^{74}Ge$ SLs measured at 10 K with $\lambda_l = 5145$ Å. The *arrows* on top of the figure and the *dotted lines* indicate the bulk phonon frequencies. The *arrows* at the spectra denote the calculated phonon frequencies. Some spectral ranges were enlarged by the factors indicated

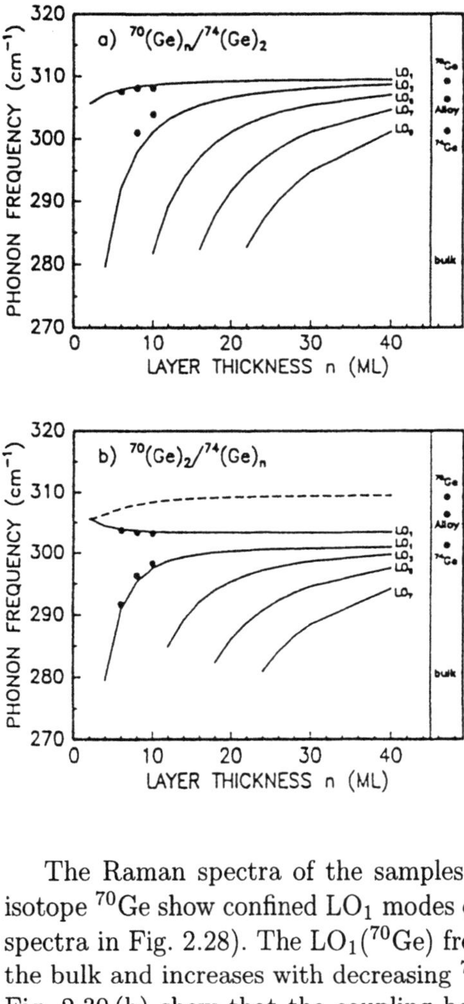

Fig. 2.29. Measured phonon frequencies (•) of asymmetric ^{70}Ge/^{74}Ge SLs compared to theory (*solid lines*). The *dashed line* in **(b)** shows the frequency of LO_1(^{70}Ge) vs. layer thickness for symmetric samples from Fig. 2.25

The Raman spectra of the samples with two monolayers of the lighter isotope ^{70}Ge show confined LO_1 modes of both materials (see the lower three spectra in Fig. 2.28). The LO_1(^{70}Ge) frequency is significantly lower than in the bulk and increases with decreasing ^{74}Ge thickness. The displacements in Fig. 2.30 (b) show that the coupling between the ^{70}Ge modes increases for thinner ^{74}Ge thicknesses, and consequently their confinement decreases. This reduces their effective wavevector and increases their frequency when backfolded onto the bulk dispersion. This tendency continues until, for quite thin layers, the value of the $^{70}Ge_{0.5}{}^{74}Ge_{0.5}$ alloy is reached. For comparison, the dashed line in Fig. 2.29 (b) shows the reduction of the LO_1(^{70}Ge) frequency in the symmetric SLs of Fig. 2.25 with decreasing layer thickness from the value of ^{70}Ge to that of the alloy. For the thicker ^{74}Ge layers Fig. 2.30 (b) shows a series of confined ^{74}Ge phonons whose frequencies decrease with decreasing number of monolayers. The LO_1(^{74}Ge) modes are experimentally observed (see Fig. 2.28) and their frequencies are in good agreement with the theory (see Fig. 2.29).

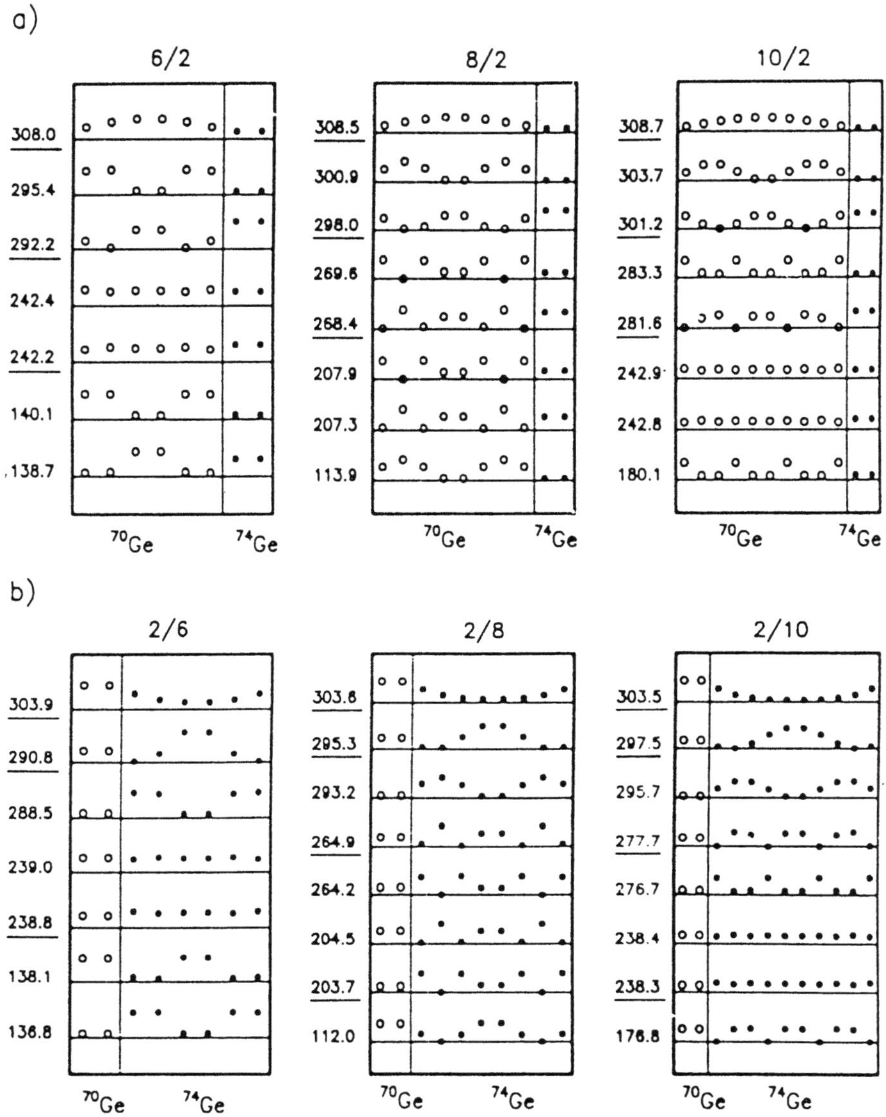

Fig. 2.30. Calculated squared displacements of phonons in asymmetric $^{70}Ge/^{74}Ge$ SLs. The mode frequencies are given next to each pattern (in cm^{-1}); Raman-active vibrations are underlined

2.3.4 Compound Semiconductors

The majority of investigations on isotopic semiconductors has so far been performed on bulk crystals of Ge [2.48, 2.49] or diamond [2.55, 2.91–2.95]. Apart from some earlier self-diffusion experiments [2.96] measurements on isotopic compound semiconductors and SLs, however, have only recently gained more attention [2.97–2.109].

The controlled variation of the average atomic mass of individual constituents by isotope substitution in crystals containing different elements allows one to investigate the vibrational amplitudes and the contributions of different atoms to a phonon mode. This makes more detailed comparisons between theory and experiment possible than can be performed on the basis of the vibrational frequencies alone. These concepts have recently been applied not only to the well-known III–V, II–VI, or I–VII semiconductors but also to other classes of substances. In the $YBa_2Cu_3O_7$-like high-temperature superconductors, for example, the substitution of ^{16}O by ^{18}O is well established and has been used in Raman investigations of quite a broad range of topics, reaching from phonon anomalies [2.110, 2.111], defect structures and laser annealing [2.112] to magneto-elastic effects [2.113, 2.114]. Recently, the mode mixing of the Ba- and Cu-related zone-center vibrations in $YBa_2Cu_3O_7$ could be clarified by Ba and Cu isotope substitution [2.56, 2.115]. The isotope measurements demonstrated the almost pure origin of these modes in either Ba or Cu displacements whereas the theory overestimated their coupling. By comparing the Raman spectrum of $^{154}Sm_2CuO_4$, originally prepared for neutron measurements with the nonabsorbing isotope ^{154}Sm, with that of natural $^{150.36}Sm_2CuO_4$, it was recently possible to unambiguously identify the low-energy rare-earth E_g-vibration in this class of materials which has an unusual temperature dependence and is often masked by crystal-field excitations [2.116, 2.117].

Apart from phonon-frequency shifts and self-energy effects due to isotope disorder, another interesting topic which can be investigated in isotopically modified compound semiconductors are changes of the band gaps due to isotope effects on the lattice constant and the influence of the electron–phonon interaction. Early investigations of these effects were performed in Cu_2O, a model substance for exciton effects, both, for oxygen and Cu substitution [2.118–2.121]. Different shifts of the various exciton series and the phonon frequencies were found. The Fröhlich electron–phonon coupling, present in polar compound semiconductors, should also offer new interesting possibilities for Raman spectroscopy in isotopically modified materials.

Presently available bulk samples include GaAs and GaP made from ^{69}Ga and ^{71}Ga isotopes. In GaAs the E_0 gap shifts by 0.39 meV/amu, in good agreement with theory [2.105, 2.106]. Single crystals of the copper halides CuCl and CuBr were grown using small amounts of ^{59}Cu and ^{63}Cu as well as ^{35}Cl, ^{37}Cl, ^{79}Br and ^{81}Br as source materials [2.97, 2.98]. Raman measurements confirmed the origin of the anomalous TO-phonon structure [2.99, 2.100]. Different signs in the gap shifts for Cu and Cl substitution lead to the understanding of the anomalous increase of the CuCl band gap with temperature, and a new contribution which is due to $p-d$-mixing in the valence band could be identified [2.101–2.103]. CdS, another strong neutron absorber due to the 12.2% abundance of ^{113}Cd in the natural element, was prepared from ^{110}Cd, ^{112}Cd, ^{114}Cd, and ^{116}Cd [2.104]. Due to mode coupling and

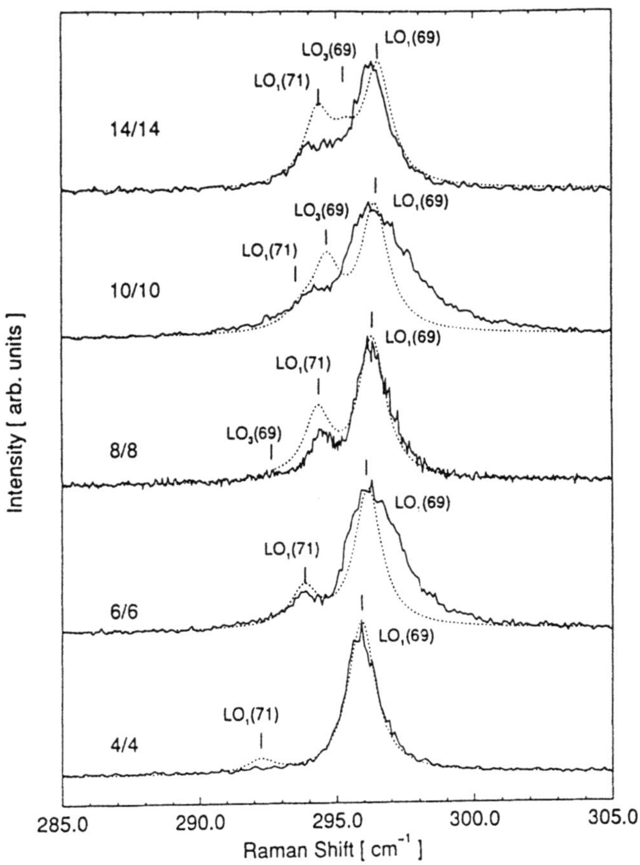

Fig. 2.31. Raman spectra of symmetric $(^{69}\mathrm{GaAs})_n/(^{71}\mathrm{GaAs})_n$ isotopic SLs for $n = 4$ to 14. The *solid lines* are spectra measured at 10 K and λ_l = 5145 Å; the *dashed lines* were calculated with the planar bond-charge model and bond polarizabilities. Confined phonons in ^{69}GaAs (^{71}GaAs) are labeled as $\mathrm{LO}_m(69)$ ($\mathrm{LO}_m(71)$)

mixed eigenvectors the isotope shifts of the two nonpolar E_2 modes deviate significantly from the reduced-mass behavior expected for optic zone-center modes [2.104].

A first result for LO phonons in [001]-oriented $(^{69}\mathrm{GaAs})_n/(^{71}\mathrm{GaAs})_n$ isotopic SLs grown by MBE is shown in Fig. 2.31 [2.57]. In this series of experimental (solid lines) and calculated (dashed lines) Raman spectra for $n = 4$ to 14 the strongest feature is the confined LO_1 phonon of ^{69}GaAs. A weaker structure whose intensity is strongly reduced with decreasing period corresponds to the LO_1 mode in ^{71}GaAs. Calculations of the bulk ^{69}GaAs and ^{71}GaAs dispersions with the planar bond-charge model show that the range of vanishing overlap is strongly reduced as compared to Ge [2.57]. The observation of the higher confined modes which are theoretically predicted is therefore more difficult. For $n = 10$, one expects a resonant coupling between $\mathrm{LO}_3(^{69}\mathrm{GaAs})$ and $\mathrm{LO}_1(^{71}\mathrm{GaAs})$ to occur. In this spectrum the $\mathrm{LO}_1(^{71}\mathrm{GaAs})$ phonon is weaker than in the measurements for $n = 8$ and 14 and it partially overlaps with a stronger contribution from $\mathrm{LO}_3(^{69}\mathrm{GaAs})$. This is in good agrement with the calculations. The broadening of the modes in some spectra (n = 6, 10) reflects a bad sample quality due to chemical impurities of the source material. This effect is not included in the calculations where all modes are uniformly broadened by Lorentzians with a width of 1.5 cm^{-1}. Isotopic SLs are also discussed in [2.75].

In some cases even the natural isotope composition of the elements in intrinsic semiconductors causes lines in the spectra which can be distinguished experimentally. Examples for such phenomena are Raman measurements on C_{60}, where the ^{13}C isotopes, or infrared experiments on α-sulphur (S_8), where the ^{34}S nuclei are responsible for well-defined vibrations [2.122, 2.123]. Isotope effects are also observed with great sensitivity in infrared absorption measurements on samples containing defects, such as, e. g., the $^{28}Si_{Ga}$ local vibrational mode found in GaAs δ-doped with Si [2.124] or lines due to different Ti isotopes and even residual Mg atoms in Ti-doped ZnTe [2.59, 2.125].

2.4 Acoustic Phonons in GaAs/AlAs Superlattices Grown along High-Index Directions

Most of the GaAs/AlAs SLs and QWs made by MBE are grown on substrates oriented along the [001]-direction. Progress in growth technology, however, has made it possible to obtain systems also for other directions. During the search for new material properties which might be beneficial for device applications the growth of high-quality epitaxial layers along different crystallographic directions was investigated and optimized. During this process it was found, e. g., that silicon on a GaAs (311)-surface can be implemented on Ga or As sites, depending on the growth conditions. It may thus act either as a donor or an acceptor [2.126]. This allows one to make modulation-doped samples or p–n-junctions using only one kind of dopant. Reports on epitaxy for [110]- [2.127,2.128], [012]- [2.129,2.130], [111]- [2.131,2.132], [310]- [2.132], [311]- [2.133] or [211]-oriented substrates [2.134] are found in the literature. An especially interesting property of the epitaxial growth along these unusual directions is the observed reordering of some surfaces to form periodic facet-like structures whose geometry is determined by the surface energy and not so much by the growth conditions [2.135–2.137]. Along the [311]-direction one finds a strong lateral corrugation which leads to the formation of quantum wires with rather small dimensions. Optical anisotropies, luminescence line sifts and anisotropic transport properties confirm the one-dimensional character of these systems [2.135–2.137].

Superlattices with growth directions other than [001] are also interesting for the investigation of lattice-dynamical properties. In a series of studies the confined optic and folded acoustic phonons were analyzed by Raman spectroscopy and compared to the bulk dispersions [1.18, 2.28, 2.129, 2.130, 2.138–2.143]. In the acoustic-phonon regime, [001]-oriented GaAs/AlAs SLs only allow one to measure the longitudinal-acoustic (LA) modes in backscattering geometry, while the transverse-acoustic (TA) vibrations are not Raman-active [1.11, 1.13]. The investigation of the transverse dispersion branches which are important in order to check theoretical predictions for the sound velocities and scattering intensities is thus only possible in samples grown along other directions. However, the TA phonons for the higher-symmetric directions [110] and [111] are also either not Raman-active ([110]) [2.130] or doubly degenerate ([111]) [2.28]. The TA modes for [110]-oriented samples were nevertheless observed in [2.130], a fact which was ascribed to deviations from the exact backscattering geometry, to reflection-induced forward scattering or to piezo-electric couplings. In [111]-oriented GaAs/AlAs SLs the doubly degenerate TA lines were only observed when better samples became available [2.144], while they were reported earlier in [111]-oriented $Si/Si_{0.5}Ge_{0.5}$ SLs [2.145, 2.146]. A complete analysis of all three dispersion branches and thus the largest possible amount of information about linear combinations of the elastic and elasto-optic constants is obtained from lower-symmetry growth directions such as [211] or [311]. Along these directions the LA and TA modes partially couple and mixed quasi-longitudinal (QL), quasi-transverse (QT), and pure transverse (T) modes are observed. In the following we present Raman investigations of acoustic phonons in [111]-, [211]- and [311]-oriented GaAs/AlAs SLs [1.18,2.144]. Before that, however, it seems appropriate to summarize the fundamentals of the acoustic phonon propagation in SLs and their Raman activities.

2.4.1 Dispersions and Raman Intensities

In the elastic continuum model the propagation of acoustic waves is described by the Christoffel equation [2.147]:

$$\begin{pmatrix} c_{11}l_x^2 + c_{44}(1-l_x^2) & (c_{12}+c_{44})l_xl_y & (c_{12}+c_{44})l_xl_z \\ (c_{12}+c_{44})l_xl_y & c_{11}l_y^2 + c_{44}(1-l_y^2) & (c_{12}+c_{44})l_yl_z \\ (c_{12}+c_{44})l_xl_z & (c_{12}+c_{44})l_yl_z & c_{11}l_z^2 + c_{44}(1-l_z^2) \end{pmatrix} e = \frac{\rho\omega^2}{q^2} e. \tag{2.24}$$

The unit vector $\boldsymbol{l} = (l_x, l_y, l_z)$ describes the propagation direction of a phonon with wavevector $\boldsymbol{q} = q\boldsymbol{l}$. In a cubic crystal the relevant material parameters are the three independent elastic constants c_{11}, c_{12}, c_{44}, and the density ρ. The Christoffel equation is an eigenvalue problem for the sound velocities ω/q and polarization vectors $\boldsymbol{e} = (e_x, e_y, e_z)$ of the acoustic modes. For the [001]-direction ($\boldsymbol{l} = (0,0,1)$) one obtains the sound velocites $\sqrt{c_{11}/\rho}$ for the LA mode and $\sqrt{c_{44}/\rho}$ for the doubly degenerate TA branches which were mentioned already in Sect. 2.1.

In the [311]-direction the eigenvalue problem reads:

$$\left| \begin{pmatrix} 9c_{11} + 2c_{44} & 3(c_{12}+c_{44}) & 3(c_{12}+c_{44}) \\ 3(c_{12}+c_{44}) & c_{11} + 10c_{44} & (c_{12}+c_{44}) \\ 3(c_{12}+c_{44}) & (c_{12}+c_{44}) & c_{11}+10c_{44} \end{pmatrix} - 11\frac{\rho\omega^2}{q^2} 1 \right| = 0. \tag{2.25}$$

After diagonalization one obtains the expressions for the sound velocities and mode polarization vectors given in Table 2.2 together with the results for other propagation directions. With these sound velocities for the two materials of a SL one obtains the folded phonon dispersion from (2.11). However, the couplings of modes which lead to the mixed character of the QL and QT branches are not yet considered. Internal band gaps, i. e., the removal of degeneracies at points of the mini-Brillouin zone where these modes interact with each other are obtained by solving the Christoffel equation with generalized boundary conditions in analogy to (2.10) [2.149–2.152]. Microscopic lattice-dynamical calculations yield further couplings of modes which are orthogonal in the elastic continuum model, e. g. for the [110]-direction [2.152].

The Raman scattering intensities for acoustic SL modes $\omega_{ph} = v \cdot q$ are obtained by transfering the Brillouin tensors of the bulk semiconductors [1.11, 1.13, 2.6]:

$$I = \frac{\epsilon_l^4 \omega_s^4 V \hbar \omega_{ph}}{2\epsilon_0^4 (4\pi)^2 c^4 \rho v^2} \left| \hat{e}_s^* \left(\tilde{\mathbf{p}} : \boldsymbol{q}\boldsymbol{e} \right) \hat{e}_l \right|^2 \left(n_{\omega_{ph}} + \frac{1}{2} \pm \frac{1}{2} \right) , \tag{2.26}$$

where $\tilde{\mathbf{p}} = p_{ijkl}(\omega)$ is the elasto-optic tensor of fourth rank which has to be contracted with the unit vector along the propagation direction $\boldsymbol{q}$ and the phonon polarization $\boldsymbol{e}$. The product $\boldsymbol{q}\boldsymbol{e}$ is proportional to the amplitude of the deformation wave caused by the phonon. The resulting second-rank tensor $\mathbf{R} = (\tilde{\mathbf{p}} : \boldsymbol{q}\boldsymbol{e})$ is treated with respect to the polarizations of incident ($\hat{e}_l$) and scattered ($\hat{e}_s$) photons like for the optic modes (see (2.6)). This photoelastic mechanism has its origin in the modulation of the dielectric function of the medium $\epsilon_{ij}(\omega)$ by the stress X_{kl} due to the phonon:

$$\Delta\epsilon_{ij}(\omega) = \pi_{ijkl}(\omega)\, X_{kl} \; . \tag{2.27}$$

Changes of $\epsilon_{ij}(\omega)$ are mediated by the piezo-optic tensor $\pi_{ijkl}(\omega)$ whose components are related to those of the elasto-optic tensor by the elastic constants c_{ijkl}. In the reduced-index notation the following relations hold for linear combinations of independent tensor elements in the cubic case [1.11]:

$$\begin{aligned} p_{11} - p_{12} &= (c_{11} - c_{12})(\pi_{11} - \pi_{12}) \\ p_{11} + 2p_{12} &= (c_{11} + 2c_{12})(\pi_{11} + 2\pi_{12}) \\ p_{44} &= c_{44}\pi_{44} \; . \end{aligned} \tag{2.28}$$

Table 2.3 summarizes the Brillouin tensors for the propagation directions $\hat{q}$ of Table 2.2. In Table 2.4 we give the selection rules and scattering intensities for different backscattering geometries. Only the factors proportional to linear combinations of the elasto-optic constants, as indicated by the Brillouin tensors of Table 2.3, were considered. An analysis of the nonvanishing scattering intensities in Table 2.4 shows that for the given directions modes from all three dispersion branches can only be observed in [111]-, [211]- or [311]-oriented samples.

Table 2.2. Polarization $\boldsymbol{e}$ and sound velocity (in 10^5 cm/s) of acoustic phonons for different propagation directions $\boldsymbol{q}$. The numbers given for GaAs and AlAs are at room temperature

$\boldsymbol{q}$	Mode	e	sound velocity v^2	v_{GaAs}	v_{AlAs}	Ref.
[001]	LA	[001]	c_{11}/ρ	4.73	5.65	[1.11]
	TA	[010]	c_{44}/ρ	3.35	3.96	
	TA	[100]	c_{44}/ρ	3.35	3.96	
[011]	LA	[011]	$(c_{11}+c_{12}+2c_{44})/2\rho$	5.24	6.26	[2.130]
	TA	$[0\bar{1}1]$	$(c_{11}-c_{12})/2\rho$	2.48	2.90	
	TA	[100]	c_{44}/ρ	3.35	3.96	
[012]	QL	[012]	$(c_{11}+c_{44}+D)/2\rho$	5.09	6.09	[2.139]
	QT	$[0\bar{2}1]$	$(c_{11}+c_{44}-D)/2\rho$	2.76	3.25	
	T	[100]	c_{44}/ρ	3.35	3.96	
			$D=\left(\frac{9}{25}(c_{11}-c_{44})^2+\frac{16}{25}(c_{12}+c_{44})^2\right)^{1/2}$			
[111]	LA	[111]	$(c_{11}+2c_{12}+4c_{44})/3\rho$	5.40	6.45	[2.28]
	TA	$[\bar{2}11]$	$(c_{11}-c_{12}+c_{44})/3\rho$	2.80	3.29	
	TA	$[0\bar{1}1]$	$(c_{11}-c_{12}+c_{44})/3\rho$	2.80	3.29	
[211]	QL	[211]	$(A+D)/12\rho$	5.27	6.30	[2.148]
	QT	$[\bar{1}11]$	$(A-D)/12\rho$	2.74	3.23	
	T	$[0\bar{1}1]$	$(c_{11}-c_{12}+4c_{44})/6\rho$	3.08	3.64	
			$A=5c_{11}+c_{12}+8c_{44}$			
			$D=\left((3c_{11}-c_{12}-4c_{44})^2+32(c_{12}+c_{44})^2\right)^{1/2}$			
[311]	QL	[311]	$(A+D)/22\rho$	5.10	6.10	[1.18]
	QT	$[\bar{2}33]$	$(A-D)/22\rho$	2.91	3.43	
	T	$[0\bar{1}1]$	$(c_{11}-c_{12}+9c_{44})/11\rho$	3.20	3.79	
			$A=10c_{11}+c_{12}+13c_{44}$			
			$D=\left((8c_{11}-c_{12}-9c_{44})^2+72(c_{12}+c_{44})^2\right)^{1/2}$			
GaAs:	$\rho=5.317\,\mathrm{g/cm^3}$,					[2.153]
	$c_{11}=119.0\,\mathrm{GPa}$, $c_{12}=53.8\,\mathrm{GPa}$, $c_{44}=59.5\,\mathrm{GPa}$					
AlAs:	$\rho=3.760\,\mathrm{g/cm^3}$,					[2.153]
	$c_{11}=120.2\,\mathrm{GPa}$, $c_{12}=57.0\,\mathrm{GPa}$, $c_{44}=58.9\,\mathrm{GPa}$					

Table 2.3. Brillouin tensors of acoustic modes, with respect to the crystal axes, for the different orientations of Table 2.2. Prefactors are given below each tensor

	Brillouin tensor		
[001]	LA [001]	TA [010]	TA [100]
	$\begin{pmatrix} p_{12} & 0 & 0 \\ 0 & p_{12} & 0 \\ 0 & 0 & p_{11} \end{pmatrix}$	$\begin{pmatrix} 0 & 0 & 0 \\ 0 & 0 & p_{44} \\ 0 & p_{44} & 0 \end{pmatrix}$	$\begin{pmatrix} 0 & 0 & p_{44} \\ 0 & 0 & 0 \\ p_{44} & 0 & 0 \end{pmatrix}$
[011]	LA [011]	TA [01$\bar{1}$]	TA [100]
	$\begin{pmatrix} 2p_{12} & 0 & 0 \\ 0 & p_+ & 2p_{44} \\ 0 & 2p_{44} & p_+ \end{pmatrix}$	$\begin{pmatrix} 0 & 0 & 0 \\ 0 & p_- & 0 \\ 0 & 0 & -p_- \end{pmatrix}$	$\begin{pmatrix} 0 & p_{44} & p_{44} \\ p_{44} & 0 & 0 \\ p_{44} & 0 & 0 \end{pmatrix}$
	$1/2$	$1/2$	$1/\sqrt{2}$
$p_+ = p_{11} + p_{12}$, $p_- = p_{11} - p_{12}$			
[012]	QL [012]	QT [0$\bar{2}$1]	TA [100]
	$\begin{pmatrix} 5p_{12} & 0 & 0 \\ 0 & p_+^{(1)} & 4p_{44} \\ 0 & 4p_{44} & p_+^{(2)} \end{pmatrix}$	$\begin{pmatrix} 0 & 0 & 0 \\ 0 & -2p_- & -3p_{44} \\ 0 & -3p_{44} & 2p_- \end{pmatrix}$	$\begin{pmatrix} 0 & p_{44} & 2p_{44} \\ p_{44} & 0 & 0 \\ 2p_{44} & 0 & 0 \end{pmatrix}$
	$1/5$	$1/5$	$1/\sqrt{5}$
$p_+^{(1)} = p_{11} + 4p_{12}$, $p_+^{(2)} = 4p_{11} + p_{12}$, $p_- = p_{11} - p_{12}$			
[111]	LA [111]	TA [$\bar{2}$11]	TA [0$\bar{1}$1]
	$\begin{pmatrix} p_+ & 2p_{44} & 2p_{44} \\ 2p_{44} & p_+ & 2p_{44} \\ 2p_{44} & 2p_{44} & p_+ \end{pmatrix}$	$\begin{pmatrix} -2p_- & -p_{44} & -p_{44} \\ -p_{44} & p_- & 2p_{44} \\ -p_{44} & 2p_{44} & p_- \end{pmatrix}$	$\begin{pmatrix} 0 & -p_{44} & p_{44} \\ -p_{44} & -p_- & 0 \\ p_{44} & 0 & p_- \end{pmatrix}$
	$1/3$	$1/\sqrt{18}$	$1/\sqrt{6}$
$p_+ = p_{11} + 2p_{12}$, $p_- = p_{11} - p_{12}$			
[211]	QL [211]	QT [$\bar{1}$11]	TA [0$\bar{1}$1]
	$\begin{pmatrix} 2p_+^{(1)} & 4p_{44} & 4p_{44} \\ 4p_{44} & p_+^{(2)} & 2p_{44} \\ 4p_{44} & 2p_{44} & p_+^{(2)} \end{pmatrix}$	$\begin{pmatrix} -2p_- & p_{44} & p_{44} \\ p_{44} & p_- & 2p_{44} \\ p_{44} & 2p_{44} & p_- \end{pmatrix}$	$\begin{pmatrix} 0 & -2p_{44} & 2p_{44} \\ -2p_{44} & -p_- & 0 \\ 2p_{44} & 0 & p_- \end{pmatrix}$
	$1/6$	$1/\sqrt{18}$	$1/\sqrt{12}$
$p_+^{(1)} = 2p_{11} + p_{12}$, $p_+^{(2)} = p_{11} + 5p_{12}$, $p_- = p_{11} - p_{12}$			
[311]	QL [311]	QT [$\bar{2}$33]	TA [0$\bar{1}$1]
	$\begin{pmatrix} p_+^{(1)} & 6p_{44} & 6p_{44} \\ 6p_{44} & p_+^{(2)} & 2p_{44} \\ 6p_{44} & 2p_{44} & p_+^{(2)} \end{pmatrix}$	$\begin{pmatrix} -6p_- & 7p_{44} & 7p_{44} \\ 7p_{44} & 3p_- & 6p_{44} \\ 7p_{44} & 6p_{44} & 3p_- \end{pmatrix}$	$\begin{pmatrix} 0 & -3p_{44} & 3p_{44} \\ -3p_{44} & -p_- & 0 \\ 3p_{44} & 0 & p_- \end{pmatrix}$
	$1/11$	$1/\sqrt{242}$	$1/\sqrt{22}$
$p_+^{(1)} = 9p_{11} + 2p_{12}$, $p_+^{(2)} = p_{11} + 10p_{12}$, $p_- = p_{11} - p_{12}$			

Table 2.4. Scattering intensities and selection rules of the acoustic phonons from Table 2.2 for backscattering in different polarization geometries

$\hat{q}$	$\hat{e}_l$	$\hat{e}_s$	Mode		
[001]			LA [001]	TA [010]	TA [100]
	[100]	[100]	p_{12}^2	0	0
	[100]	[010]	0	0	0
	[010]	[010]	p_{12}^2	0	0
[011]			LA [011]	TA [0$\bar{1}$1]	TA [100]
	[100]	[100]	p_{12}^2	0	0
	[100]	[0$\bar{1}$1]	0	0	0
	[0$\bar{1}$1]	[0$\bar{1}$1]	$A_+^2/4$	0	0
[012]			QL [012]	QT [0$\bar{2}$1]	TA [100]
	[100]	[100]	p_{12}^2	0	0
	[100]	[0$\bar{2}$1]	0	0	0
	[0$\bar{2}$1]	[0$\bar{2}$1]	$\frac{1}{625}(8p_{11}+17p_{12}-16p_{44})^2$	$\frac{36}{625}A_-^2$	0
[111]			LA [111]	TA [$\bar{2}$11]	TA [0$\bar{1}$1]
	[0$\bar{1}$1]	[0$\bar{1}$1]	$(p_{11}+2p_{12}-2p_{44})^2/9$	$A_-^2/18$	0
	[0$\bar{1}$1]	[$\bar{2}$11]	0	0	$A_-^2/18$
	[$\bar{2}$11]	[$\bar{2}$11]	$(p_{11}+2p_{12}-2p_{44})^2/9$	$A_-^2/18$	0
[211]			QL [211]	QT [$\bar{1}$11]	TA [0$\bar{1}$1]
	[0$\bar{1}$1]	[0$\bar{1}$1]	$(p_{11}+5p_{12}-2p_{44})^2/36$	$A_-^2/18$	0
	[0$\bar{1}$1]	[$\bar{1}$11]	0	0	$A_-^2/18$
	[$\bar{1}$11]	[$\bar{1}$11]	$(p_{11}+2p_{12}-2p_{44})^2/9$	0	0
[311]			QL [311]	QT [$\bar{2}$33]	TA [0$\bar{1}$1]
	[0$\bar{1}$1]	[0$\bar{1}$1]	$\frac{1}{121}(p_{11}+10p_{12}-2p_{44})^2$	$9A_-^2/242$	0
	[0$\bar{1}$1]	[$\bar{2}$33]	0	0	$9A_-^2/242$
	[$\bar{2}$33]	[$\bar{2}$33]	$\frac{1}{121^2}(27p_{11}+94p_{12}-54p_{44})^2$	$\frac{225}{2\cdot 121^2}A_-^2$	0
			$A_\pm = p_{11} \pm p_{12} - 2p_{44}$		

2.4.2 Experimental Results

Figure 2.32 shows Raman spectra of a (111)-GaAs/AlAs SL with layer thicknesses of (17/20) Å for different polarizations and a comparison of the measured lines with the backfolded phonon dispersions which were calculated from the values of Table 2.2 [2.144]. In contrast to previous measurements [2.28] sharp lines of folded TA phonons are observed in this sample. For parallel polarizations they correspond to the TA $[\bar{2}11]$, in crossed ones to the TA $[0\bar{1}1]$ modes. The intensities of the lines are almost equal for all three cases, in agreement with the selection rules of Table 2.4. The LA [111] modes are only observed for parallel polarizations.

Raman spectra of a (211)-GaAs/AlAs SL with (28/26) Å layer thicknesses are shown in Fig. 2.33 and compared to the calculated dispersions [2.144]. The QL modes appear in parallel polarizations, the QT phonons are observable only for parallel polarizations along $[0\bar{1}1]$, and the T lines are seen only in

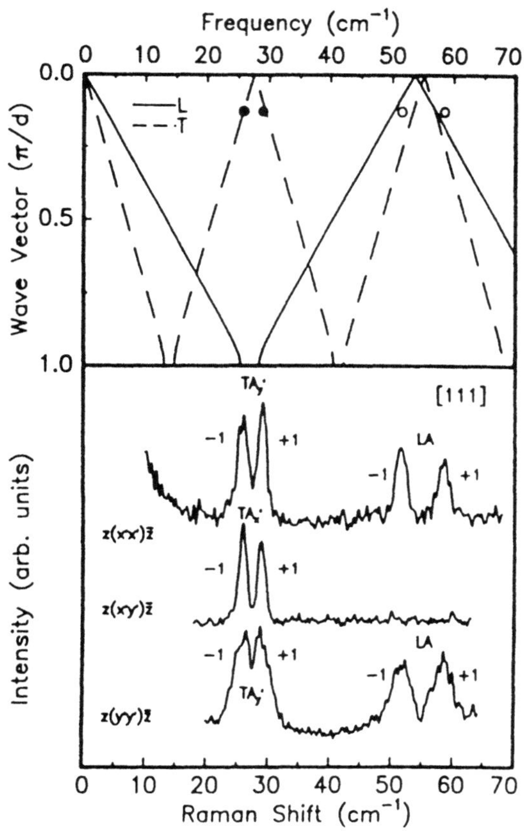

Fig. 2.32. Raman spectra of a [111]-oriented (17/20) Å GaAs/AlAs SL, T = 300 K and $\lambda_l = 4579$ Å, in comparison with the calculated dispersions. The degenerate TA $[\bar{2}11]$ and TA $[0\bar{1}1]$ modes as well as the LA [111] phonons are observed ($x' = [0\bar{1}1]$, $y' = [\bar{2}11]$)

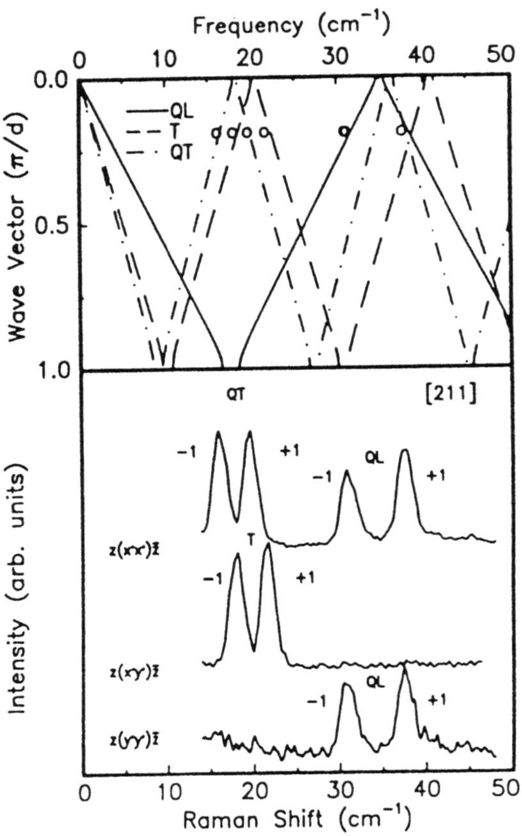

Fig. 2.33. Raman spectra of a [211]-oriented (28/26) Å GaAs/AlAs SL ($T = 300\,K$, $\lambda_l = 4579$ Å) in comparison with the calculated dispersions. The quasi-transverse QT $[\bar{1}11]$, quasi-longitudinal QL [211], and pure transverse TA $[0\bar{1}1]$ modes are observed ($x' = [0\bar{1}1]$, $y' = [\bar{1}11]$)

crossed polarizations. For this crystallographic direction the degeneracy of the transverse branches is removed. The QT and T phonons therefore appear at different frequencies with the T modes having the larger sound velocity and thus the larger Raman shift. According to the selection rules of Table 2.4 their intensities should be equal.

Figure 2.34 shows Raman spectra of a (24/22) Å GaAs/AlAs SL oriented along the [311]-direction [1.18]. Again, all three types of modes are observed according to the selection rules. The Raman shifts are in good agreement with the calculated dispersions, a fact which is further confirmed by measurements at three different excitation wavelengths. From the selection rules of Table 2.4 one expects the same scattering intensity for the QT modes in parallel polarizations along $[0\bar{1}1]$ (x') and the T phonons in crossed polarizations. While this is experimentally observed, a deviation is found when comparing these intensities with the QT signal for parallel polarizations along $[\bar{2}33]$ (y') which should be weaker by a factor of 4.8.

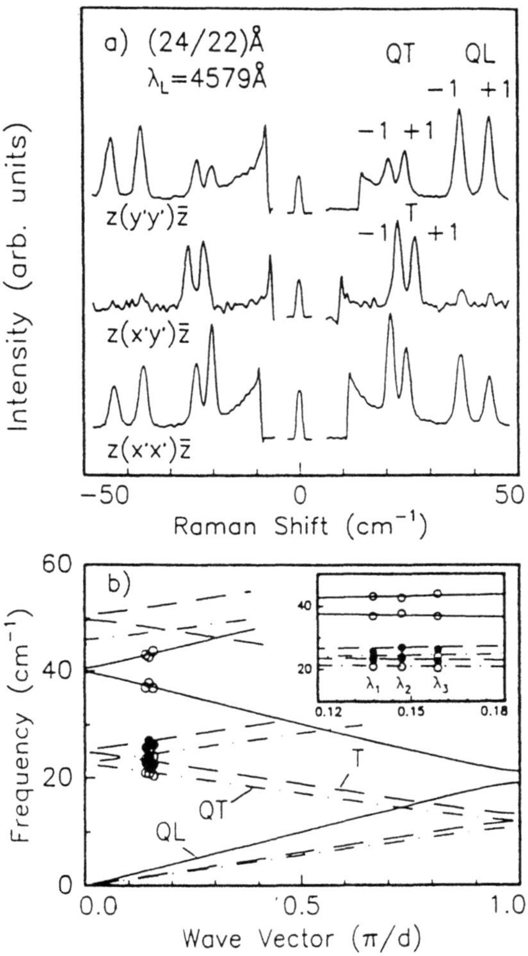

Fig. 2.34. (a) Raman spectra of a [311]-oriented (24/22) Å GaAs/AlAs SL. The quasi-longitudinal (QL) and quasi-transverse (QT) modes appear in parallel, the pure transverse (T) modes in crossed polarizations ($x' = [0\bar{1}1]$, $y' = [\bar{2}33]$). Positive (negative) Raman shifts correspond to Stokes (anti-Stokes) scattering. **(b)** Measured doublet frequencies (*dots*) compared to the calculated dispersion curves. The inset shows the dispersions of the modes for three excitation wavelengths: $\lambda_1 = 5145$ Å, $\lambda_2 = 4880$ Å, and $\lambda_3 = 4579$ Å

This observation leads us to a more detailed consideration of the scattering intensities. In the photoelastic model of Brillouin scattering, which can also be applied to folded acoustic phonons in SLs, the intensity I_m of modes from a backfolded dispersion branch m ($m = 0, \pm1, \pm2, ...$) is proportional to the squared magnitude of the corresponding Fourier component of the spatially-varying elasto-optic constant $p(z)$ [2.6]:

$$I_m \propto \omega_m(n_m + 1)\,|p_m|^2 \; . \qquad (2.29)$$

Here n_m is the Bose–Einstein factor and ω_m is the mode frequency. We neglect changes of the phonon amplitudes due to SL effects [1.13, 2.6]. Under the simplifying assumption of equal layer thicknesses and sound velocities in both materials the coefficients p_m are defined by

$$p(z) = \sum_m p_m \exp\left(i2\pi mz/d\right) . \tag{2.30}$$

This approximation holds for wavevectors away from the band gaps where the dispersion $\omega_m(q) = v_{SL}|q+2\pi m/d|$ is given by the average sound velocity

$$v_{SL} = d/\sqrt{\frac{d_A^2}{v_A^2} + \frac{d_B^2}{v_B^2} + (\kappa + \frac{1}{\kappa})\frac{d_A d_B}{v_A v_B}} \tag{2.31}$$

with $\kappa = \rho_B v_B/(\rho_A v_A)$ [2.6]. In the case of piecewise constant elasto-optic coefficients p_A and p_B in the respective layers one finds for the Fourier components of $p(z)$:

$$\begin{aligned} p_0 &= (d_A p_A + d_B p_B)/d \\ p_m &= \frac{i(p_B - p_A)}{2\pi m}(1 - \exp\left(-i2\pi m d_A/d\right)), \quad m \neq 0 . \end{aligned} \tag{2.32}$$

For Stokes processes this leads to the following ratio of the folded-phonon scattering intensities ($m \neq 0$) with respect to the Brillouin line ($m = 0$) [2.6]:

$$\frac{I_m}{I_0} = \left|\frac{p_B - p_A}{p_0}\right|^2 \frac{\sin^2(\pi m\, d_A/d)}{(\pi m)^2} \frac{\omega_m(n_m+1)}{\omega_0(n_0+1)} . \tag{2.33}$$

When comparing the average intensities $(I_m + I_{-m})/2$ of different modes with the same index m, which is possible in Raman measurements on [111]-, [211]- or [311]-oriented samples, the dependence of the scattering intensity on the SL sound velocity v_{SL} has to be taken into account (see (2.26)) [1.18]. In this case one obtains for $\hbar\omega_{ph} \ll kT$

$$\frac{I_m^{(i)}}{I_m^{(j)}} = \left|\frac{p_B^{(i)} - p_A^{(i)}}{p_B^{(j)} - p_A^{(j)}}\right|^2 \left|\frac{v_{SL}^{(j)}}{v_{SL}^{(i)}}\right|^2 = \left|\frac{p_B^{(i)} - p_A^{(i)}}{p_B^{(j)} - p_A^{(j)}}\right|^2 \left|\frac{\Delta\omega^{(j)}}{\Delta\omega^{(i)}}\right|^2 , \tag{2.34}$$

and the theoretical values of v_{SL} can be expressed by the experimentally accessible frequency splitting $\Delta\omega = \omega_m - \omega_{-m}$ between two modes with the same magnitude of m. In the photoelastic model no SL effects on the acoustic phonons are considered. Near dispersion gaps the mode mixing causes changes in the phonon amplitudes which lead to different intensities for the two lines of a doublet [1.13, 2.154–2.157]. These asymmetries strongly depend on the material parameters such as the layer thicknesses or the Al concentration in the barriers and can be used to study the interface properties [2.154]. A generalization of (2.29) and (2.32) is obtained when the impedance differences between the SL layers are neglected ($\rho_A v_A = \rho_B v_B$). In this case different sound velocities are permitted for each of the bulk constituents, and the scattering intensity I_m becomes proportional to [2.156]

$$I_m \propto \sin^2\left[\frac{d_A}{d}\left(\frac{\omega_{ph}d_B}{2}\left(\frac{1}{v_A}-\frac{1}{v_B}\right)-\pi m\right)\right]\times$$
$$\left|\frac{p_A/v_A}{\frac{\omega_{ph}d_B}{d}\left(\frac{1}{v_A}-\frac{1}{v_B}\right)-\frac{2\pi m}{d}}-\frac{p_B/v_B}{\frac{\omega_{ph}d_A}{d}\left(\frac{1}{v_B}-\frac{1}{v_A}\right)-\frac{2\pi m}{d}}\right|^2 . \qquad (2.35)$$

While the intensity of the Brillouin line ($m = 0$) in the photoelastic model depends on a weighted average of the elasto-optic constants in both materials, the scattering of folded phonons ($m \neq 0$) is determined by the difference between these parameters (see (2.32)). In higher-symmetric GaAs/$Al_xGa_{1-x}As$- and $Si/Si_{0.5}Ge_{0.5}$ SLs it was possible to determine the magnitude ratio of the elasto-optic constants p_{12} by measuring the relative scattering intensities $I_{\pm1}/I_0$ [2.155, 2.157] or I_1/I_{-1} [2.154] of LA phonons. For GaAs/AlAs SLs one finds values of $|p_{12}^{\mathrm{AlAs}}/p_{12}^{\mathrm{GaAs}}| = 0.1$ ($\lambda_l = 6764$ Å), 0.15 ($\lambda_l = 4880$ Å), and 0.2 ($\lambda_l = 4579$ Å) which exhibit significant dispersion [2.154, 2.155]. As can be seen from Table 2.4, the scattered intensity is determined by a single elasto-optic constant only in the most simple cases. For lower-symmetry directions it depends on linear combinations of all three parameters. Another complication arises from the fact that the elasto-optic constants have both, a real and an imaginary part, above the band gap where Raman spectra are usually excited in thin SLs. For GaAs the dispersion of the piezo-optic tensor components between 1.5 and 5.5 eV was determined by ellipsometry under uniaxial stress [2.158]. No direct measurements exist for AlAs. Calculations confirm the ratios of p_{12} found in Raman experiments [2.6, 2.159]. There is, however, no detailed information on the complex tensor elements or their dispersion. For these reasons we first analyze the intensity ratios of Fig. 2.34 under the assumption that the AlAs layers can be neglected. With the complex elasto-optic constants of GaAs for the excitation wavelength of 4579 Å from [2.158] ($p_{11} = 115.7 - 35.6\,\mathrm{i}$, $p_{12} = 40.7 + 199.1\,\mathrm{i}$, $p_{44} = 99.9 + 13.7\,\mathrm{i}$) the linear combinations of the coefficients for the allowed scattering geometries according to Table 2.4 can be calculated [2.160]. Using these numbers and the measured doublet splittings of $\Delta\omega_{QL} = 6.6\,\mathrm{cm}^{-1}$ and $\Delta\omega_{QT} = 3.7\,\mathrm{cm}^{-1}$ from Fig. 2.34 one finds with (2.34) intensity ratios of $I_{QT}/I_{QL} = 0.32$ (0.10) for the scattering geometries with parallel polarizations along $[0\bar{1}1]$ ($[\bar{2}33]$). The measured averages of these intensity ratios are 1.4 (0.4) and thus about four times as large as expected.

Figure 2.35 shows Raman spectra of a [311]-oriented GaAs/AlAs SL with layer thicknesses of (66/61) Å [1.18]. These spectra were measured at room temperature with $\lambda_l = 5145$ Å using a double monochromator with a focal length of 2.12 m. The quite good stray-light rejection and resolution of such an instrument allow one, in addition to the folded-phonon doublets QL [311], QT $[\bar{2}33]$ and T $[0\bar{1}1]$, mentioned already in Fig. 2.34, to observe the corresponding Brillouin lines ($m = 0$) as well. All Raman shifts are in good agreement with the calculated dispersion relations. The larger period of this sample also allows one to observe the ($m = \pm3$)-doublet of the QT branch. The QT

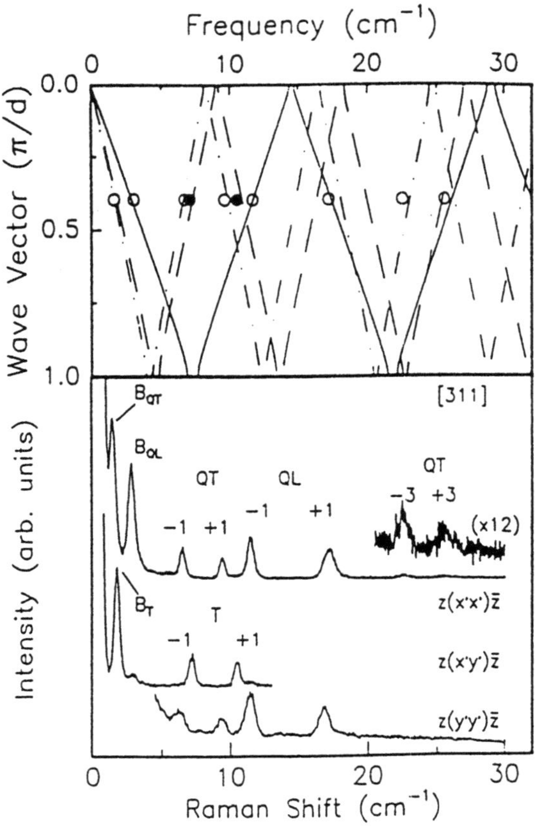

Fig. 2.35. Raman spectra of a (66/61) Å [311]-GaAs/AlAs SL and the comparison with the calculated dispersion relations. In addition to the folded acoustic modes QL, QT, and T, the Brillouin lines B_{QL}, B_{QT} and B_T are also observed ($x' = [0\bar{1}1]$, $y' = [\bar{2}33]$)

modes with $m = \pm 2$ do not appear due to their almost vanishing scattering intensity for equal layer thicknesses of GaAs and AlAs [1.11, 2.154]. In order to analyze the intensities we calculated the appropriate linear combinations of elasto-optic constants for the allowed scattering geometries according to Table 2.4 using the parameters from [2.158] for 5145 Å ($p_{11} = 67.2 - 28.7\,\mathrm{i}$, $p_{12} = 11.7 + 105.0\,\mathrm{i}$, $p_{44} = 78.0 - 5.4\,\mathrm{i}$) [2.160]. Together with the measured doublet splittings of $\Delta\omega_{QL} = 5.5\,\mathrm{cm}^{-1}$ and $\Delta\omega_{QT} = 3.0\,\mathrm{cm}^{-1}$ from Fig. 2.35 one obtains with (2.34) $I_{QT}/I_{QL} = 0.39$ (0.10) for parallel polarizations along $[0\bar{1}1]$ ($[\bar{2}33]$). The measured averages of these intensity ratios are 0.8 (0.35). Their disagreement with the calculation is comparable to that for the (24/22) Å sample. The observation of Brillouin lines allows one to determine the ratios $I_{\pm 1}/I_0$ according to (2.33) or (2.35). Using again only the elasto-optic constants of GaAs and taking into account a small correction due to the different occupation numbers one calculates from (2.35) intensity ratios of $I_{QL}/I_{B_{QL}} = 0.39$ (0.36), $I_{QT}/I_{B_{QT}} = 0.38$ (0.36), and $I_T/I_{B_T} = 0.38$ (0.33). The average values of the measured ratios for each doublet are given in brackets. Within the experimental accuracy of ± 0.05 they are in good agreement with the calculations. This result is expected since the dependence on the photoelastic constants in the expression for the intensity ratio cancels when only GaAs is taken into account. The relative scattering intensities are then determined by the structural parameters of the SL and the sound velocities.

In order to include the contribution of the barriers in the photoelastic model, it has been suggested that the unknown constants of AlAs should be approximated by those of GaP which has a similar electronic structure [1.18, 2.144]. With values for GaP determined from pressure-dependent measurements one obtains the following ratios $|p^{\mathrm{GaP}}/p^{\mathrm{GaAs}}|$ for the magnitudes of the elasto-optic constants at 4579 Å (5145 Å): p_{11}: 25% (19%), p_{12}: 14% (19%), p_{44}: 3% (7%) [2.161, 2.162]. The ratio for p_{12} is consistent with the results of Raman measurements in GaAs/AlAs SLs [2.154, 2.155] which lends some credibility to this ansatz. Considering these contributions in the calculation of the scattering intensities, however, does not significantly reduce the discrepancies in the I_{QT}/I_{QL} ratios [2.163].

The discrepancy between the calculated and measured I_{QT}/I_{QL} ratios for the two samples investigated indicates that additional effects may play a role. In [311]-oriented samples it was found that resonances with higher subbands and the vicinity of the E_1 band gap lead to interference effects in the scattering intensities of optic phonons [2.142, 2.143]. Changes of the electronic structure as compared to the bulk should also influence the elasto-optic constants and their dispersion. This may cause resonance effects which are not included in the simple calculations presented. Other effects like miniband formation and the indirect character of the lowest optical transitions, i. e., the X wave functions of the AlAs barriers, have to be taken into account in the short-period (24/22) Å sample. Calculations of the SL band structure under uniaxial stress could lead to a more precise determination of the elasto-optic constants and thus to a better understanding of the observed intensity ratios and their resonance behavior.

3. Continuous Emission of Acoustic Phonons

In addition to the sharp folded-phonon lines from crystal-momentum conserving scattering processes, the acoustic-phonon Raman spectra of MQWs and SLs also exhibit a continuous emission background where phonons from the whole vibrational density of states participate because crystal momentum is not conserved. In the following sections we discuss the basic features of this effect and demonstrate how it can be applied to determine material and fundamental parameters such as electronic linewidths, effective masses, and the electron–phonon interaction.

3.1 Phenomenology of the Continuous Emission

As has already been mentioned in Sect. 2.1 resonant Raman scattering is widely used in semiconductor physics. In many cases weak signals in absorbing materials can only be observed due to the enhancement of the scattering cross-section in the vicinity of electronic resonances. Beyond the measurement of phonon frequencies and other lattice-dynamical properties this resonance behavior allows one to investigate electronic phenomena, and Raman scattering can therefore be regarded as a complementary technique to methods like reflectivity, transmission or the different varieties of modulation spectroscopy [1.12].

Raman scattering is especially suitable for investigations of semiconductor SLs since it requires no modifications of the samples such as etching of the substrate material or the evaporation of contacts. The technique is rather material-specific and also allows one to perform measurements with high spatial resolution. The possibilities which resonant Raman scattering offers to materials research are highlighted, e. g., by measurements on single InAs/AlSb QWs. Excitation in the vicinity of the $E_1/E_1 + \Delta_1$ critical point of InAs allows one to observe an InSb-like interface mode which originates from an area with a width of less than four monolayers between the InAs and AlSb layers [3.1]. In Si/Ge SLs resonant micro-Raman measurements on the sample edge yield the components of the dielectric function perpendicular and parallel to the growth direction [3.2].

Resonant Raman scattering is also used to investigate the electronic structure of SLs or bulk semiconductors under the influence of external parameters. Under resonant excitation the Landau quantization of electrons and holes in a strong magnetic field causes pronounced magneto-Raman oscillations of the LO phonon intensity which can be used for a detailed analysis of the electronic structure, electron–phonon interaction effects and different scattering mechanisms [1.28–1.39]. Various examples of magneto-Raman spectroscopy as well as the basics of the technique are presented in Chap. 4. In GaAs/$Al_xGa_{1-x}As$ SLs under the influence of an electric field doubly and triply resonant Raman scattering by optic phonons at Stark-quantized levels was observed. This effect occurs whenever the separation of two energy levels participating in a scattering process just equals the phonon energy [3.3, 3.4]. The intensities of confined optic modes in GaAs/AlAs SLs are modified by an electric field due to changes of the electron and hole wave functions [3.5]. Uniaxial stress allows one to continuously shift the electronic states and to tune a system into resonance. In bulk GaAs and GaP doubly and triply resonant Raman scattering of optic phonons between heavy- and light-hole bands was realized this way [3.6–3.8]. In GaAs/$Al_xGa_{1-x}As$ SLs and QWs similar resonance effects were observed with the hole energies being tuned by the well width [3.9–3.12].

As can be seen from these examples, the resonance behavior of the optic phonons in SLs has been intensively studied during the past decade, and it is still an active field of research. However, much less was known until recently about the behavior of the acoustic modes in the vicinity of electronic resonances. This was due to the fact that a strong background signal, tentatively interpreted as luminescence, appeared in many experiments and made the observation of acoustic-phonon lines impossible. This interpretation, however, is dubious in view of the fact that the scattering intensities of optic phonons under conditions of strong, e. g., double resonance in QWs were sometimes larger than the recombination emission between electrons and holes and thus appeared superimposed on the luminescence background [3.9–3.11]. For the acoustic phonons one expects the same tendency since, due to their smaller energies, they nearly fulfill the double resonance condition (see (2.3)) even in processes where only two energy levels participate. Superimposed on the apparent luminescence, intensity anomalies were also observed and preliminarily attributed to Fano resonances of the acoustic modes with an electronic or acoustic continuum or to dispersion effects of the elasto-optic constants [2.148, 3.13]. These open questions led to detailed investigations on the resonance behavior of acoustic-phonon Raman scattering which are summarized in this chapter.

Figure 3.1 shows Raman spectra of a [311]-oriented (24/22) Å GaAs/AlAs SL in the acoustic-phonon regime which were measured under different resonance conditions [1.24]. The parameter Δ in Fig. 3.1 (a) gives the detuning of the laser excitation energy with respect to the lowest direct transition between electron and heavy-hole subbands which for this sample is located at 1.772 eV ($T = 300$ K). For excitation far away from resonance ($|\Delta| \gg 0$)

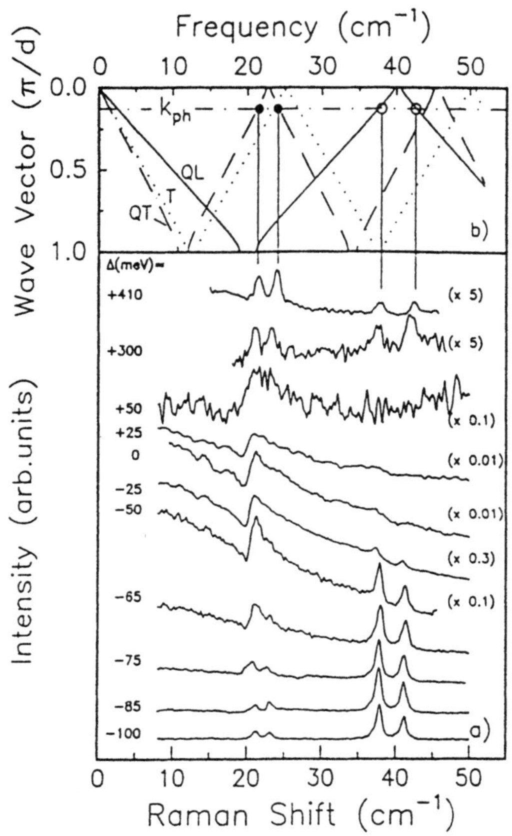

Fig. 3.1. **(a)** Acoustic-phonon Raman spectra of a [311]-oriented (24/22) Å GaAs/AlAs SL for different excitation energies shifted from exact resonance by the given amounts Δ ($T = 300\,\mathrm{K}$). Some of the spectra were scaled with the factors indicated. **(b)** The mode dispersion according to the continuum model

one observes only doublets of folded acoustic phonons whose Raman shifts, with the crystal-momentum transfer k_{ph} being determined by the photon energy, coincide well with the calculated dispersion of QL and QT modes shown in Fig. 3.1 (b). When approaching resonance ($\Delta \simeq 0$) a continuous background signal appears which has a maximum at small frequencies and disappears for larger Raman shifts. At first, this background is comparable to the folded-phonon signal. In resonance, however, it becomes much stronger and no doublets are observed any more. In the vicinity of the QT doublet one observes for small Δ an intensity anomaly which has the form of a pronounced anti-resonance. It seems to evolve out of the QT doublet which is observed away from resonance, hence the above-mentioned interpretation as a Fano line due to a coupling of the discrete doublet states with the continuum of acoustic modes. One also notices, however, that this anomaly and another one near the QL doublet are located at frequencies, where the QL dispersion has band gaps at the edge and center of the SL mini-Brillouin zone.

Another example for the observation of this so-called continuous emission is shown in Fig. 3.2 for a [001]-oriented GaAs/$Al_{0.35}Ga_{0.65}As$ MQW structure with layer thicknesses of (100/103) Å [1.19]. In this case Landau quantization

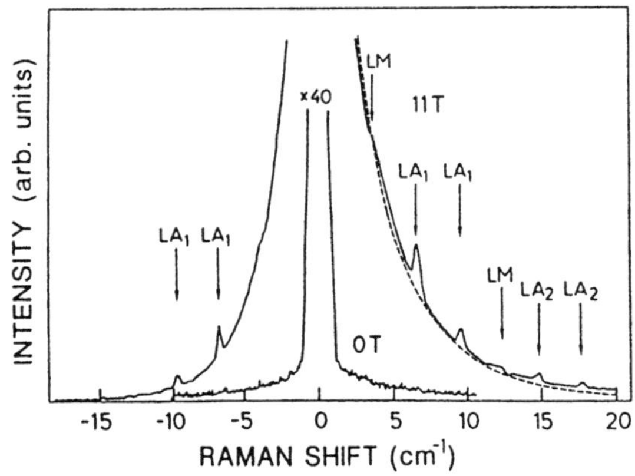

Fig. 3.2. Raman spectra of a (100/103) Å [001]-oriented GaAs/$Al_{0.35}Ga_{0.65}As$ MQW measured in resonance with an interband transition between Landau levels at 11 T and off resonance at 0 T. Note the enlargement of the 0 T spectrum by a factor of 40. In resonance one observes folded phonons (LA_n), continuous emission, and intensity anomalies (LM)

of the electronic states in a strong magnetic field is used to create resonances. The spectrum at 11 T was measured in backscattering parallel to the field (Faraday geometry) with equal circular polarizations (σ^+) for incident and scattered photons. The electronic states involved were the $n = 1$ Landau levels of the lowest electron and hole subbands ($\hbar\omega_l = 1.585\,\text{eV}$, $T = 6\,\text{K}$). In addition to the expected folded-acoustic-phonon doublets, labelled LA_n, one also observes continuous emission and superimposed intensity anomalies, denoted by LM. In comparison to the 0 T spectrum, which was measured under identical conditions, the enormous enhancement of the scattering signal in resonance is evident. In fact, the observation of spectral features under these conditions is only possible due to resonant excitation. In Fig. 3.2 both, Stokes (positive Raman shifts) and anti-Stokes spectra (negative Raman shifts) are shown. Note that the background signal does not, as one would expect from a luminescence tail, increase continuously towards lower absolute energies. Instead, the intensity is centered around the laser line. By varying the excitation energy and the magnetic field it has been shown that this signal is indeed due to inelastically scattered light since the spectrum shifts with the laser line. To ensure that the continuous emission signal does not accidentally arise from insufficient stray-light rejection of the spectrometer, Fig. 3.3 shows a series of magneto-Raman spectra (Stokes scattering) measured at different detection energies Δ relative to a constant excitation line ($\hbar\omega_l = 1.588\,\text{eV}$, $T = 6\,\text{K}$) [1.19]. These spectra are intensity profiles of the continuous emission vs. magnetic field at a given Raman shift with the spectrometer being used as a bandpass filter. Note that the detection energies of $\Delta = 4$, 16, 24 and $32\,\text{cm}^{-1}$ correspond to Raman shifts at which no folded-phonon doublets occur. All spectra show pronounced magneto-oscillations which reflect the Landau quantization of the electronic structure in the potential wells. Two types of resonances can be distinguished: The maxima, marked by vertical dashed lines, do not move with Δ. Therefore they correspond to *incoming resonances*, as known from optic-phonon measurements, which occur whenever the laser energy matches an interband transition between

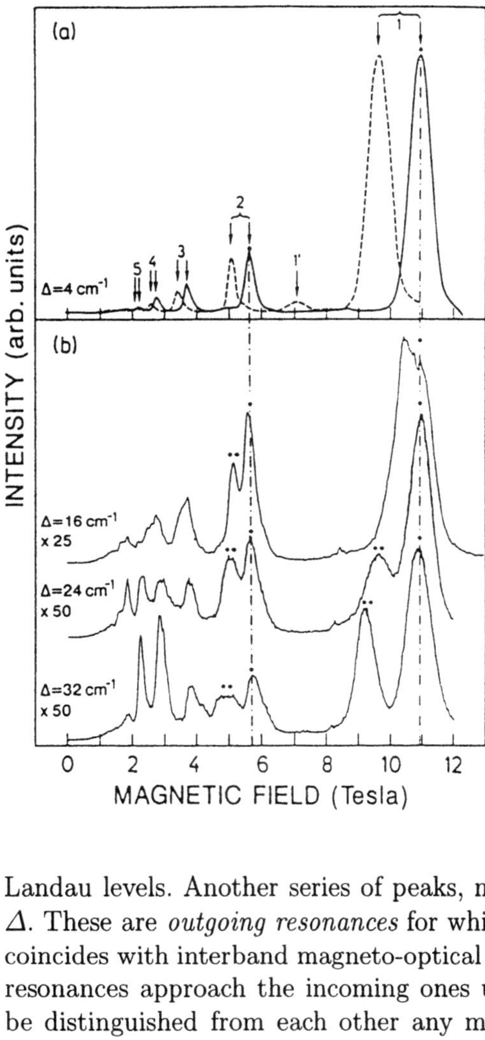

Fig. 3.3. Magneto-oscillations of the continuous emission for resonant laser excitation of a (100/103) Å GaAs/$Al_{0.35}Ga_{0.65}As$ MQW at interband transitions between Landau levels. Δ is the shift between laser and detection energy. **(a)** *solid* (*dashed*) line: $\bar{z}(\sigma^+,\sigma^+)z$ ($\bar{z}(\sigma^-,\sigma^-)z$) scattering geometry; **(b)** $\bar{z}(\sigma^+,\sigma^+)z$ configuration for different Δ

Landau levels. Another series of peaks, marked by double stars, shifts with Δ. These are *outgoing resonances* for which the energy of scattered photons coincides with interband magneto-optical transitions. For decreasing Δ these resonances approach the incoming ones until, at $\Delta = 4\,\mathrm{cm}^{-1}$, they cannot be distinguished from each other any more. In this case *double resonance* conditions are fulfilled to a very good approximation, a fact which results in a further enhancement of the signal (note the scaling factors with which the spectra in Fig. 3.3 were multiplied). For different circular polarizations, as illustrated for $\Delta = 4\,\mathrm{cm}^{-1}$ in Fig. 3.3 (a), characteristic shifts of the resonances appear which reflect the g factors of electrons and holes. In the scattering configurations with complementary polarizations ($\bar{z}(\sigma^+,\sigma^-)z$, $\bar{z}(\sigma^-,\sigma^+)z$) no folded-phonon doublets are observed and the continuous emission is reduced by a factor of about 100. The resonance behavior of the continuous emission and the folded-phonon doublets is compared in Fig. 3.4 [1.19]. To obtain these curves the LA phonon intensity at $9.7\,\mathrm{cm}^{-1}$ was determined from individual Raman spectra after background subtraction (1.585 eV, $T = 6\,\mathrm{K}$). The continuous emission signal was measured as an intensity profile for the same excitation energy at a shift of $4\,\mathrm{cm}^{-1}$. In both cases one obtains basically the same resonance behavior. The LA phonon resonance, however, is

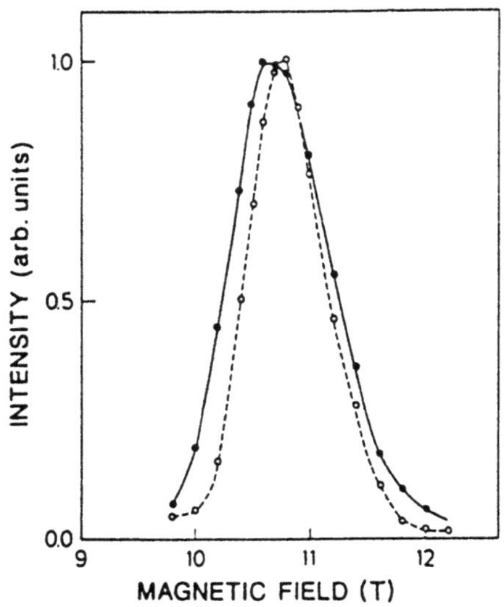

Fig. 3.4. Resonance profiles of the scattering intensities in the sample of Fig. 3.1 for the folded-phonon line at $9.7\,\text{cm}^{-1}$ (*solid line, filled circles*) and the continuous emission signal at a Raman shift of $4\,\text{cm}^{-1}$ (*dashed line, open circles*). Both curves were normalized to one

somewhat broader due to the larger value of Δ. Finally, it should be pointed out that this type of continuous emission only appears in QWs and has so far not been observed in bulk materials.

3.2 Single-Quantum-Well Effects

After the general description and introduction of the continuous emission in the previous section we now address the question of its origin. In Sect. 3.1 we showed that many properties of the continuous emission are analogous to those of Raman scattering by acoustic SL phonons. For both effects the same polarization selection rules and resonance behavior were observed. The continuous emission thus appears as a resonance phenomenon of light scattering in QWs, independently of whether these resonances are due to the size-quantization of the electronic states in the wells or an additional cyclotron motion of the carriers in a strong magnetic field (see Fig. 3.2). The question therefore arises whether the continuous emission can be interpreted as Raman scattering by acoustic phonons in SLs. Considering the law of energy conservation which governs a Raman process and the quasi-continuous character of the folded phonon dispersion it seems possible, at least in principle, that modes from a broad range of energies participate in the scattering. However, the crystal-momentum conservation law, which also has to be fulfilled, appears as a limiting factor. It is due to the coherent superposition of the scattering contributions of all QWs in a sample (or within the penetration depth of light, whichever is smaller). Consequently, modes with a well-defined crystal-momentum transfer appear as sharp lines in the spectrum. This situation is analogous to bulk crystals where crystal-momentum conservation in Raman processes arises from a coherent sum over the unit cells [1.11]. Scattering for a continuum of modes would thus be possible if the requirement for crystal-momentum conservation was weakened.

3.2.1 Continuous Emission in a Single Quantum Well

Theoretical Model. In order to test these ideas we consider a simple model for Raman scattering from acoustic phonons in a *single*, isolated QW for which crystal momentum is not conserved because of the missing SL periodicity [1.19, 3.14]. In a second step this model is generalized to *multiple* QWs, which are mostly used in experiments, and its applicability is discussed. Approximating the displacement field of an LA phonon propagating along the growth direction z of a (001) SL by a plane wave

$$u(z) \sim e^{iq_z z} , \tag{3.1}$$

as in a bulk crystal, the associated strain field is given by

$$\epsilon_{zz} = \frac{\partial u(z)}{\partial z} \sim iq_z e^{iq_z z} . \tag{3.2}$$

The changes of the electronic structure due to ϵ_{zz} constitute a mechanism for electron–phonon interaction which is described by a deformation potential D [1.10]. The perturbation operator for scattering of electrons by acoustic phonons is thus

$$H_{DP} = D_{ij}\,\epsilon_{ij} \sim D\, iq_z e^{iq_z z} . \tag{3.3}$$

For the confined electrons and holes in the QW we assume infinite barriers, i. e., no penetration into the adjacent layers, and describe them by simple sine or cosine wave functions. Perpendicular to the growth direction the states are approximated by plane waves. The envelopes of the wave functions along the growth direction are thus

$$\begin{aligned} \phi_N(z) &= \sqrt{\frac{2}{a}} \cos\frac{\pi N z}{a}, \quad N = 1, 3, 5... \\ \phi_N(z) &= \sqrt{\frac{2}{a}} \sin\frac{\pi N z}{a}, \quad N = 2, 4, 6... \end{aligned} \tag{3.4}$$

and extend over the well width a with $z \in [-a/2, a/2]$. For odd (even) subband indices N they have even (odd) parity with respect to the well center. With these wave functions the electron–phonon interaction matrix element, which enters into the expression for the Raman intensity (see (2.3)), can be calculated. One finds [1.19]

$$\langle \phi_N(z) | H_{DP} | \phi_N(z) \rangle \sim \frac{q_z}{\sqrt{\omega_{q_z}}} D \frac{\sin(aq_z/2)}{aq_z/2} \frac{4N^2}{4N^2 - (aq_z/\pi)^2} , \tag{3.5}$$

where ω_{q_z} is the frequency of a mode with wavevector q_z. A prefactor $1/\sqrt{\omega_{q_z}}$ which arises from the normalization of the phonon displacement $u(z)$ is also

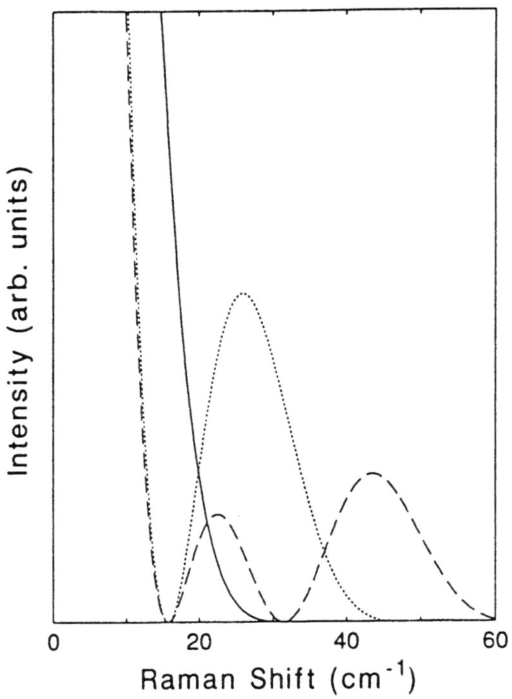

Fig. 3.5. Theoretical continuous emission spectra of a 100 Å GaAs QW for different subband indices N. The *solid* (*dotted, dashed*) *lines* are for $N = 1$ (2, 3)

included [1.19]. The phonon dispersion in this simple approximation is given by $\omega_{q_z} = v_{SL} \cdot q_z$ with an average SL sound velocity v_{SL}, which can be calculated from the elastic constants of the bulk constituents according to Table 2.2 and (2.31).

According to (3.5) scattering for a continuum of acoustic phonons is possible in an isolated QW. The scattering intensity for a mode ω_{q_z} depends on two factors. The oscillatory dependence of the electron–phonon matrix element on q_z, expressed by the term $\sin(aq_z/2)/(aq_z/2)$, causes the signal to vanish for wavevectors which are integer multiples of $2\pi/a$. The decay of this term for large q_z reflects the fact that crystal-momentum conservation in a QW is only relaxed to about π/a. For very wide wells one retrieves the ($q_z = 0$)-selection rule of the bulk. The term in (3.5) which contains the dependence on the subband index N has poles at $q_z = 2\pi N/a$. For $q_z \to 0$ it becomes one and does not influence the spectrum. Above the poles, however, this term rapidly vanishes and the continuous emission intensity is further suppressed. For each N the poles at $q_z = 2\pi N/a$ coincide with a zero of the $\sin(aq_z/2)/(aq_z/2)$-term and the expected divergence is canceled. For these q_z one therefore expects a nonvanishing scattering intensity and a smooth spectrum. Figure 3.5 shows the theoretically expected continuous emission intensity in a single 100 Å GaAs QW for different N. The calculated spectra are proportional to the squared magnitudes of the two terms discussed. They are characterized by a series of zeros at multiples of $2\pi/a \simeq 16\,\text{cm}^{-1}$ which are canceled, however, for $N = 1$ (2, 3) at Raman shifts of about $16\,\mathrm{cm}^{-1}$ ($32\,\mathrm{cm}^{-1}$, $48\,\mathrm{cm}^{-1}$). The continuous emission spectra for $\omega_{q_z} \to 0$ are, in addition to the squared magnitudes of (3.5) ($\sim q_z$), further influenced by the Bose–Einstein factor which takes the thermal population of the modes into account. For small energies it diverges as $kT/(\hbar\omega_{q_z}) \sim 1/q_z$. For the simple case of a one-dimensional linear phonon dispersion one thus obtains a finite, nonvanishing scattering intensity at $\omega_{q_z} = 0$. If one considers, however, the three-dimensional character of the dispersion, as it is necessary for the scattering of modes with wavevectors $q_{\|} \neq 0$ in the QW plane, the continuous emission intensity vanishes for $\omega_{q_z} \to 0$.

Experimental Verification. The dashed line in Fig. 3.2 is a continuous emission spectrum calculated from (3.5) which was fitted to the experiment by adjusting only a linear scaling factor. Otherwise experimental parameters for the resonance condition ($\Delta = 0$), the temperature ($T = 6\,\mathrm{K}$) and the linewidth of the electronic transition ($\Gamma \simeq 0.7\,\mathrm{meV}$) were used. Due to resonant excitation only subbands with $N = 1$ contribute to the scattering, and the spectrum does not show any oscillations. The continuous emission observed in this 40-period MQW sample is apparently well described by the simple single-QW model for acoustic-phonon Raman scattering.

Figure 3.6 shows magneto-oscillations of the continuous emission calculated with the single-QW model [1.19]. Landau quantization of the electronic structure was taken into account. Only optical transitions for which both, the Landau index n and the subband index N of electrons and holes, are conserved appear in the spectra. In these qualitative considerations we used a parabolic band structure with effective masses adapted to reproduce the experimental resonance positions. For comparison with the spectra in Fig. 3.3 we chose a Raman shift of $16\,\mathrm{cm}^{-1}$ between the excitation and the detec-

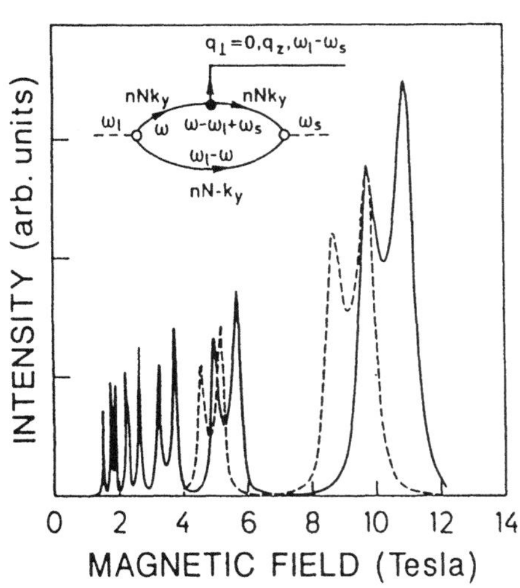

Fig. 3.6. Theoretical magneto-Raman spectra of the continuous emission for a single QW according to the experimental conditions of Fig. 3.2. The Raman shift between the excitation and the detection energy is $\Delta = 16\,\mathrm{cm}^{-1}$. The *solid* (*dashed*) *line* is for the $\bar{z}(\sigma^+,\sigma^+)z$ ($\bar{z}(\sigma^-,\sigma^-)z$) polarization. At each Landau level transition incoming and outgoing resonances occur. The inset shows a Feynman diagram of the process considered: the phonon is created by crystal-momentum non-conserving scattering of electrons

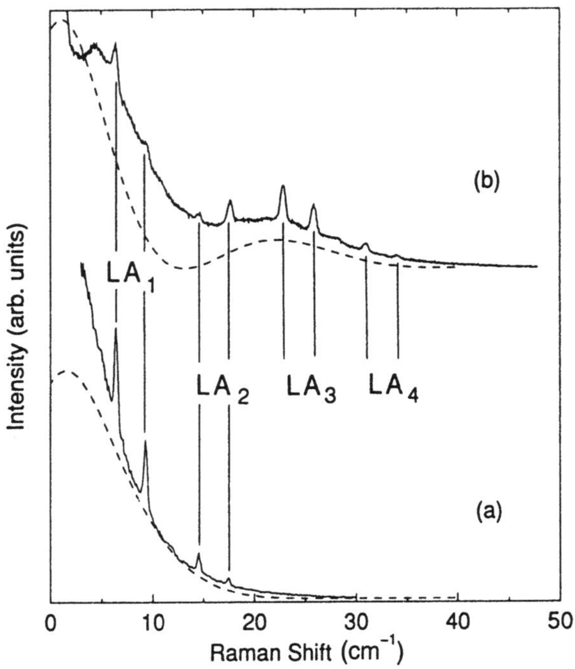

Fig. 3.7. Raman spectra of a [001]-oriented (100/100) Å GaAs/$Al_{0.34}Ga_{0.66}As$ MQW measured for resonant excitation at the ($n = 1$)-Landau level of the ($N = 1$)- **(a)** and ($N = 2$)-transitions **(b)** between heavy holes and electrons (*solid lines*). The *dashed lines* are calculated spectra. Note the analogous intensity variations of the folded-phonon doublets LA_n superimposed on the continuous emission

tion energy. At each Landau level transition the spectra show peaks due to incoming and outgoing resonances which are also found experimentally.

Another possibility to test the single-QW model for the continuous emission arises from the predicted dependence of the electron–phonon interaction matrix element on the subband index N (see (3.5)). As shown in Fig. 3.5 the continuous emission spectra should vary significantly with N. Resonant excitation at different interband transitions provides an experimental possibility to test these predictions. Figure 3.7 shows Raman spectra from a [001]-oriented (100/100) Å GaAs/$Al_{0.34}Ga_{0.66}As$ MQW for excitation at $N = 1$ (a) and $N = 2$ (b) [1.20,1.24]. In these experiments tuneable resonances were obtained by excitation at Landau levels in a magnetic field. With this approach the continuous emission can be investigated far away from the much stronger fundamental MQW luminescence. The measurements (solid lines) were performed at 12.4 T in resonance with the ($n = 1$)-Landau level of each interband transition using excitation energies of $\hbar\omega_l = 1.585\,\text{eV}$ ($N = 1$) and 1.682 eV ($N = 2$), respectively ($T = 10\,\text{K}$). The dashed spectra were calculated with the theory for single QWs. For $N = 1$ one observes a monotonic decrease of the continuous emission with increasing Raman shift while the ($N = 2$)-

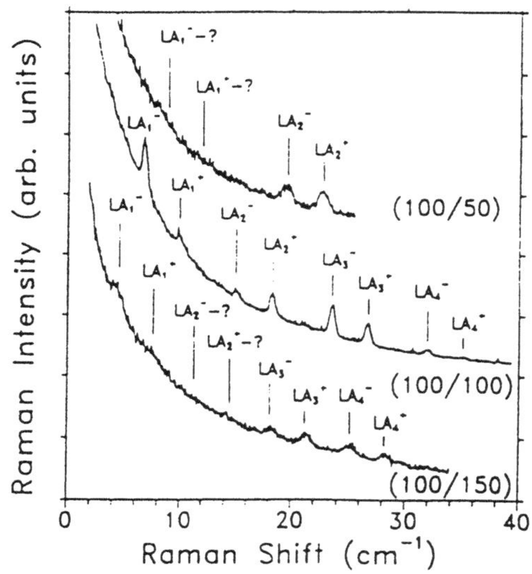

Fig. 3.8. Raman spectra of [001]-oriented GaAs/$Al_{0.34}Ga_{0.66}As$ MQWs with 100 Å wells and different barrier widths (50, 100, 150 Å) measured at the ($N = 2$)-resonance. Between 8 and 15 cm^{-1} the folded-phonon intensity is suppressed

spectrum shows a pronounced minimum around 15 cm^{-1} which is followed by another intensity maximum. Both calculated spectra are in good agreement with the experiment. Note that, in addition to the continuous emission, described by crystal-momentum *nonconserving* scattering from single QWs, crystal-momentum *conserving* scattering with characteristic doublet peaks of the SL dispersion is also observed. Even more important, the scattering intensity of these lines, denoted by LA_n in Fig. 3.7, varies in a similar way as the continuous emission. In the ($N = 1$)-spectrum it decreases monotonically for increasing frequencies. For excitation at $N = 2$ a minimum is observed for the higher-energy LA_1 and the lower-energy LA_2 mode, while the LA_3 lines have maximum intensity.

The dependence of the continuous emission minimum for $N = 2$ on the MQW parameters was further investigated in a series of samples with constant well widths and different barrier thicknesses. These systems have different folded phonon dispersions but the same electronic structure. Figure 3.8 shows Raman spectra measured without magnetic field in resonance with the ($N = 2$)-transition between heavy holes and electrons ($\hbar\omega_l = 1.666$ eV, $T = 6$ K, $\bar{z}(\sigma^-,\sigma^-)z$) [1.20]. In this case the strong background in each spectrum originates from resonant luminescence of the ($N = 2$)-transition. It makes it impossible to separate the underlying continuous emission. However, the folded-phonon doublets LA_n exhibit a uniform intensity modulation which is independent of the MQW period and can thus be attributed to the common electronic structure of the three samples. Between 8 and 15 cm^{-1} the folded-phonon intensities are strongly suppressed while a maximum occurs around 25 cm^{-1}. This behavior corresponds to that of the continuous emission spectra for $N = 2$ in Figs. 3.5 and 3.7.

3.2.2 Continuous Emission in Multiple Quantum Wells

The Influence of an Energy-Level Distribution. By comparison with the experiment the examples just discussed show that the model of acoustic-phonon scattering in a single, isolated QW correctly describes the basic properties of the continuous emission. It can therefore be considered as a crystal-momentum nonconserving Raman process. The question therefore arises how the *single*-QW effects, considered theoretically, are related to the *multiple*-QW systems, on which the experiments were performed. One would also like to understand what determines the simultaneous appearance of folded-phonon doublets and continuous emission in the spectra. As will be shown in the following, these problems are strongly connected with the resonance behavior of the scattering processes. It has already been mentioned that crystal-momentum conservation in SLs is due to the coherent superposition of the scattering contributions from the individual layers. The Raman intensity is thus proportional to a sum over all potential wells m in a sample or those within the penetration depth of the light, respectively:

$$I \sim \left| \sum_m R(m) e^{iq_z md} \right|^2 . \tag{3.6}$$

In this expression factors which are common to all wells, such as the electron–phonon interaction matrix element of (3.5), were neglected. $R(m)$ is the resonance contribution to the scattering (see (2.3)) for which we admit an explicit dependence on the layer index m:

$$R(m) = \frac{1}{(\Delta(m) + i\Gamma_{\mathrm{hom}})(\Delta(m) - \hbar\omega_l + \hbar\omega_s + i\Gamma_{\mathrm{hom}})} . \tag{3.7}$$

The homogeneous broadening of electronic resonances is described phenomenologically by the constant Γ_{hom}. $\Delta(m)$ is the detuning of the exciting laser energy $\hbar\omega_l$ from the resonance,

$$\Delta(m) = \hbar\omega_l - \hbar\omega_{\mathrm{gap}}(m, E_0, N, n, B) , \tag{3.8}$$

which, in principle, may also depend on m. The energies $\hbar\omega_{\mathrm{gap}}$ of critical points in the electronic structure are composed of the direct band gap E_0 as well as contributions due to the size quantization of electrons and holes in different subbands N and, in the case of an applied magnetic field B, the cyclotron energy in Landau level n.

If $\hbar\omega_{\mathrm{gap}}$ is equal for all QWs which contribute to the scattering, i. e., independent of the layer index m, $R(m)$ can be taken out of the sum in (3.6). For $m \to \infty$ the summation over the remaining exponentials leads to a delta function $\delta(q_z)$ and thus to crystal-momentum conservation. Zero wavevector transfer has been assumed here for the sake of simplicity. This limit describes pure SL effects, and the Raman spectrum consists only of folded-phonon doublets. Admitting, however, a fluctuation of the critical points around an average value via the dependence of $\hbar\omega_{gap}$ on m, i. e., assuming that these energies vary from well to well, the summation in (3.6) has to be performed with the explicit consideration of $R(m)$. For excitation around a distribution of energy levels (small $\Delta(m)$) the scattering contributions of those QWs which are exactly in resonance can become very strong, while others hardly participate at all. This effect is even more pronounced for the acoustic phonons as compared to the optic ones since their small frequencies cause both denominators in (3.6) to vanish almost simultaneously and double resonance occurs. In this limit single-QW effects dominate and the continuous emission is observed in the Raman spectrum.

Homogeneous and Inhomogeneous Linewidths, Resonance Effects. The interplay between the folded-phonon doublets and the continuous emission is basically influenced by three parameters. The *homogeneous linewidth* Γ_{hom} determines the strength of the scattering contributions from single QWs in resonance. As a measure of the fluctuations of $\hbar\omega_{gap}$ around a mean value we introduce the *inhomogeneous broadening* Γ_{inh}. If the energy-level distribution is caused by variations of the layer thickness from well to well and by interface roughness, as is usually the case in SLs, Γ_{inh} can be regarded as the width of a Gaussian distribution of critical points. Finally, the difference between the excitation energy and the electronic resonances enters the Raman cross-section via the *detuning parameter* $\Delta(m)$. For small inhomogeneous broadening compared to the homogeneous linewidth ($\Gamma_{inh} \ll \Gamma_{hom}$), i. e., in very good samples or for excitation far away from resonance ($\Delta \gg \Gamma_{inh}$), the fluctuations of the energy levels can be neglected. Coherence effects dominate and the folded-phonon doublets are observed. For strong inhomogeneous broadening compared to the homogeneous linewidth ($\Gamma_{inh} \gg \Gamma_{hom}$) or for excitation within the inhomogeneously broadened resonance ($\Delta \lesssim \Gamma_{inh}$) scattering from individual QWs is most pronounced and one obtains the continuous emission. These considerations explain the resonance behavior of the spectra in Fig. 3.1.

The appearance of the continuous emission in nonideal MQWs as compared to perfect samples is illustrated in Fig. 3.9 which shows Raman spectra calculated according to (3.6) for an eleven-period sequence of (100/100) Å GaAs/AlAs [3.15]. In the ideal case (solid line) we assumed that all QWs have identical critical points $\hbar\omega_{gap}$ with a broadening of $\Gamma_{hom} = 5\,\mathrm{cm}^{-1}$. The excitation is $10\,\mathrm{cm}^{-1}$ above this resonance. For simplicity the wavevector transfer onto the sample was taken to be zero. The sharp lines at multiples of about $8\,\mathrm{cm}^{-1}$ thus correspond to acoustic modes at the zone center of the SL mini-Brillouin zone. The finite number of periods in the calculation is reflected by the nonvanishing oscillating intensity between the main maxima. In $Si/Si_{1-x}Ge_x$ SLs with 15 periods such side peaks of the folded phonons

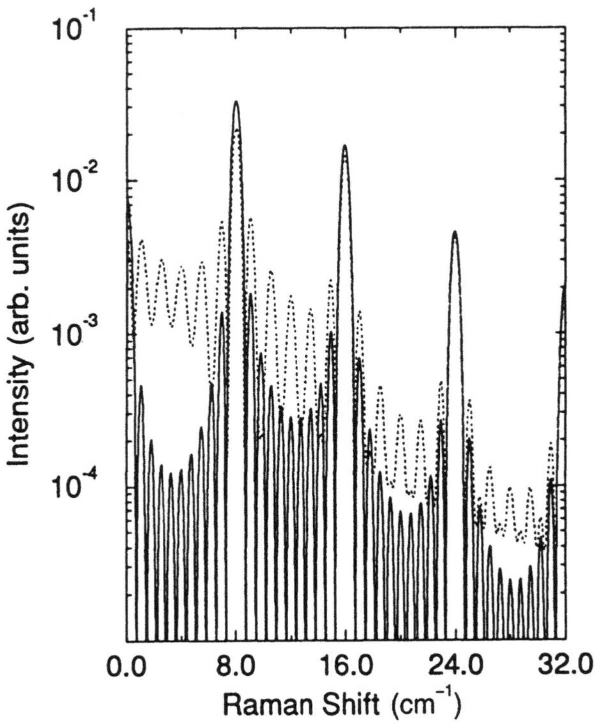

Fig. 3.9. Influence of disorder on the acoustic-phonon Raman spectrum in a sample with eleven periods of (100/100) Å GaAs/AlAs. The excitation energy is $10\,\mathrm{cm}^{-1}$ above an electronic resonance. *Solid line*: spectrum for identical energy levels in each well. *Dashed line*: spectrum assuming that the energy of the center well is exactly in resonance, i.e., $10\,\mathrm{cm}^{-1}$ above the others

have been observed in high-resolution Raman experiments [3.16, 3.17]. A coherence-induced enhancement of these side peaks and line splittings of the main doublet components occur in samples with a modified periodicity, such as mirror-plane SLs like $(\mathrm{SL})_{10}/(\mathrm{LS})_{10}$, where SL = $(\mathrm{GaAs})_{13}/(\mathrm{AlAS})_{18}$ and LS = $(\mathrm{AlAS})_{18}/(\mathrm{GaAs})_{13}$ [3.18]. The dashed spectrum in Fig. 3.9 is obtained if the transition energy of one QW in the center of the sample exceeds the other ones by $10\,\mathrm{cm}^{-1}$, e.g., due to a layer thickness fluctuation. It is thus exactly in resonance with the excitation. In the spectral range between the main (folded-phonon) peaks the scattering intensity is strongly enhanced due to the resonant excitation of this particular well, and the continuous emission appears. The difference between the envelopes for the two cases in Fig. 3.9 shows the frequency dependence given by (3.5).

Coherent and Incoherent Scattering. The example just given suggests that the continuous emission in MQWs with a disorder-induced distribution of energy levels arises from those wells for which the resonance conditions are exactly fulfilled. The intensity ratio of this signal to the crystal-momentum conserving scattering is determined by the parameters discussed above. Since, according to (3.6) – (3.8), such calculations require the explicit knowledge of the resonance conditions in each layer ($\Delta(m)$), which are generally not known, it appears useful to introduce an average over a distribution of energy levels, described by a Gaussian with half-width at half maximum (HWHM) Γ_{inh}. This was done in [3.19] and from (3.6) one obtains the following expressions for the continuous emission intensity I_B and the folded-phonon doublet signal I_F:

$$I_B \sim (\langle |R| \rangle^2 - |\langle R \rangle|^2)/(2\pi v/d) \tag{3.9a}$$

$$I_F \sim |\langle R \rangle|^2 \, \delta(\hbar\omega - \hbar\omega_{q_z}) \, . \tag{3.9b}$$

Here R is the resonance term of the Raman cross section given in (3.7). The brackets $\langle\rangle$ indicate the average of the included quantity over the inhomogeneous broadening. I_B is basically determined by the first term which is an *incoherent sum* over the scattering contributions from all layers. The average is performed with the absolute values of R before the result, which contains no phase information anymore, is squared. In the calculation of I_F the average is a *coherent sum*, i. e., the phase of the complex expression for R is taken into account before the squared magnitude is evaluated. While I_B appears at all Raman shifts, values of $I_F \neq 0$ are obtained only for crystal-momentum conservation, hence the δ-function. Analogous considerations for the coherent and incoherent superposition of scattering contributions are also relevant for neutron or X-ray diffraction [3.20]. In Sects. 3.5 and 3.6 the connection between the experimental quantities I_B and I_F and the broadening parameters Γ_{hom} and Γ_{inh}, which is constituted by these equations, is exploited in order to determine the natural lifetimes of electronic states and to investigate effects of the electron–phonon interaction in QWs and SLs. The measurement of these macroscopic quantities in Raman spectra thus allows one to make a connection to microscopic parameters.

3.3 Structures at Phonon Dispersion Gaps

The model of disorder-induced acoustic phonon Raman scattering from single QWs presented in the previous section describes the experimentally observed continuous emission. It has been pointed out already in Figs. 3.1 and 3.2 that characteristic *intensity anomalies* are superimposed on these "smooth" background spectra. These features will be discussed further in the following.

3.3.1 Types of Intensity Anomalies

Figure 3.10 compares the Raman spectrum of a [001]-oriented (45/45) Å GaAs/AlAs MQW with the backfolded SL dispersion [1.19]. The spectrum was measured at 11 T and 6 K with $\hbar\omega_l = 1.727\,\text{eV}$ in resonance with the ($n = 1$)-Landau level transition between the ($N = 1$)-heavy hole and electron subbands in $\bar{z}(\sigma^-, \sigma^-)z$ backscattering geometry. The folded-phonon

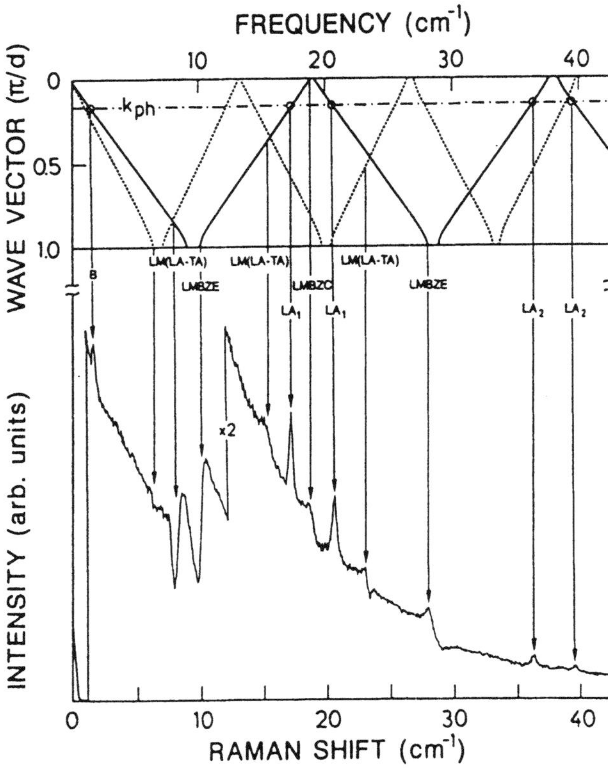

Fig. 3.10. Intensity anomalies and folded-phonon doublets in the Raman spectrum of a [001]-oriented (45/45) Å GaAs/AlAs MQW compared to the acoustic-mode dispersion. See text for experimental conditions and further details

doublets at the crystal-momentum transfer k_{ph} expected for an ideal SL are denoted by LA_{n}. The strong resonant enhancement of the scattering intensity allows one to use rather narrow spectrometer slits. Therefore the spectrum, which was measured with a conventional double monochromator of 85 cm focal length, also shows the Brillouin line (B) of the LA phonon dispersion branch. These "allowed" signals are in good agreement with the calculated dispersion shown in the upper part of the figure for the LA (solid lines) and TA branches (dotted lines). Superimposed on the continuous emission, which, according to Fig. 3.5 and (3.5), decreases monotonically ($N = 1$) for larger Raman shifts, are various intensity anomalies. These spectral features are connected to the folded phonon dispersion by vertical arrows. Two anomalies (LMBZE) occur at energies which correspond to band gaps of the LA dispersion at the SL mini-Brillouin zone edge ($q_z = \pi/d$). Another one (LMBZC) is located at a zone center gap ($q_z = 0$). Three structures (LM(LA-TA)) are observed at frequencies where the LA and TA branches cross each other inside the Brillouin zone. A weak step at about 6 cm^{-1} can be associated with a band gap of the TA dispersion at $q_z = \pi/d$.

3.3.2 Mini-Brillouin Zone Edge and Center

Theory. The location of many intensity anomalies in Fig. 3.10 at band gaps of the SL dispersion suggests that changes in the phonon displacement patterns at these points might be responsible for their appearance. The continuous emission spectra in the previous section were derived from bulk-like three-dimensional modes (plane waves), and their SL character was neglected (see (3.1)). It is well-known, however, that the intensities of crystal-momentum conserving folded-phonon lines exhibit characteristic changes in the vicinity of dispersion gaps [2.155, 2.156] which depend strongly on the sample parameters such as the relative layer thicknesses [1.13, 2.151, 2.154]. This suggests that the continuous emission is influenced by SL effects as well. In a first approach we extend the model used so far ((3.1) to (3.5)) by considering the solutions of the wave equation for the SL acoustic modes (2.9) and the phonon displacements $u(z)$. An ansatz of counter-propagating plane waves along the z-direction yields for the displacement of a mode ω_{q_z}

$$u(z) \sim A_+(q_z)e^{ik_z z} + A_-(q_z)e^{-ik_z z} , \tag{3.10}$$

where the wavevectors $k_z = \omega(q_z)/v$ in each material are given by the appropriate bulk sound velocities v. The amplitudes $A_\pm(q_z)$ are determined by the boundary conditions (2.1) for the transmission and reflection of waves at the SL interfaces. Taking this expression into account when calculating the strain field (3.2) and the matrix element for the electron–phonon interaction (3.5) one obtains, in addition to the squared magnitude of (3.5), a new term which determines the continuous emission intensity I:

$$I \sim |A_+(q_z) - A_-(q_z)|^2 . \tag{3.11}$$

The scattered signal thus depends on the amplitude difference of the two plane waves, and changes of $A_\pm(q_z)$ with the SL wavevector q_z have an influence on the continuous emission spectrum. The appearance of intensity anomalies can be easily understood qualitatively at the dispersion band gaps. At these points Bragg reflection of the modes leads to standing waves. The factors $A_\pm(q_z)$ are therefore equal in magnitude. They may, however, differ in their sign. The solution on one side of a gap therefore has even and the other one odd parity with respect to the well center. The strain field of these modes which enters in the electron–phonon coupling (see (3.5)) has the opposite parity, respectively. Scattering within subbands $\Phi_N(z)$ (intraband processes), which was assumed in the derivation of the matrix element in (3.5), only leads to nonvanishing values for operators H_{DP} with even parity. Thus the continuous emission disappears on that side of a gap where $u(z)$ is even while it has a finite nonzero value on the other side.

Experiment. Figure 3.11 shows the spectrum of Fig. 3.10 compared to the folded LA dispersion and the intensity factor of (3.11) [1.22]. One recognizes the behavior at the dispersion gaps just explained, and good agreement with the experiment is found. On the lower-energy side of the lowest band

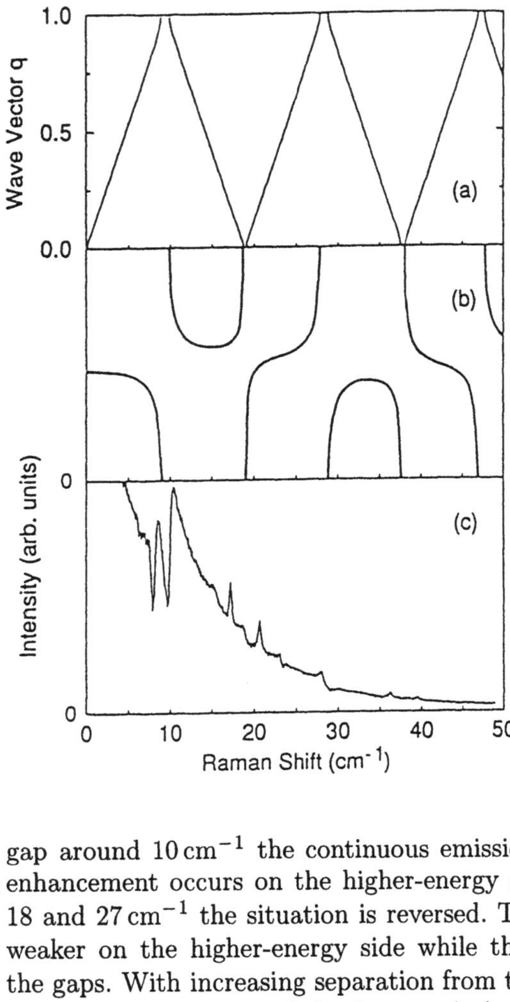

Fig. 3.11. LA phonon dispersion **(a)** and intensity factor from (3.11) **(b)** in comparison to the spectrum of Fig. 3.10 **(c)**. The measured anomalies at LA dispersion gaps have the shape expected from the theory

gap around $10\,\mathrm{cm}^{-1}$ the continuous emission signal is suppressed while an enhancement occurs on the higher-energy side. At the higher gaps around 18 and $27\,\mathrm{cm}^{-1}$ the situation is reversed. There the intensity anomalies are weaker on the higher-energy side while the intensities are stronger below the gaps. With increasing separation from the gaps the modes become more propagating, and one of the factors $A_{\pm}(q_z)$ can be neglected. In this case $|A_+(q_z) - A_-(q_z)|^2$ has a nearly constant average value and does not influence the continuous emission.

Dependence on Superlattice Parameters. The magnitudes of the SL dispersion gaps and the intensities of the doublets depend on the relative layer thicknesses of the constituents [1.13,2.151,2.154]. Defining the parameter $\alpha = d_{\mathrm{AlAs}}/d$ for GaAs/AlAs systems, where d is the SL period and d_{AlAs} is the AlAs layer thickness, one finds a maximum splitting of the lowest gap for $\alpha \simeq 0.55$, while it vanishes for $\alpha = 0$ and $\alpha = 1$. For the higher band gaps this splitting oscillates with α and has further zeros. At each of these zeros a change occurs in the side on which the Raman-active component of a folded-phonon doublet is located. Note that, due to symmetry, scattering at $q_z = 0$ or π/d is allowed only for one component of a folded-phonon doublet. This property of the crystal-momentum-conserving signal is also reflected in the intensity factor of the continuous emission (see (3.11)). Figure 3.12 shows

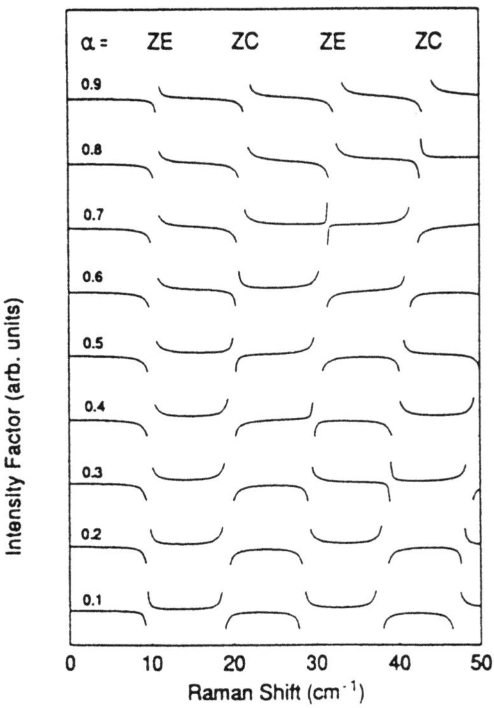

Fig. 3.12. Intensity factor according to (3.11) for a series of GaAs/AlAs SLs with a period of 30 monolayers. The parameter α indicates the relative layer thickness of AlAs with respect to the period. ZE and ZC denote LA dispersion gaps at the edge and the center of the Brillouin zone. Note the changing character of the intensity anomalies near zeroes of the band gaps

the dependence of $|A_+(q_z) - A_-(q_z)|^2$ on α for a series of GaAs/AlAs SLs with a constant period of 30 monolayers [1.22]. The first zone-edge band gap (ZE, $\approx 10\,\mathrm{cm}^{-1}$) has no zeroes. The character of the associated intensity anomalies is independent of α. In all spectra the intensity vanishes on the lower-energy side while it is enhanced on the higher-energy one. The symmetric and antisymmetric modes of the first zone-center gap (ZC, $\approx 20\,\mathrm{cm}^{-1}$) change their relative positions once at $\alpha = 0.55$. This causes a "flip" in the character of this intensity anomaly between $\alpha = 0.5$ and 0.6. The next higher dispersion gaps have two (ZE, $\approx 30\,\mathrm{cm}^{-1}$) and three (ZC, $\approx 40\,\mathrm{cm}^{-1}$) such changes, respectively, which correspond to zero gaps at $\alpha \simeq 0.4$ and 0.7 as well as $\alpha \simeq 0.3$, 0.55 and 0.8. Note that these oscillations depend only on α and not on the SL period itself [1.13, 2.154].

3.3.3 Internal Gaps

Theory. In addition to the intensity anomalies at the LA dispersion gaps Fig. 3.10 also shows features where the LA and TA branches cross each other. The agreement between the shape of the LA anomalies and the extended model, which takes SL effects on the phonon displacements into account, makes it probable that these structures can be treated in a similar way. One has to consider, however, that the LA and TA branches are orthogonal for the [001]-direction. Thus there are no gaps due to inter-mode Bragg reflection which could be responsible for intensity anomalies. Only for other than high-symmetry directions, e. g., for phonons with arbitrary crystal-momentum components $q_{\parallel} \neq 0$, one finds, as was discussed in Sect. 2.4, a mixing of the modes which then have quasi-longitudinal (QL), quasi-transverse (QT) or purely transverse (T) character. The dispersions of such mixed QL and QT phonons are split at internal band gaps. Their experimental observation in Fig. 3.10 thus indicates that phonons with $q_{\parallel} \neq 0$ participate in the continous emission. This is a fundamentally new aspect since we have considered so far only crystal-momentum nonconservation *along* the SL growth direction, i. e., for q_z. A single QW, however, has translational invariance in the plane. For a Raman backscattering geometry this requires crystal-momentum conservation and $q_{\parallel} \equiv 0$. A model for continuous emission which considers intensity anomalies at internal band gaps therefore cannot be described by first-order Raman scattering. It is thus necessary to consider higher-order processes which make the scattering by modes with $q_{\parallel} \neq 0$ possible.

However, before such processes are further investigated, it is necessary to determine the acoustic SL phonon dispersion for arbitrary wavevectors $\boldsymbol{q} = (q_{\parallel}, q_z)$. This is possible by solving the Christoffel equation with generalized boundary conditions for the displacement and strain fields [2.149, 2.150]. In analogy to (3.10) the displacement field $\boldsymbol{u}_{nqi}(\boldsymbol{r})$ of a mode i with wavevector $\boldsymbol{q}$ in the SL unit cell n made from the materials a and b is given by

$$
\begin{aligned}
\boldsymbol{u}_{nqi}(\boldsymbol{r}) = C_{iq}\, & e^{(iq_z^i nd + i\boldsymbol{q}_{\parallel}\boldsymbol{r}_{\parallel})} \times \\
& \times \left[\sum_{j=1}^{6} \hat{e}_{qj}^{ai} A_j^i(\boldsymbol{q}) e^{ik_{zj}^{ai}(\boldsymbol{q})z}\, \Theta(\frac{a}{2}+z)\Theta(\frac{a}{2}-z) + \right. \\
& \left. + \sum_{j=1}^{6} \hat{e}_{qj}^{bi} B_j^i(\boldsymbol{q}) e^{ik_{zj}^{bi}(\boldsymbol{q})(z-\frac{d}{2})}\, \Theta(z-\frac{a}{2})\Theta(b+\frac{a}{2}-z) \right] . \qquad (3.12)
\end{aligned}
$$

Here $C_{iq} = L^{-3/2}\left(\hbar/(2\rho\omega_{qi})\right)^{1/2}$ is a normalization constant with the volume L^3. The step functions $\Theta(z)$ limit the range of validity of the two terms in brackets to the different material layers. The SL period is $d = a + b$. $A_j^i(\boldsymbol{q})$ and $B_j^i(\boldsymbol{q})$ are the amplitudes of six plane waves j counterpropagating in each layer (two waves per bulk dispersion branch); $\hat{e}_{\boldsymbol{q}j}^{ai,bi}$ are their polarization vector components. The values of $k_{zj}^{ai,bi}(\boldsymbol{q})$ and the polarization vectors are related to the bulk dispersions and can be determined by solving the Christoffel equation (see (2.24)). For this purpose it is practical to chose a fixed mode frequency ω and wavevector component $\boldsymbol{q}_{\parallel}$ since these quantities, just as $\boldsymbol{k}_{\parallel}$ in the layers, are conserved at each interface. Applying the boundary and periodicity conditions one obtains six solutions for each frequency which correspond to the SL phonons at $\pm q_z$. For propagation along the [001]-direction these modes are orthogonal. There are no internal gaps at the crossing points of their dispersions. However, for $q_{\parallel} \neq 0$ the modes

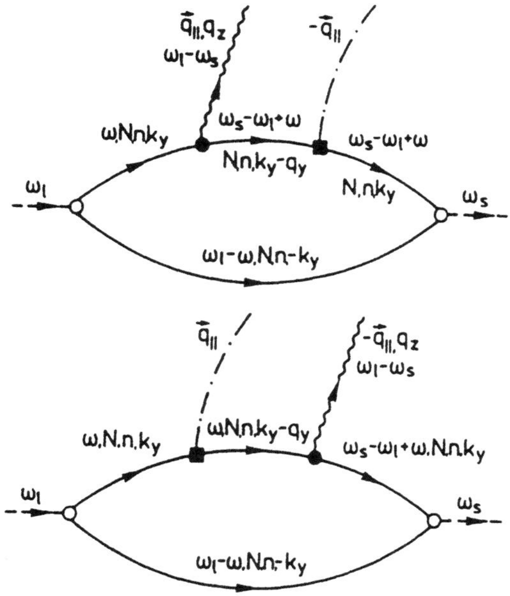

Fig. 3.13. Feynman-diagrams of second-order Raman processes (fourth-order perturbation theory) used for the description of the continuous emission at internal band gaps. ω_l and ω_s denote the frequencies of absorbed and emitted photons; *wavy lines* and *round vertices* label the emission of acoustic phonons; *dashed-dotted lines* with *square vertices* are steps of elastic scattering at potential fluctuations in the QW plane

couple and inter-mode Bragg reflection occurs. Band gaps appear at points of the Brillouin zone which are connected by integer multiples of reciprocal lattice vectors [2.149, 2.150]. From the displacement $\boldsymbol{u}_{nqi}(\boldsymbol{r})$ one obtains the strain field ϵ for each mode from which the continuous emission intensity can be calculated.

As has already been mentioned, the crystal-momentum component $q_{\|}$ is conserved for backscattering along the growth direction even in a single QW. It is therefore necessary to consider higher-order Raman processes and suitable intermediate steps which make it possible that phonons with $q_{\|} \neq 0$ participate. As a starting point we consider disorder-induced Raman processes in bulk crystals where elastic scattering of electronic intermediate states at defects causes large changes of the wavevector. This leads to a relaxation of the crystal-momentum conservation law for one-phonon scattering [3.21]. Analogously, potential fluctuations in the planes of QWs may act as scattering centers and transfer states created by photon absorption at $q_{\|} = 0$ into others with $q_{\|} \neq 0$. Such elastic scattering processes only change the crystal momentum of the excitation, its energy is conserved. Recombination of the scattered states, however, is only possible if they can get rid of the acquired crystal momentum. This is possible, e. g., by interaction with acoustic phonons with wavevector $\boldsymbol{q} = (-\boldsymbol{q}_{\|}, q_z)$. After this step the crystal-momentum component in the plane is again zero, while its conservation along the growth direction (single QWs) is not required. Such scattering processes can be described in fourth-order perturbation theory. Typical diagrams which are used for the calculation of continuous emission spectra are shown in Fig. 3.13 [1.22]. The emission of acoustic modes is labeled by wavy lines and round vertices, while the elastic scattering at potential fluctuations is given by dashed-dotted lines and square vertices. In these processes we assume that an electron–hole pair is created in a strong magnetic field in subband N and Landau state n and that recombination takes place from the same state. The description of these processes in one higher order of perturbation theory does not necessarily mean that their intensity can be neglected compared to one-phonon scattering, especially since we are dealing with acoustic phonons. Despite the additional factor from the scattering probability for the elastic step which should lead to a reduction of the signal, the Raman intensity is likely to be enhanced by the additional resonance denominator which arises form the higher order in perturbation theory. Due to the small mode frequencies one obtains almost a triple resonance which is generally stronger than the double-resonant emission of a phonon in a first-order process. The question of the relative importance of the two processes could be clarified by measurements of the absolute scattering intensities. The experimental observation of structures at internal dispersion gaps, however, can only be explained by higher-order processes with that presented here being the simplest one. Admitting elastic scattering at potential fluctuations whose specific features are addressed in the following section, an integration over $q_{\|}$ in the well plane has to be performed in the calculation of the continuous emission spectra.

Experiment. Figure 3.14 shows the spectrum of Fig. 3.10 in comparison to theoretical continuous emission spectra and phonon dispersions [1.22]. For a suitable choice of the integration interval over $q_{\|}$, which will be discussed further in the following section, one obtains a very good description of the measured spectrum (a) by the theory (b). Figure 3.14 (c) shows calculated dispersion curves (sample parameters as given above) for a constant value of $q_{\|} = 0.4\,\pi/d$. The QL, QT and T branches are shown by solid, long- and short-dashed lines. One readily recognizes the internal band gaps where the QL and QT modes interact. A series of spectra with different values of $q_{\|}$ is shown in Fig. 3.14 (d). For $q_{\|} = 0$, as mentioned above, the continuous emission spectrum has only intensity anomalies due to gaps of the LA dispersion. These are marked by the asterisks in Fig. 3.14 (b). For $q_{\|} \neq 0$ further anomalies occur at internal gaps (triangles in Fig. 3.14 (b)). According to changes of the dispersion they shift to larger frequencies for increasing $q_{\|}$ and become more pronounced. From this series of spectra one can easily imagine how the integration over $q_{\|}$ leads to the "broadened" spectra of Fig. 3.14 (b). Another consequence of this integration is that the dispersion considered is no longer one-dimensional but rather three-dimensional in character. This change is responsible for the vanishing continuous emission as $\omega_{q_z} \to 0$ which can be seen in spectrum (b). Experimentally this behavior is not observed here due to insufficient stray-light rejection at very small Raman shifts. It was, however, verified on other samples. Mixing with the QL branch also lends some Raman activity to the QT modes in the vicinity of internal gaps. This causes a weak

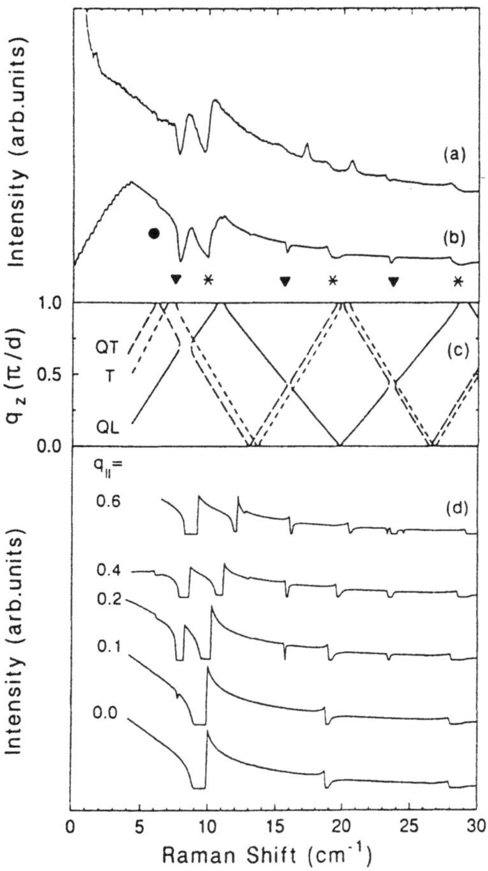

Fig. 3.14. **(a)** Experimental Raman spectrum from Fig. 3.10. **(b)** Theoretical continuous emission spectrum after integration over $q_{\parallel}$. The *stars, triangles* and the *filled circle* label the intensity anomalies at band gaps of the QL dispersion, at internal QL-QT gaps and at a QT gap. **(c)** Dispersion of the QL, QT and T modes for $q_{\parallel} = 0.4\pi/d$. **(d)** Continuous emission spectra for different $q_{\parallel}$ (in units of π/d)

shoulder near $6\,\mathrm{cm}^{-1}$ which is also observed in the experiment (filled circle in spectrum (b)).

These results show that the intensity anomalies of the continuous emission arise from the wavevector dependence of crystal-momentum nonconserving scattering processes. Not only the dispersion along the growth direction [001] but also modes with wavevectors $q_{\parallel} \neq 0$ are important. It should be added that intensity anomalies were observed already in earlier investigations of acoustic phonons in GaAs/$Al_xGa_{1-x}As$ systems [2.6, 3.22]. They were also reported in other materials such as Si/$Si_{0.5}Ge_{0.5}$ SLs [2.145,2.146,3.23]. However, the origin of these features remained elusive. They were attributed to anti-resonances of the acoustic phonons at the Brillouin zone edge [2.6], to disorder-induced q_z-non conserving scattering processes [2.145,3.22] or to local phonon modes inside the dispersion gaps [1.19]. Another recent model leads to intensity anomalies due to the scattering from modes with vanishing displacement near the SL surface for which the electron–phonon matrix element also vanishes [3.24]. This theory predicts intensity anomalies at dispersion band gaps, however, it does not explain their asymmetric character. In quasi-periodic SLs with the layers ordered, e. g., according to a Fibonacci sequence one also observes density-of-states effects under resonant excitation. The continuous emission and the superimposed intensity anomalies reflect gaps of the acoustic-phonon dispersion along the growth direction of these structures [3.25–3.27]. Reduced scattering intensities occur at frequencies which are given by powers of the "golden mean" $\tau = (1 + \sqrt{5})/2$ [3.26].

3.4 Interface Roughness

In Sect. 3.3 we explained the intensity anomalies at folded-phonon dispersion gaps with the dependence of the scattering intensity on the SL wavevector q_z. It was further shown that intensity anomalies at internal band gaps of [001]-oriented SLs can only be understood by scattering processes involving phonons with a crystal-momentum component $q_{\parallel} \neq 0$. In the following we discuss the microscopic origin for the elastic scattering of electronic intermediate states in the QW plane which occurs during such processes. This allows us to make a quantitative comparison between theory and experiment and to determine the lateral extent of the potential fluctuations due to interface roughness.

3.4.1 The Model

In analogy to the impurity-induced Raman scattering in bulk semiconductors [3.21] mentioned above, the interface roughness may cause elastic scattering of electronic intermediate states and crystal-momentum components $q_{\parallel} \neq 0$. Such perturbations of the periodicity in the plane, as well as deviations from the nominal layer thickness along the SL direction, occur during the epitaxial growth and one obtains samples with nonideal transitions from one material to the other. In order to model the impurity-induced scattering we assume that the potential fluctuations $\Phi(r_{\parallel})$ occur at otherwise ideal interfaces of a single QW in the form of randomly distributed islands with a certain lateral extent. We further assume that these perturbations do not interact, i. e., they are well separated from each other. The coupling strength of the electronic intermediate states in a Raman process and therefore the elastic scattering at the crystal-momentum transfer $q_{\parallel}$ is determined by the Fourier component $\chi(q_{\parallel})$ of $\Phi(r_{\parallel})$. Since $\Phi(r_{\parallel})$ is generally not known for a given sample one has to make suitable assumptions. On the other hand, the comparison with the experiment should allow one to estimate this otherwise hardly accessible quantity. Taking one of the simplest approximations we express $\chi(q_{\parallel})$ in the

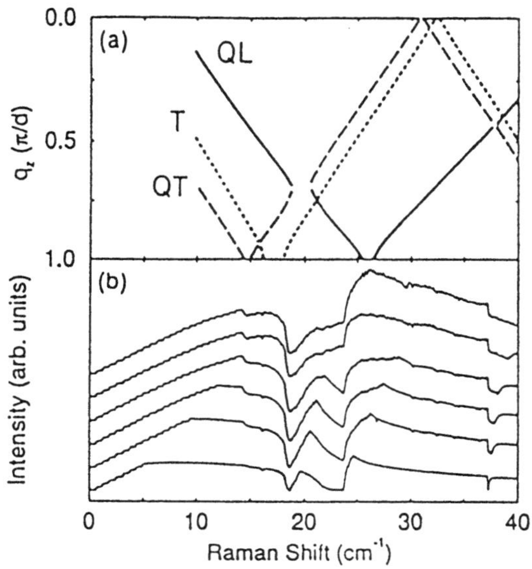

Fig. 3.15. (a) Dispersion of the folded acoustic phonons in a symmetric [001]-oriented GaAs/AlAs SL with 6.7 ml layer thickness for $q_{\|} = 0.4\pi/d$. The *solid*, *dashed*, and *dotted lines* are the quasi-longitudinal (QL), quasi-transverse (QT), and pure transverse (T) branches, respectively. **(b)** Continuous emission spectra for integration over $q_{\|}$ with different upper limits. From bottom to top: $0 \leq q_{\|} \leq 0.2$, 0.4, 0.5, 0.6, 0.8, $1.0\,\pi/d$

following by a step function with a constant value $\chi_0 \neq 0$ up to a certain upper limit $q_{\|,\max}$ and zero beyond. In real space this distribution for $\chi(q_{\|})$ corresponds to a $\sin(q_{\|,\max}r_{\|})/(q_{\|,\max}r_{\|})$-like function whose first zeros can be used to define the lateral extent λ of the potential fluctuations which is then given by $\lambda = 2\pi/q_{\|,\max}$.

Figure 3.15 shows theoretical continuous emission spectra of a symmetric GaAs/AlAs SL with a layer thickness of 6.7 ml (1 ml = 1 monolayer = 2.83 Å) and the dispersions of the different acoustic branches for $q_{\|} = 0.4\,\pi/d$ [1.21–1.24]. The spectra in Fig. 3.15 (b) were calculated by integrating the second-order scattering processes described in Sect. 3.3 (see Fig. 3.13) over $q_{\|}$ using the step function for $\chi(q_{\|})$ just mentioned. In order to illustrate the influence of the upper integration limit on the spectra different values were chosen for $q_{\|,\max}$. From bottom to top we show $0 \leq q_{\|} \leq 0.2$, 0.4, 0.5, 0.6, 0.8 and $1.0\,\pi/d$. The strength of the potential fluctuations χ_0 does not enter since we are not interested in the absolute scattering intensities. By comparison with the mode dispersion in Fig. 3.15 (a) one finds that the spectrum for the smallest $q_{\|,\max}$ is dominated by the intensity anomaly at the zone-edge QL gap which causes a broad minimum at about $23\,\mathrm{cm}^{-1}$. In addition to that there is a weaker structure at the internal gap between the QL and QT dispersions around $19\,^{-1}$. For larger values of the upper integration limit, as can be seen in the other spectra, the structure at the internal gap becomes stronger, and, in the region just above the QL anomaly, the spectrum rounds off as compared to the sharp edge found for the smallest value of $q_{\parallel,\max}$. At the same time another edge-shaped structure appears below the QL-QT anomaly at about $14\,\mathrm{cm}^{-1}$. It can be traced back to the QT branch which becomes Raman-active due to its coupling with the QL modes at the nearby internal gap. For the two upper spectra in Fig. 3.15 (b) the value of $q_{\parallel,\max}$ is so large that the QL and internal gaps partially overlap within the integration interval. This causes a single broad structure instead of the two separate minima found for smaller $q_{\parallel,\max}$. The extension of the previous one-dimensional model to include nonzero values of $q_{\parallel}$ leads to a quasi-three-dimensional phonon density of states being considered in these calculations. As a consequence the scattering intensity vanishes at small Raman shifts. Since the continuous emission is also suppressed at larger frequencies one obtains a characteristic maximum in its intensity which depends on the upper integration limit. With increasing $q_{\parallel,\max}$ it shifts towards larger frequencies. These model calculations show that, due to the strong sensitivity of the QL, QL-QT and QT anomalies as well as the position of the intensity maximum on $q_{\parallel,\max}$, the fits of the theoretical continuous emission spectra to the experiment should allow one to accurately determine this parameter and thus λ, the lateral extent of the growth islands.

3.4.2 Comparison to the Experiment

Figure 3.16 shows fitted continuous emission spectra (lower curves) in comparison with the experiment (upper curves) [1.21–1.24]. The general character of the continuous emission is highlighted by the fact that these spectra were measured under different conditions. Those of Fig. 3.16 (a) and (b) were excited at $T = 77\,\mathrm{K}$ and $10\,\mathrm{K}$, respectively, at zero magnetic field. The laser energy was in resonance with the lowest direct transition between the heavy-hole and electron subbands at the Γ point of the Brillouin zone ($\Gamma - \Gamma$ transition). In short-period type-II SLs another, spatially indirect, transition occurs at a lower energy between the electron states derived from the AlAs X point and the GaAs holes at Γ. Via the so-called Γ-X transfer photoexcited carriers rapidly reach this level and the strongest recombination occurs at this energy ($\Gamma - X$ transition) [2.46]. While the short-period indirect-gap SLs thus allow one to observe the continuous emission without the additional presence of a magnetic field, the spectrum of the wider sample in Fig. 3.16 (c) was measured at $11\,\mathrm{T}$ in resonance with the ($n = 1$)-Landau levels of the ($N = 1$)-electron and heavy-hole subbands. In samples with a direct band gap the much stronger luminescence at the $\Gamma - \Gamma$ transition allows one to observe the continuous emission only at higher-lying resonances which can be conveniently created and tuned by a magnetic field. It has been shown that the theoretical model of Sect. 3.4.1, which for computational reasons has been developed for Landau levels, can be extended to the zero-field case [1.22]. Differences in the electronic structure were found to influence the

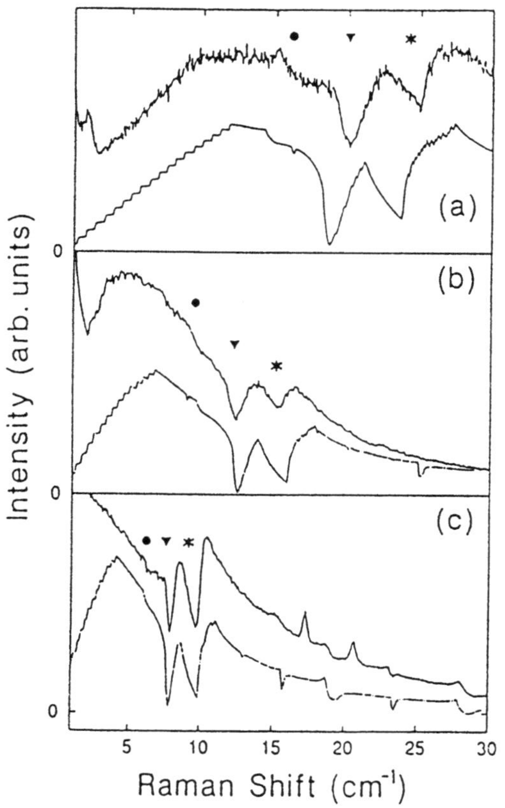

Fig. 3.16. Experimental (upper curves) and theoretical (lower curves) continuous emission spectra for three symmetric GaAs/AlAs SLs with 6.7 **(a)**, 10 **(b)** and 16 ml **(c)** layer thickness. The *circles*, *triangles*, and *asterisks* denote the intensity anomalies at band gaps of the QT, QL-QT, and QL dispersions

resonance behavior, however, the shape of the continuous emission spectra, which is of importance here, remains unaffected [1.22].

The calculated spectra in Fig. 3.16 were fitted to the experiment by varying $q_{\|,max.}$ such that best agreement was obtained for the above-mentioned spectral features. From the upper integration limits determined this way we find values for λ of about 150 Å (a), 280 Å (b) and 450 Å (c) with estimated errors of 20 %. However, before further conclusions concerning the interface roughness are made, one has to consider that the smallest length scale to which this method is sensitive is given by the exciton diameter. Electron–hole pairs act as intermediate electronic states in the Raman process and therefore also participate in the elastic scattering by potential fluctuations in the QW plane. The typical lateral exciton diameter in GaAs/AlAs QWs is about 250 Å [3.28]. The value of $\lambda = 150$ Å found for the sample with a layer thickness of 6.7 ml is below this limit and has to be discarded. In this case roughness might well exist on still smaller length scales, however, it would not have an influence on the intensity anomalies of the continuous emission spectrum. The values obtained for the two samples with larger layer thicknesses (Fig. 3.16 (b) and (c)) are above the sensitivity limit. They are therefore interpreted within the model as the lateral extensions of potential fluctuations such as those caused by interface roughness.

3.4.3 Other Methods

In order to discuss these results, obtained by Raman spectroscopy, and to evaluate their significance within the general context we now describe other investigations of interface properties. The determination of the interface quality in heterostructures and methods which allow one to make quantitative statements about their deviations from the ideal case are an active field of research, not least driven by the expectation of obtaining important information for the improvement of growth processes and device properties. In a simple picture the interface roughness is assumed to consist of island-like perturbations (growth islands) in the composition of otherwise ideally sharp transitions between different materials [3.28–3.31]. The quantities which characterize these defects are their lateral extent, their height in terms of atomic monolayers or potential fluctuations and the fraction to which they cover an interface. Most optical investigations of the interface roughness use methods of photoluminescence, partially with temporal or spatial resolution. Modern micropositioning technology allows one to record intensity profiles over large areas up to whole wafers [3.28, 3.32, 3.33]. Cathodoluminescence, excited by the strongly focused beam of an electron microscope, represents a complementary method with the advantage of still higher spatial resolution [3.28]. Other methods to determine the interface roughness include "chemical lattice imaging", a combination of transmission electron microscopy and pattern recognition [3.34], electron diffraction (RHEED) [3.35] and electronic Raman scattering of hole inter-subband excitations which allows one to determine precisely the height of the growth islands and roughness power spectra [3.36]. Scanning tunneling microscopy in combination with MBE yields an insight into the development of islands during their growth and therefore detailed information about individual interfaces [3.37].

From the different results obtained by optical techniques a comprehensive description of the semiconductor heterostructure interface roughness has been derived [3.38]. Since excitons are also involved as intermediate states in luminescene processes, their diameter, as in Raman experiments, is the relevant quantity to which the lateral potential fluctuations have to be compared to in order to estimate their influence on a spectrum. In samples with "rough interfaces" [3.38] the size of the growth islands is comparable with the exciton diameter. The excitonic recombination energy has a strong spatial dependence, and the luminescence spectrum is dominated by an inhomogeneously broadened line around that transition which corresponds to the average layer thickness of the potential well [3.29]. Spectra of samples with "large atomically flat areas" compared to the exciton diameter [3.38] exhibit different discrete lines at the transition energies which correspond to an integer number of monolayers. In samples with "pseudo-flat" or "micro-rough interfaces" [3.38] the growth islands are much smaller than the exciton diameter. The exciton energy is an average over many of these perturbations [3.30,3.31], and sharp luminescence lines are observed in the spectra. For micro-roughness superimposed on larger, flat areas on the scale of the exciton diameter one observes discrete luminescence lines which, however, cannot be attributed to transitions corresponding to an integer number of monolayers but are somewhat shifted. In order to decide whether the excitons created at a growth island also contribute to the luminescence with an energy which corresponds to that particular layer thickness one has to take into account their diffusion during the time until the recombination occurs [3.28]. Discrete transitions due to local potential fluctuations are observed when their extension is larger than the exciton diffusion length. It has been noted that this effect limits the spatial resolution of luminescence and cathodoluminescence experiments to those areas of an interface which are "flat", i. e., ideally sharp or, more likely, micro-rough over at least several micrometers [3.28].

The growth islands with diameters between 300 and 500 Å, as found by continuous emission, therefore do not appear in conventional optical experiments. As an example, the cathodoluminescence image of the sample with 16 ml GaAs and AlAs thickness in Fig. 3.16 (c) shows no steps which could be attributed to islands or large-scale terraces and one would thus conclude that it has excellent interface properties. Also, the luminescence spectra of all three samples in Fig. 3.16 show no satellites at other thicknesses but only a transition at the energy corresponding to the well width as determined by X-ray measurements. This observation also hints to large areas with ideally flat interfaces. Structures at the internal band gaps which are observed in the acoustic-phonon Raman spectra due to scattering processes with $q_{\|} \neq 0$, however, show that even in the best samples roughness exists on a smaller length scale which cannot be probed by ordinary luminescence. Further support for the existence of potential fluctuations with this lateral extent comes from RHEED measurements where the width of the diffraction peaks also reflects Fourier components of the interface roughness [3.35]. Recent spatially-resolved luminescence experiments on GaAs/$Al_xGa_{1-x}As$ single QWs with a very small focus of the excitation (diameter less than 2 μm) show line splittings and a fine structure of the inhomogeneously broadened luminescence which can be attributed to excitons localized at interface roughness [3.33]. The energy of these lines depends rather sensitively on the size of the growth islands, and one obtains similar values for λ as in continuous emission [3.33].

3.5 Homogeneous and Inhomogeneous Linewidths

In Sect. 3.2 we have shown that the continuous emission is due to the disorder-induced, crystal-momentum nonconserving Raman scattering of acoustic phonons in SLs. In contrast to crystal-momentum conserving, coherent processes which lead to sharp doublets of folded acoustic phonons in the spectrum the continuous emission arises from the incoherent superposition of the scattering contributions of individual layers. The signal intensities of these two phenomena, which are often observed simultaneously, reflect the homogeneous linewidth Γ_{hom}, a measure of the sharpness of electronic transitions, and the inhomogeneous broadening Γ_{inh} due to the disorder of a sample. In the following we determine these parameters from the resonances of the continuous emission and the crystal-momentum conserving scattering [1.23].

3.5.1 Model Calculations

To illustrate the expected behavior Fig. 3.17 shows calculated continuous emission spectra of a GaAs QW with 10 ml thickness for various resonance conditions at a single electronic transition without (a) and with (b) inhomogeneous broadening [1.23]. The separation of the excitation energy from the electronic resonance is given by Δ (see (3.8)). In contrast to the previous section, only the general shape of the continuous emission spectra is of interest

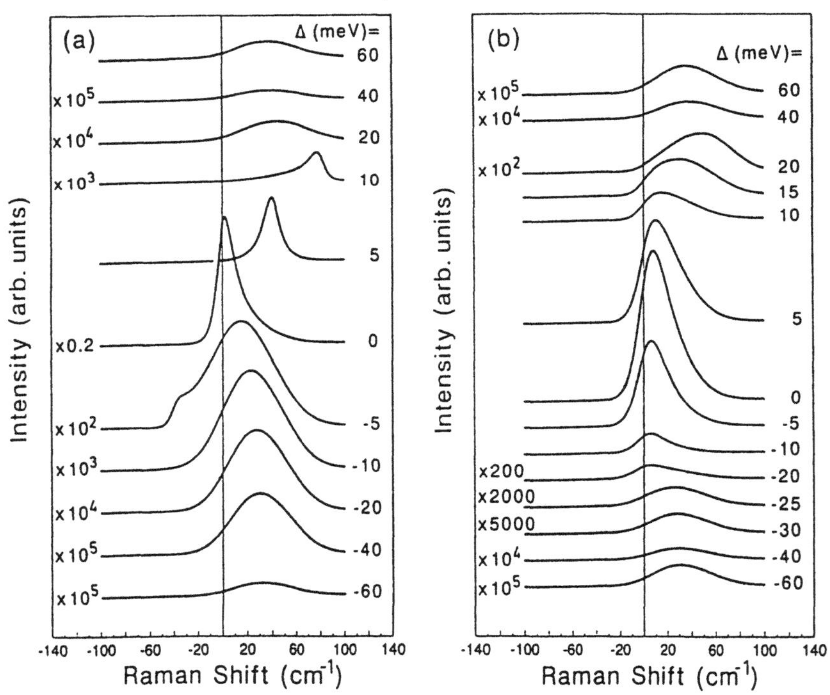

Fig. 3.17. Theoretical continuous emission spectra of a 10 ml GaAs QW for different detunings Δ of the excitation energy from resonance: **(a)** $\Gamma_{\text{hom}} = 1\,\text{meV}$, $\Gamma_{\text{inh}} = 0\,\text{meV}$, **(b)** $\Gamma_{\text{hom}} = 1\,\text{meV}$, $\Gamma_{\text{inh}} = 5\,\text{meV}$, (T = 10 K). Some spectra were scaled by the factors indicated. Positive (negative) Raman shifts correspond to Stokes (anti-Stokes) scattering

here. The calculations were thus performed for bulklike acoustic modes and no intensity anomalies are obtained.

For larger values of $|\Delta|$, far away from resonance, one finds in all cases the same spectrum which is characterized by the shift of its intensity maximum from the excitation energy and its width (FWHM). These parameters depend on the QW thickness and increase according to (3.5) if the wells become thinner. The maximum intensity of the spectra is also of interest for the analysis of the resonance behavior. Approaching resonance from either direction the shape of the spectrum does not change at first. However, characteristic modifications occur within a small range around $\Delta = 0$. For the single homogeneously broadened resonance with $\Gamma_{\mathrm{hom}} = 1\,\mathrm{meV}$ in Fig. 3.17 (a) one observes for decreasing $\Delta > 0$ a strong peak around $\hbar\omega = \Delta$ which shifts linearly with the excitation energy thereby gaining in intensity. The continuous emission becomes rather strong for these phonons since their energy matches the difference between the excitation and the electronic transition which leads to an outgoing resonance in the Raman process. For $\Delta = 0$ double resonance occurs for small phonon frequencies. This causes a further enhancement of the continuous emission. For negative Δ the outgoing resonance appears in the anti-Stokes region until, for larger values of $|\Delta|$, the nonresonant spectrum with its characteristic shape reappears. Note the large intensity differences which make the observation of the continuous emission away from the resonance practically impossible. The series of spectra in Fig. 3.17 (b) was calculated under the same conditions as in (a), however, an inhomogeneous broadening of $\Gamma_{\mathrm{inh}} = 5\,\mathrm{meV}$ was considered. While the shape of the spectra does not change appreciably away from the resonance, one observes for small $|\Delta|$ a decrease in their width and a shift of the intensity maximum towards smaller Raman frequencies.

This different behavior compared to the case of pure homogeneous broadening in Fig. 3.17 (a) arises from the interplay of two contributions to the signal and can be easily understood in the nonresonant case ($|\Delta| \gg \Gamma_{\mathrm{inh}}$). Within a Gaussian distribution of electronic transitions those in the vicinity of the laser line cause strong resonances even for large $|\Delta|$. Their contribution to the continuous emission intensity is proportional to $\Gamma_{\mathrm{hom}}^{-4} \exp(-\Delta^2/(2\Gamma_{\mathrm{inh}}^2))$. Despite the small resonance denominators in the expression for the Raman efficiency, which would lead to strong scattering ($\sim \Gamma_{\mathrm{hom}}^{-4}$), the signal is exceedingly weakened by the exponential factor due to the inhomogeneous broadening. Another possibility is the nonresonant excitation of transitions in the center of the inhomogeneously broadened Gaussian distribution. Due to the larger energy denominators the signal is weaker compared to resonant excitation. On the other hand, these contributions are not suppressed by the exponential factor. The scattering intensity in this case is proportional to Δ^{-4} and dominates the spectrum. Considering a MQW structure instead of a single layer the coherence of the scattering contributions is restored and folded-phonon doublets appear in the spectrum. For $|\Delta|$ within several Γ_{inh}

the scattering contributions from closeby resonances are no longer suppressed by the exponential factor, and the signal is determined by the first mechanism. The Gaussian distribution of transitions additionally centers the strong continuous emission signal around the excitation energy and thus reduces the width of the spectrum. In both limits the shape of the spectra hardly depends on Δ, and only the maximum intensity shows a pronounced resonance. The transition between these two regimes, where scattering is dominated by either one or the other mechanism, occurs at a certain value of Δ which depends on the homogeneous and inhomogeneous linewidths. It is marked by sharp and pronounced changes in the spectra.

3.5.2 Indirect-Gap Superlattices

Figure 3.18 shows continuous emission spectra of a (10/10) ml GaAs/AlAs SL for different detunings Δ of the excitation energy with respect to the resonance at the direct $\Gamma - \Gamma$ transition between the lowest electron and the highest hole states [1.23]. The spectra were measured at a temperature of 16 K. Points close to the laser, where the finite stray-light rejection of the spectrometer becomes apparent, were omitted. Positive Raman shifts are for Stokes, negative ones for anti-Stokes scattering. For all Δ the continuous emission is so strong that no folded-phonon doublets are observed. The luminescence spectrum of this sample shows an emission maximum due to the $\Gamma - \Gamma$ transition at 1.825 eV from which an inhomogeneous broadening of $\Gamma_{\text{inh}} = 15$ meV is determined. The lower-lying indirect $\Gamma - X$ transition leads to luminescence with a maximum at 1.764 eV. At higher energies this recombination causes a linear background which was subtracted in the spectra of Fig. 3.18 prior to fitting. The luminescence energies are in good agreement with earlier measurements [3.39]. The intensity of the continuous emission has a maximum 15 meV above the $\Gamma - \Gamma$ luminescence. This shift is analogous to the energy difference found for luminescence and luminescence excitation spectra and can be attributed to exciton localization [3.29]. It has been shown that the absorption edge for optical transitions is located at this higher energy [3.29] which we therefore use to define $\Delta = 0$. The values of Δ given in Fig. 3.18 refer to this origin. At Raman shifts of about 15 and 45 cm^{-1} the spectra show intensity anomalies which, however, shall not be considered here. The lines in Fig. 3.18 are theoretical spectra which were fitted to the data by only varying Γ_{hom} and a scaling factor for the overall intensity. The inhomogeneous broadening of the resonance was taken into account by averaging over a Gaussian distribution of transitions with a width Γ_{inh} determined from luminescence. For negative Δ the maximum of the continuous emission spectra is close to the excitation energy (zero Raman shift) and their width is constant. For larger positive Δ the intensity maximum moves towards larger Raman shifts on the Stokes-side of the spectra. At the same time their width increases.

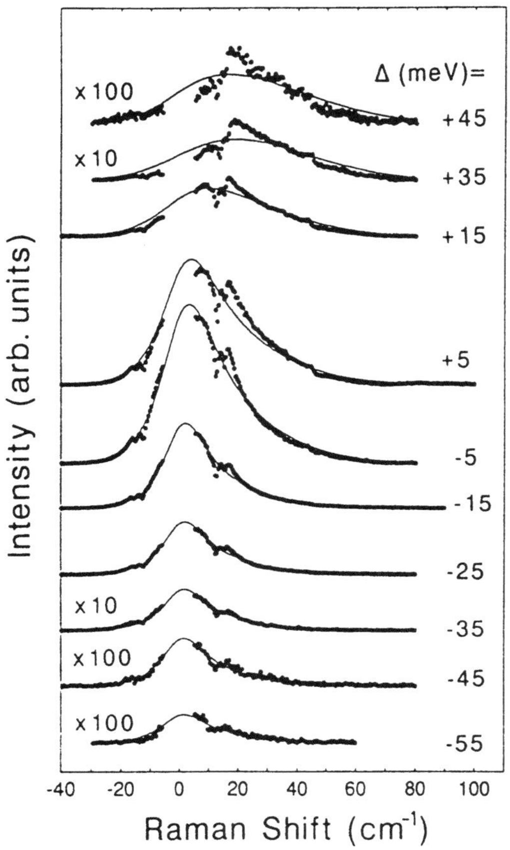

Fig. 3.18. Experimental (*points*) and theoretical (*lines*) continuous emission spectra of a (10/10) ml GaAs/AlAs SL for different detunings $\varDelta$ of the excitation from resonance. The calculated spectra were fitted to the experiment by only adjusting the homogeneous linewidth Γ_{hom} and a scaling factor for the overall intensity. Some spectra were magnified by the factors indicated

Values of Γ_{hom} vs. $\varDelta$ obtained from the fits to the data in Fig. 3.18 are shown in Fig. 3.19 [1.23]. For excitation below resonance ($\varDelta < 0$) the homogeneous linewidth is constant. As an average in this region we obtain $\Gamma_{\text{hom}} = (0.63 \pm 0.1)$ meV. For positive $\varDelta$ the value of Γ_{hom} increases significantly. This behavior reflects the increasing number of relaxation channels for electrons excited in higher-energy states of the SL dispersion along and perpendicular to the growth direction. Therefore electronic intermediate states with larger Γ_{hom} participate in resonant Raman processes. In contrast to that scattering processes for laser energies below the critical point ($\varDelta < 0$) involve virtually excited direct transitions with the closest ones yielding the largest contribution. Thus we interpret the value of Γ_{hom} in this region as the homogeneous broadening of an average $\Gamma - \Gamma$ transition. In SLs with indirect band gap which include the symmetric GaAs/AlAs systems with layer thicknesses

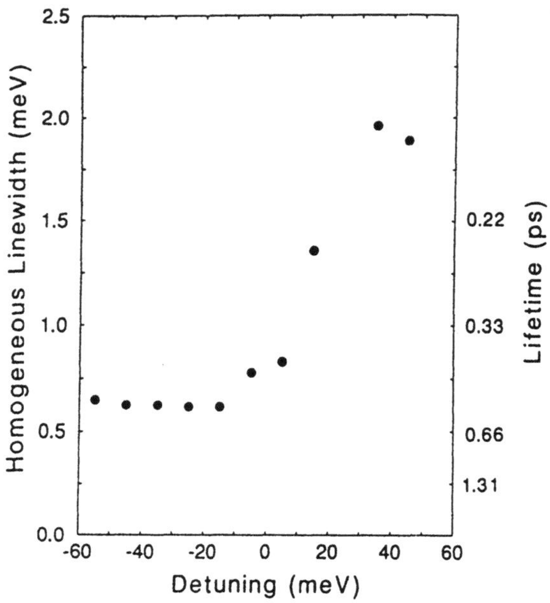

Fig. 3.19. Homogeneous linewidths Γ_{hom} for different detunings Δ of the excitation energy from resonance as determined from fits to the spectra in Fig. 3.18

below 12 monolayers [2.46] the homogeneous linewidth at the $\Gamma - \Gamma$ transition is determined by the transfer time of carriers from the GaAs Γ to the lower-lying X states of the AlAs barriers and does not depend on temperature [3.40]. Taking into account the fact that the values of Γ_{hom} in the expressions of the Raman intensity (see (3.6) – (3.8)) are half widths of Lorentzians at half maximum (HWHM) and defining a life time τ via the 1/e-decay of the electric field intensity [3.41], the carrier transfer times can be obtained from the experimental values of Γ_{hom} (in meV) using $\tau = 0.66/(2\Gamma_{\mathrm{hom}})$ ps. For the (10/10) ml GaAs/AlAs sample investigated here one thus obtains a $\Gamma - X$ transfer time of $\tau = (524 \pm 90)$ fs. This is rather close to the average value of $\tau = (435 \pm 50)$ fs found for a series of SLs with 10 ml GaAs thickness and different barrier widths by energy- and time-resolved differential transmission experiments [3.40]. In that work it was also observed that the transfer rates increase for nonresonant excitation above the $\Gamma - \Gamma$ transition where hot carriers are excited [3.40]. This effect is due to the increasing delocalization of the envelopes of Γ and X state wave functions above the $\Gamma - \Gamma$ transition [3.40]. Another contribution to Γ_{hom} comes from the coupling of the carriers to polar optical phonons which leads to intraband relaxation. In investigations of the hot electron luminescence in wide GaAs quantum wells with direct optical transitions the corresponding scattering time was determined to be about 150 fs [3.42, 3.43]. It is therefore of the same order as that of the $\Gamma - X$ transfer [3.40]. The analysis of the results in Fig. 3.19 with a theoretical model for both processes should allow one to determine their relative importance in short-period SLs. Similar investigations were also performed for a sample with (6.7/6.7) ml GaAs/AlAs. At T=15 K and $\Delta \simeq 0$

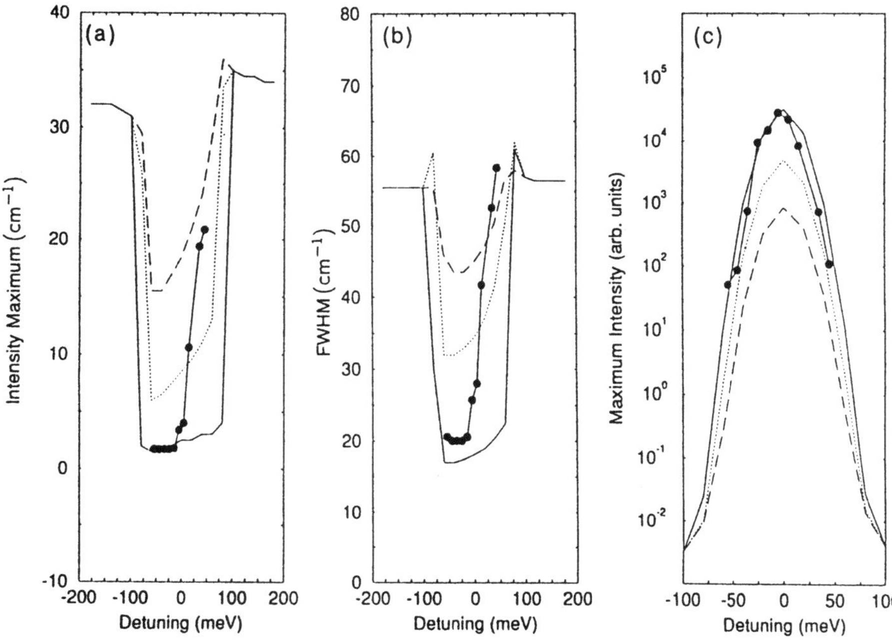

Fig. 3.20. Characteristic parameters of the continuous emission spectra in Fig. 3.18 vs. the detuning from resonance Δ. **(a)** Raman shift of the intensity maximum; **(b)** full spectral width at half maximum (FWHM); **(c)** maximum Raman intensity. The values obtained from fits to the data (*dots*) are connected by *solid lines.* The *solid, dotted,* and *dashed lines* are calculations with T=10 K, $\Gamma_{\rm inh} = 15\,$meV, and $\Gamma_{\rm hom} = 0.5$, 1, and 2 meV, respectively

the continuous emission spectra give a value of $\Gamma_{\rm hom} = (1.6 \pm 0.1)\,$meV which corresponds to a $\Gamma - X$ transfer time of about 200 fs.

The dots in Fig. 3.20 show further parameters of the continuous emission spectra in Fig. 3.18 which result from fits of the theory, such as the position of the intensity maximum (a), the width (b), and the maximum signal intensity (c) [1.23]. The experimental results are connected by lines. Theoretical values of these parameters for different homogeneous linewidths are also given. The calculations in Fig. 3.20 (a) and (b) underline the qualitative arguments for the resonance behavior of the continuous emission given above. For excitation away from resonance the intensity maximum occurs at a large Stokes shift from the excitation energy, and its width is independent of $\Gamma_{\rm hom}$. The same holds for the spectral width. Approaching resonance from either side one finds a characteristic transition into a regime where the spectra depend on $\Gamma_{\rm hom}$. Spectra excited close to resonance have an intensity maximum which is significantly shifted towards smaller frequencies. Their width is reduced. These changes are more pronounced for small $\Gamma_{\rm hom}$ and decrease with increasing homogeneous linewidth. The calculations in Fig. 3.20 (c) show the decrease of the maximum Raman intensity with increasing $\Gamma_{\rm hom}$, as expected from the resonance denominators of (3.7). For negative Δ the values obtained from the fits agree well with the model calculations. Due to strong $\Gamma - X$ luminescence the transition regime for $\Delta \ll 0$ could not be investigated. For positive detunings the parameters in Fig. 3.20 (a) and (b) increase already at smaller values of Δ than expected theoretically for constant Γ_{hom}. These differences thus reflect the increase of the homogeneous linewidth above the critical point. The variation of the maximum intensity in Fig. 3.20 (c) does not show a dependence on Γ_{hom} as pronounced as the other two parameters. This is partially due to the larger uncertainties connected with the comparison of the scattering intensities obtained in series of spectra excited with different laser wavelengths. One observes, however, a slight asymmetry of the resonance curve on the higher-energy side ($\Delta > 0$) due to the increasing linewidth.

3.5.3 Direct-Gap Samples

In the example just illustrated the very strong crystal-momentum nonconserving background signal makes it impossible to compare the continuous emission intensity I_B with the coherent scattering by folded phonons I_F and to investigate the changing influence of the homogeneous and inhomogeneous linewidths. The reason for this is the much smaller value of Γ_{hom} in comparison to Γ_{inh} which in this case was determined from luminescence measurements. In Figure 3.21 we show a series of spectra of a (16/16) ml GaAs/AlAs sample in which the resonance detuning Δ was varied continuously in a magnetic field [1.23]. The continuous emission and the folded-phonon doublets are simultaneously observed in this sample which makes an analysis of the scattering intensities according to (3.9a) and (3.9b) possible. The spectra were excited at T=10 K with a constant laser energy of $\hbar\omega_l = 1.727\,\text{eV}$ in $\bar{z}(\sigma^-, \sigma^-)z$ backscattering geometry. A resonance of $\hbar\omega_l$ with the ($n = 1$)-Landau level transition between the first subbands of heavy holes and electrons occurs at a magnetic field of $B_0 = 11\,\text{T}$ which is used to define $\Delta = 0$. From the experimentally determined slope $\beta_1 = 2.5\,\text{meV/T}$ of the Landau level transition one obtains values of Δ, tunable by the magnetic field, via $\Delta(B) = \beta_1(B_0 - B)$. For magnetic fields above 9 T the neighboring transitions are more than 20 meV away. Thus only a single transition has to be considered. In luminescence measurements we find an inhomogeneous broadening of the resonance by 3.3 meV, much less than the separation from neighboring transitions. Some spectra are superimposed on a linear luminescence background which was subtracted before analyzing the signal intensities. In resonance the ratio of the scattering intensities I_B/I_F of the first LA doublet at 17.1 and $20.5\,\text{cm}^{-1}$ is large and amounts to about four. At the strongest and smallest fields, away from resonance, the continuous emission decreases and I_B/I_F is reduced to about one. The resonance behavior of I_B/I_F with magnetic field for the two components of the first folded-phonon doublet is shown in Fig. 3.22 for Stokes scattering [1.23]. The solid lines are

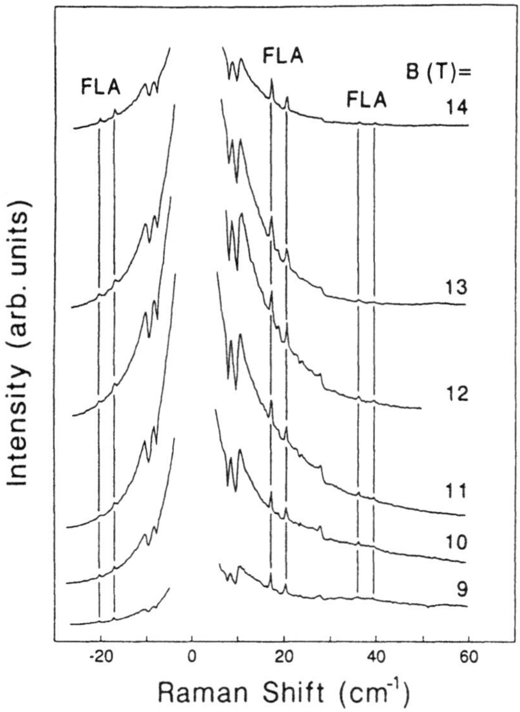

Fig. 3.21. Raman spectra of a (16/16) ml GaAs/AlAs MQW for different magnetic fields. The continuous emission and the folded-phonon doublets (FLA, *vertical lines*) are simultaneously observed. The spectra were excited with a constant laser energy at the $(n = 1)$-Landau level transition between states of heavy holes and electrons. Resonance $(\Delta = 0)$ occurs at 11 T

calculated intensity ratios according to (3.9a) and (3.9b) which were fitted to the experiment using a least-squares method thereby determining Γ_{hom} and Γ_{inh}. For the $17.1\,\text{cm}^{-1}$ mode (a) one obtains values of $\Gamma_{\text{hom}} = 0.64\,\text{meV}$ and $\Gamma_{\text{inh}} = 2.2\,\text{meV}$. The resonance at the $20.5\,\text{cm}^{-1}$ phonon (b) is best described by $\Gamma_{\text{hom}} = 0.76\,\text{meV}$ and $\Gamma_{\text{inh}} = 2.5\,\text{meV}$. From calculations with slightly different numbers we determine an uncertainty of these parameters of $\pm 0.2\,\text{meV}$. The maximum value of I_B/I_F and the width of the resonance curves in Fig. 3.22 contain sufficient information for an independent determination of Γ_{hom} and Γ_{inh}. The ratio I_B/I_F increases with increasing $\Gamma_{\text{inh}}/\Gamma_{\text{hom}}$ and vice versa. This behavior reflects the origin of the continuous emission in Raman processes involving single QWs. Smaller values of Γ_{hom} lead to stronger electronic resonances which are responsible for an increase of the continuous emission. For larger Γ_{inh} more transitions within the Gaussian distribution of energy levels contribute to the continuous emission which also leads to stronger signals. While both parameters, Γ_{inh} and Γ_{hom}, thus have an influence on the maximum of I_B/I_F, the width of the resonance curves is only given by Γ_{inh}. These ideas are also confirmed by the spectra of Fig. 3.18. The very large value of $\Gamma_{\text{inh}}/\Gamma_{\text{hom}}$ leads to a continuous emission signal which is so strong that the superimposed weak folded-phonon doublets cannot be observed. Since the values of Γ_{hom} found for both samples are quite similar this effect has to be attributed to the much stronger influence of monolayer fluctuations on Γ_{inh} for smaller layer thicknesses.

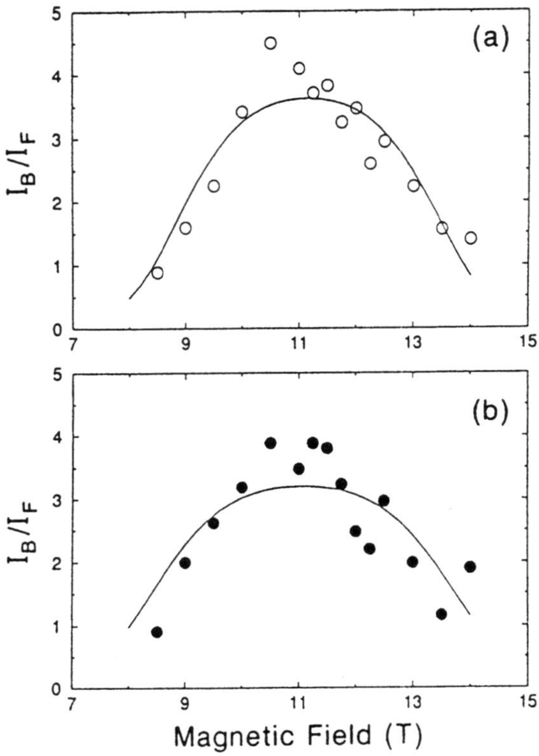

Fig. 3.22. Resonance behavior of the intensity ratio I_B/I_F of the continuous emission and the folded-phonon doublet signal for the two LA modes of Fig. 3.21 at 17.1 cm^{-1} **(a)** and 20.5 cm^{-1} **(b)**. The *solid lines* are fits to the experiment from which Γ_{hom} and Γ_{inh} can be determined

3.6 Continuous Emission and Electron–Phonon Interaction

In the previous section we showed that the continuous emission allows one to determine the homogeneous and inhomogeneous linewidths of interband transitions in QWs. In the following we use the relation between the measured quantities I_B and I_F and the electronic structure parameters Γ_{hom} and Γ_{inh} to determine the constants of the electron–phonon interaction.

3.6.1 Temperature Dependence of the Homogeneous Linewidth

Figure 3.23 shows a series of continuous emission spectra of the (16/16) ml GaAs/AlAs sample of Fig. 3.21 at different temperatures [1.23]. All spectra were excited in resonance with a detuning of $\Delta = 0$. In order to compensate shifts of the electronic transition with temperature and to maintain comparable experimental conditions, the laser energies for all spectra were chosen such that the resonance of the $(n = 1)$-Landau level transition always occured at 11 T. This condition was checked in each case by measuring the magneto-oscillations of the continuous emission close to the laser line, similar to those shown in Fig. 3.3 [1.19]. While the overall strength of the spectra

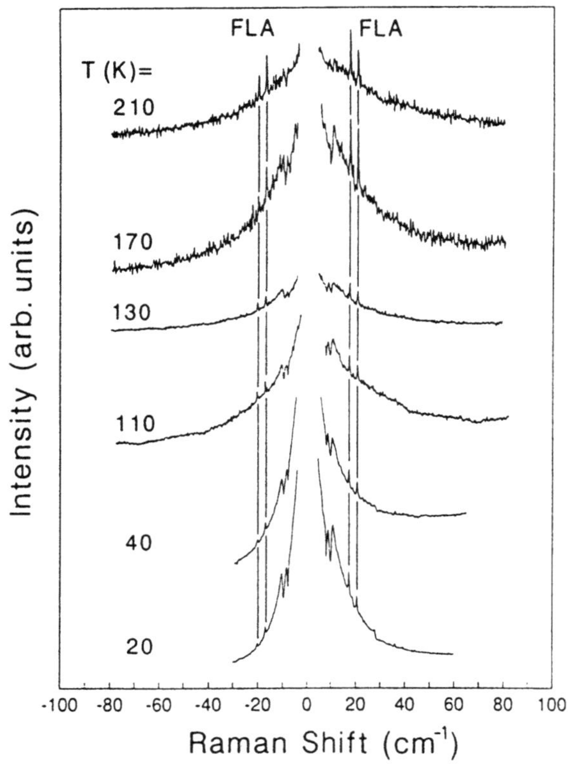

Fig. 3.23. Temperature-dependent continuous emission spectra of the (16/16) ml GaAs/AlAs MQW of Fig. 3.21. The doublets of folded acoustic phonons are marked by FLA and *vertical lines*. Positive (negative) Raman shifts denote Stokes (anti-Stokes) scattering

in Fig. 3.23 decreases with increasing temperature, the relative intensities of the folded-phonon doublets I_F and the continuous emission I_B show characteristic changes which were first observed in [3.19]. With increasing temperature the ratio I_B/I_F decreases and the folded-phonon doublets become more pronounced on the continuous emission background. With its origin in layer thickness fluctuations and interface roughness the inhomogeneous linewidth Γ_{inh} can be regarded as temperature independent. Changes of I_B/I_F are thus due to an increase of Γ_{hom} with temperature [3.19]. Figure 3.24 shows values of $\Gamma_{\text{hom}}(T)$ which were determined from the spectra of Fig. 3.23 in this way [1.23].

3.6.2 Electron–Phonon Interaction in Quantum Wells

Coupling Constants. In bulk semiconductors, QWs and SLs the most important contributions to the temperature dependence of the homogeneous linewidth of electronic resonances Γ_{hom} comes from the electron–phonon interaction [3.44–3.49]. Generally, the deformation potential and piezo-electric

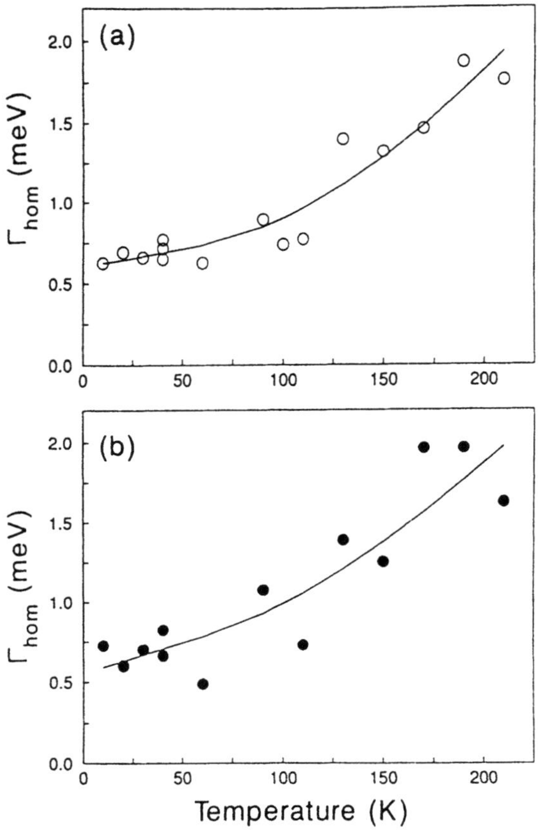

Fig. 3.24. Temperature dependence of the homogeneous linewidth $\Gamma_{\text{hom}}(T)$ of the $(n=1)$-Landau level transition in the sample of Fig. 3.21 determined from I_B/I_F in the spectra of Fig. 3.23. The *open* (**a**) and *filled* points (**b**) are measurements at LA phonon Raman shifts of $17.1\,\text{cm}^{-1}$ and $20.5\,\text{cm}^{-1}$, respectively. The *solid lines* were fitted to the data with the theoretical model explained in the text

interactions are taken into account to describe the coupling of the lowest-energy exciton state with acoustic phonons. Due to the small frequencies of these modes one obtains a linear increase of Γ_{hom} with temperature. The Fröhlich interaction mediates the scattering of electronic states by longitudinal-optic phonons. In this case the linewidth is proportional to the Bose–Einstein statistical factor $n_{\text{LO}}(T)$ and becomes linear only at higher temperatures. Scattering by ionized impurities gives an additional contribution to $\Gamma_{\text{hom}}(T)$ which has the temperature dependence of a thermally activated process. Due to the low defect concentration in the samples investigated it shall not be further considered. The temperature dependence of the homogeneous linewidth is thus given by

$$\Gamma_{\text{hom}}(T) = \Gamma_0 + \sigma T + \Gamma_{\text{ep}}\, n_{\text{LO}}(T) \,. \tag{3.13}$$

The parameters in this expression have been intensively investigated by theory and experiment [3.46–3.49]. The majority of the experiments is based on the determination of spectral linewidths or time-resolved measurements. Modulation spectroscopy, notably photoreflectance, as well as absorption and reflectivity were used to determine the width of direct transitions in bulk materials and QWs [3.50–3.56]. Luminescence linewidths were analyzed with respect to the contributions of $\Gamma_{\text{hom}}(T)$ and Γ_{inh}, and efforts to separate these two parameters have been made [3.48, 3.57–3.59]. In time-resolved four-wave mixing and energy- and time-resolved differential transmission experiments the phase relaxation time of excitons, which is directly related to Γ_{hom}, could be determined [3.60–3.62]. Tables 3.1 and 3.2 compile the experimental values of σ and Γ_{ep} for bulk GaAs and GaAs/$Al_xGa_{1-x}As$ QWs with different thicknesses which were determined by various methods.

The solid lines in Fig. 3.24 are fits of (3.13) to the experiment using a nonlinear least-squares method. With a fixed LO phonon energy of $\hbar\Omega_{\text{LO}} = 36.7\,\text{meV}$ one obtains by variation of the three other parameters

$$\Gamma_{\text{hom}}(T) = 0.60\,\text{meV} + 2.1\,\frac{\mu\text{eV}}{\text{K}}T + 5.9\,\text{meV}\, n_{\text{LO}}(T) \tag{3.14}$$

for the phonon at $17.1\,\text{cm}^{-1}$ (Fig. 3.24 (a)) and

$$\Gamma_{\text{hom}}(T) = 0.56\,\text{meV} + 3.7\,\frac{\mu\text{eV}}{\text{K}}T + 4.2\,\text{meV}\, n_{\text{LO}}(T) \tag{3.15}$$

Table 3.1. Experimental values of the coefficient σ, the contribution of acoustic phonons to Γ_{hom} in (3.13). The results are for bulk GaAs and GaAs/$Al_xGa_{1-x}As$ QWs with different thicknesses a

a (Å)	σ(μeV/K)	method	reference
bulk GaAs	12	absorption	[3.46, 3.63]
bulk GaAs	8.5 ± 0.5	phase relaxation	[3.61]
bulk GaAs	7	reflectivity	[3.55]
340	2.2	reflectivity	[3.56]
325	3	photoluminescence	[3.58]
277	4.5	phase relaxation	[3.61]
250	2.6	reflectivity	[3.56]
170	1.9	reflectivity	[3.56]
150	1.7	photoluminescence	[3.58]
135	2.5	phase relaxation	[3.61]
130	1.9	reflectivity	[3.56]
120	1.25	phase relaxation	[3.61]
45	3 ± 1	Raman	[1.23], this work

Table 3.2. Experimental values of the coefficient Γ_{ep}, the contribution of optic phonons to Γ_{hom} in (3.13), for bulk GaAs and GaAs/$Al_xGa_{1-x}As$ QWs with different thicknesses a

a (Å)	Γ_{ep} (meV)	method	reference
bulk GaAs	8.3	absorption	[3.56]
bulk GaAs	20 ± 1	photoreflectance	[3.52]
28 - 340	9 ± 1	absorption, reflectivity	[3.56], see Fig. 3.25
205	8.1	photoreflectance	[3.50]
200	7.8	absorption	[3.57]
200	4.7 ± 0.6	photoreflectance	[3.52, 3.53]
130	3.7 ± 0.5	photoreflectance	[3.52, 3.53]
120	12.3	photoluminescence	[3.57]
102	5.5	absorption	[3.64]
96	10.4	absorption	[3.57]
84	11.1 ± 1	photoluminescence	[3.65]
80	1.1 ± 0.9	photoreflectance	[3.52, 3.53]
78	12 ± 1	photoluminescence	[3.65]
60	1.7 ± 0.9	photoreflectance	[3.52, 3.53]
60	10.9	absorption	[3.57]
53	9.3	reflectivity	[3.51]
45	5 ± 1	Raman	[1.23], this work

for the mode at $20.5\,\text{cm}^{-1}$ (Fig. 3.24 (b)). The average values for the contribution of acoustic and optic phonons to the temperature dependence of the homogeneous linewidth in this 16 ml (45 Å) GaAs MQW are thus $\sigma = (3 \pm 1)\,\mu\text{eV/K}$ and $\Gamma_{ep} = (5 \pm 1)\,\text{meV}$.

Acoustic Phonons. The value obtained for σ is in good agreement with other experimental results and thus further confirms the observed reduction of this parameter in QWs as compared to the bulk and possibly a slight well width dependence. However, the predictions for σ significantly deviate from the measured values. A comprehensive theoretical investigation of the temperature dependence of exciton linewidths in bulk semiconductors is given in [3.46]. In this study acoustic-phonon scattering via the deformation potential and the piezo-electric interaction is taken into account. While the latter only gives a negligible contribution, the deformation-potential coupling of longitudinal-acoustic phonons leads to $\sigma = 0.64\,\mu\text{eV/K}$ for GaAs [3.46]. In view of the large discrepancy to the above-mentioned experiments (see Table 3.1) excitonic-polariton effects were additionally considered in the calculation [3.46]. This leads to values of $\sigma_T = 0.95\,\mu\text{eV/K}$ and $\sigma_L = 1.12\,\mu\text{eV/K}$ for the transverse and longitudinal branches of the exciton-polariton dispersion [3.46]. Despite this increase a difference to the experiments remains, and no comparable calculations for QWs have been attempted [3.56]. Further contributions to the value of σ in bulk GaAs, which arise from the shear-components of the valence band deformation potential and final-state broadening effects, are discussed in [1.23]. They should increase the exciton-acoustic phonon scattering rate and thus give better agreement with the experimental value of σ.

Optic Phonons. The experimental values for the LO-phonon contribution to $\Gamma_{\text{hom}}(T)$ in bulk GaAs span the wide range between $\Gamma_{\text{ep}} = 8.3\,\text{meV}$ [3.56] and 20 meV [3.52]. The same holds for GaAs QWs where values of $\Gamma_{\text{ep}} = 1.1\,\text{meV}$ for 80 Å layers [3.52] and 10.9 meV for a sample with 60 Å well width [3.57] have been reported. Similar discrepancies exist for the dependence of Γ_{ep} on the QW thickness. In Ref. [3.49] it is predicted that Γ_{ep} increases with decreasing well width by up to a factor of three compared to the bulk, depending on the effective mass used in the calculations. The results of a careful determination of Γ_{hom} and Γ_{inh} from absorption and luminescence linewidth measurements in [3.57] are quoted to support these calculations. In other linewidth measurements one observes a linear decrease of Γ_{ep} with the well width [3.52]. It is thus apparent that more precise experiments and additional calculations are required in order to resolve this controversy. In view of the reported decrease of Γ_{ep} with the well width [3.52] and the resulting consequences for the dependence of the electron–phonon interaction on dimensionality discussed below, more experiments and calculations are also required for the bulk.

At this point the values of Γ_{ep} obtained from the continuous emission Raman experiments help to solve the problem. At first, however, it should be noted that in most earlier determinations of Γ_{ep} the acoustic-phonon contribution to $\Gamma_{\text{hom}}(T)$ which is linear in temperature was not taken into account [3.50–3.53, 3.57]. According to our and other measurements of $\sigma \simeq 3\,\mu\text{eV/K}$ in GaAs QWs (see Table 3.1) this process gives a contribution to Γ_{hom} of almost 1 meV at 300 K. Therefore it cannot be neglected in comparison to the broadening of 2.5 meV from optic-phonon scattering for $\Gamma_{\text{ep}} = 8\,\text{meV}$. In fact, a larger value of $\Gamma_{\text{ep}} = 8.6\,\text{meV}$ is found from fits of the data in Fig. 3.24 with (3.13) and $\sigma = 0$. Our data for a 45 Å GaAs QW are thus in good agreement with most other determinations which extend over a range of layer thicknesses from 54 Å ($\Gamma_{\text{ep}} = 9.3\,\text{meV}$ [3.51]) over 60 Å ($\Gamma_{\text{ep}} = 10.4\,\text{meV}$ [3.57]) and 96 Å ($\Gamma_{\text{ep}} = 10.4\,\text{meV}$ [3.57]) up to about 200 Å ($\Gamma_{\text{ep}} = 8.1$, 7.8 meV [3.50, 3.57]) if one takes into account that the acoustic-phonon contribution was not considered in the interpretation of these measurements.

Much smaller values for Γ_{ep} were observed in a photoreflectance study of GaAs QWs with thicknesses between 60 and 200 Å in [3.52, 3.53]. Decreasing coupling constants of $\Gamma_{\text{ep}} = 4.7\,\text{meV}$ at 200 Å and 1.7 meV at 60 Å were interpreted as a reduction of the electron–phonon interaction due to phonon confinement and the different dimensionality of the electronic band structure [3.52, 3.53]. In this context the importance of a separation between the contributions of Γ_{hom} and Γ_{inh} to the width of an observed line needs to be emphasized, especially if one attempts to determine the coupling constants of the electron–phonon interaction. As explained above, this requirement is fulfilled in the continuous emission measurements and in earlier investigations which distinguish between homogeneous and inhomogeneous components of a broadened excitonic line [3.57]. It can only be circumvented in samples of excellent quality where Γ_{inh} is much smaller than Γ_{hom} [3.50, 3.51, 3.58]. For samples with $\Gamma_{\text{inh}} \gg \Gamma_{\text{hom}}$ the increase of Γ_{hom} due to electron–phonon interaction appears in fits which do not distinguish between these two parameters only at much higher temperatures. This leads to an apparently reduced increase of the linewidth. The resulting smaller values of Γ_{ep} could then be misinterpreted as a reduction of the electron–phonon coupling which is expected for confined optic modes due to the dependence of the Fröhlich interaction on crystal momentum [3.52, 3.53]. However, one also obtains a much smaller value of $\Gamma_{\text{ep}} = 2.7\,\text{meV}$ when (3.13) with $\sigma = 0$ is fitted to the temperature-dependent width of the continuous emission magneto-oscillation at the ($n = 1$)-Landau level transition (see the experimental conditions for the spectra in Fig. 3.23) which is mostly determined by Γ_{inh} and not by the electron–phonon interaction [1.23, 1.24].

From our and other measurements of Γ_{ep} in which the homogeneous and inhomogeneous broadenings are treated separately or which were performed in samples where Γ_{inh} can be neglected [3.50, 3.51, 3.57, 3.58] one finds that a value of $\Gamma_{\text{ep}} \simeq 5\,\text{meV}$ describes the LO-phonon contribution to $\Gamma_{\text{hom}}(T)$ for GaAs QWs between 45 and 200 Å well if acoustic-phonon scattering is taken into account. In reliable theoretical calculations of Γ_{ep} in GaAs and other bulk semiconductors a multitude of effects such as, e. g., exciton-polariton states, mixed valence bands or the different contributions to exciton–phonon scattering between discrete and continuous exciton states have to be considered [3.46]. It is also necessary to describe the excitonic continuum by scattering states, simple plane waves are not sufficient [3.46]. In a careful calculation considering all these effects one finds $\Gamma_{\text{ep}} = 8.7\,\text{meV}$ for GaAs [3.56]. This value, which also agrees with the theory of [3.63] ($\Gamma_{\text{ep}} = 7\,\text{meV}$ [3.66]), is close to the numbers found for GaAs QWs. It therefore does not support the reported strong reduction of Γ_{ep} in QWs from the bulk side which seemed more likely in view of earlier calculations where values of about 20 meV were obtained [3.46, 3.67]. The experimental coupling constant of $\Gamma_{\text{ep}} = 20\,\text{meV}$ for GaAs in [3.52, 3.53] is in better agreement with this value if an acoustic-phonon contribution of $\sigma = 10\,\mu\text{eV/K}$, an average of the determinations in [3.46] ([3.63]) and [3.61], is taken into account; thereby Γ_{ep} is reduced to about 12 meV. New calculations, which consider confined optical phonons as well as the appropriate changes of the electronic structure in QWs and the scattering of excitons into bound and continuum excitonic states associated with higher-lying subbands, show that Γ_{ep} is almost independent of the layer thickness and hardly different from the bulk value [3.56]. Small changes of 1–2 meV, however, are expected at well widths where the electronic inter-

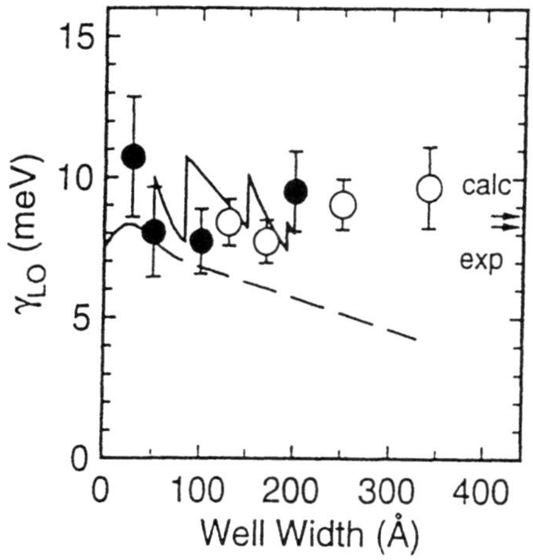

Fig. 3.25. Well-width dependence of Γ_{ep} in GaAs/$Al_{0.30}Ga_{0.70}$As QWs. The *filled (open) circles* are results from multiple (single) QWs. The *solid (dashed) line* gives the theoretical result with (without) exciton scattering into higher subbands, which is important for large well widths. Results for bulk GaAs are indicated by *arrows* (from [3.56])

subband separations are comparable to the LO phonon energy [3.56]. This behavior is also confirmed by the recent transmission and reflectivity measurements on the temperature dependence of the excitonic linewidth in [3.56]. Experimental values of Γ_{ep} from this work are compared to the theory in Fig. 3.25 [3.56]. Similar theoretical results are described in [3.68]. However, that work uses bulk phonons and plane waves for the excitonic continuum, and the three-dimensional limit for Γ_{ep} does not agree with the experimental value [3.68].

Hot-Phonon Measurements. Further evidence for a coupling constant Γ_{ep} almost independent of the QW width and deeper insight about the other arguments given above comes from time-resolved Raman measurements in which one observes the nonequilibrium population of hot phonons due to intraband carrier relaxation [3.69]. In contrast to linewidth measurements, which only allow one to observe the sum of all mechanisms, such experiments make it possible to study the electron–phonon interaction of individual optic modes and thus the different contributions to Γ_{ep}. Figure 3.26 shows the occupation numbers for several phonons in a series of GaAs/AlAs QWs with different well widths and constant barrier thickness of 70 Å [3.69]. In these experiments the electrons were excited with a constant excess energy of 200 meV above the lowest direct interband transition, respectively. During the intrasubband relaxation they create a distribution of hot phonons. This population is probed in anti-Stokes spectra at 10 K with a second (weaker) laser in outgoing resonance with the transition considered. In such experiments, however, one is only sensitive to a certain value of the crystal momentum within the whole distribution. Due to the wavevector dependence of the Fröhlich electron–phonon interaction for confined phonons LO_n their occupation indeed decreases with the well width. For smaller layer thicknesses, however, the electrons relax preferentially via the emission of interface phonons (IF1, IF2). The occupation of these modes increases with decreasing well width

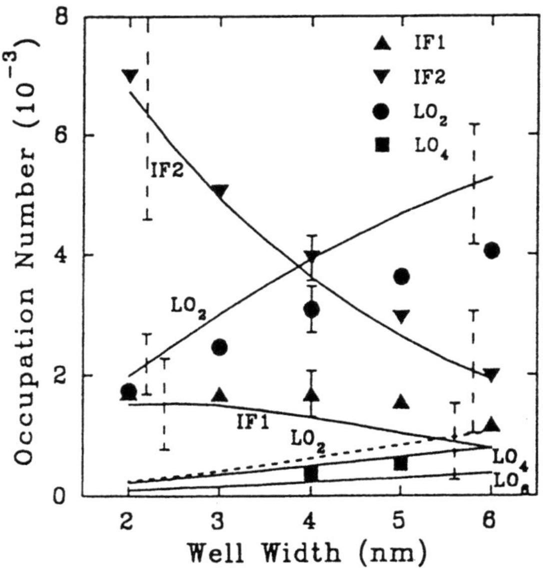

Fig. 3.26. Well-width dependence of the occupation numbers for different optic phonons in GaAs/AlAs QWs. The occupation of the confined phonons (LO_n) decreases with the well width, that of interface modes (IF1, IF2), however, increases. Note that the sum of the occupation numbers for all modes, which enters into Γ_{ep}, is almost constant. The lines result from calculations using different models for the electron–phonon interaction

and compensates the reduction in the emission of the confined modes. The resulting overall change is rather small. The total occupation, a measure of Γ_{ep}, thus remains almost unaffected.

3.6.3 Short-Period Superlattices

Linewidths from the Continuous Emission. In the measurements just discussed $\Gamma_{hom}(T)$ was obtained from the intensity ratio I_B/I_F of the continuous emision and the folded-phonon signals. In the following we present results from short-period SLs where, as mentioned above, the determination of this ratio is not possible due to the much stronger crystal-momentum nonconserving signal.

Figure 3.27 shows a series of continuous emission spectra of the (10/10) ml GaAs/AlAs sample of Fig. 3.18 for different temperatures (points) and fits to the data (solid lines) obtained by varying the homogeneous linewidth Γ_{hom} [1.23]. Positive Raman shifts denote Stokes and negative ones anti-Stokes scattering. Scaling factors for the intensities are given on the right-hand side of the curves. The laser excitation energy was shifted along with the temperature variation of the resonance such that all spectra were measured under comparable conditions with $\Delta = 0$. In all calculations we used the value of $\Gamma_{inh} = 15\,\text{meV}$ for the inhomogeneous broadening as determined from luminescence. We assume this value to be temperature-independent. With increasing temperature the spectra broaden and their maximum intensity decreases. This is indicative of an increase in the homogeneous linewidth. However, the expected shift of the intensity maximum and the increase of the broadening (FWHM) in the spectra with increasing Γ_{hom}, as discussed

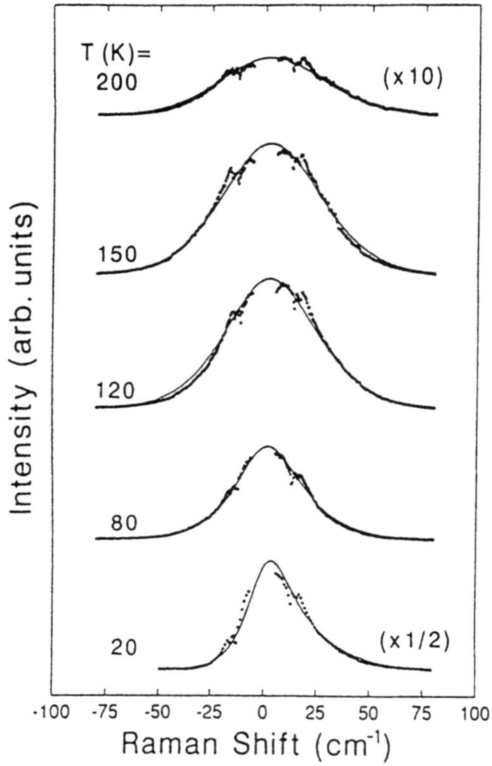

Fig. 3.27. Experimental continuous emission spectra (*points*) and fitted theoretical curves (*solid lines*) for the (10/10) ml GaAs/AlAs SL of Fig. 3.18 at different temperatures

in Sect. 3.5 (see Fig. 3.20), is partially compensated by the increasing signal on the anti-Stokes side due to the larger thermal population of the modes at higher temperatures. The homogeneous linewidths determined from these spectra are shown in Fig. 3.28 [1.23]. The solid line is a fit of (3.13) to the data. One obtains

$$\Gamma_{\mathrm{hom}}(T) = 0.66\,\mathrm{meV} + 9.3\,\frac{\mu\mathrm{eV}}{\mathrm{K}}T + 4.9\,\mathrm{meV}\,n_{\mathrm{LO}}(T)\ . \tag{3.16}$$

The constant term is mostly due to $\Gamma - X$ transfer. It has been found that this contribution is temperature independent in GaAs/AlAs SLs since it is mediated by the interface mixing potential ($\Gamma - X_z$) and interface roughness ($\Gamma - X_{x,y}$) [3.40]. The value for Γ_{ep} is comparable to the result for the 16 ml sample (see Fig. 3.24). The contribution due to acoustic-phonon scattering is larger than for the wider sample and closer to measurements on bulk GaAs (see Table 3.1). In order to estimate the accuracy of these results we performed a fit with fixed $\Gamma_{\mathrm{ep}} = 8\,\mathrm{meV}$, chosen according to [3.56]. We find a constant term of 0.77 meV and $\sigma = 7.1\,\mu\mathrm{eV/K}$. Hence, while the uncertainty in the determination of Γ_{ep} is large, the enhanced value for σ is confirmed.

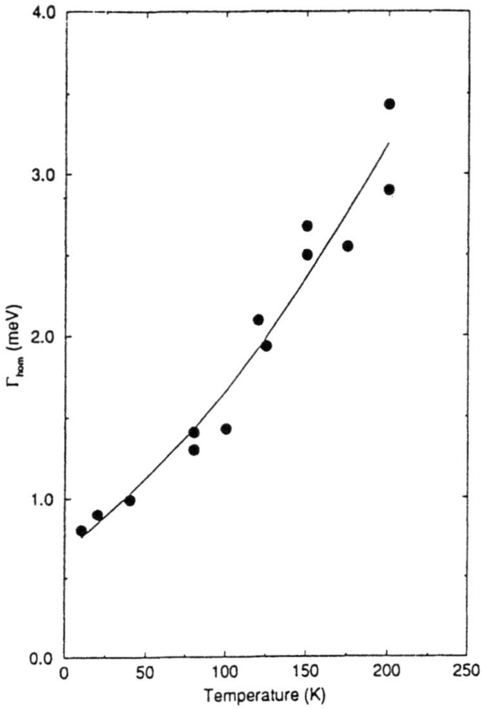

Fig. 3.28. Temperature dependence of Γ_{hom} determined from the spectra in Fig. 3.27 (*points*). The *solid line* is a fit of (3.13) to the data.

This effect could be related to the formation of minibands which has been neglected in the QWs discussed so far. One expects that the exciton–phonon scattering in short-period SLs whose character becomes more and more three-dimensional due to the increasing miniband width approaches the bulk behavior.

Sensitivity limits. Further changes of the behavior described so far for short-period SLs occur in samples with still shorter periods. Figure 3.29 shows temperature-dependent continuous emission spectra of a SL with six monolayers of GaAs and AlAs, respectively [1.25]. The values of Γ_{hom} for this sample, the one of Fig. 3.27 with (10/10) ml and another one with (8/8) ml GaAs/AlAs are summarized in Fig. 3.30 [1.25]. The (10/10) ml sample shows the linewidth increase due to contributions of acoustic and optic phonons discussed above. For the other two SLs with thinner layers, however, $\Gamma_{\text{hom}}(T)$ first increases but then saturates at higher temperatures. For the 8 ml sample this saturation appears at higher temperatures than for the one with 6 ml thickness. This behavior of $\Gamma_{\text{hom}}(T)$ cannot be described by (3.13), and the lines drawn for these samples only underline the observed trend. From calculated continuous emission spectra one obtains an upper sensitivity limit of $\Gamma_{\text{hom}} \simeq 5\,\text{meV}$ for temperatures up to 300 K and layer thicknesses below 10 ml, determined from changes of the maximum position and the width (FWHM) of the spectra at different temperatures, layer thicknesses and homogeneous linewidths. The observations for the samples with 6 and 8 ml in

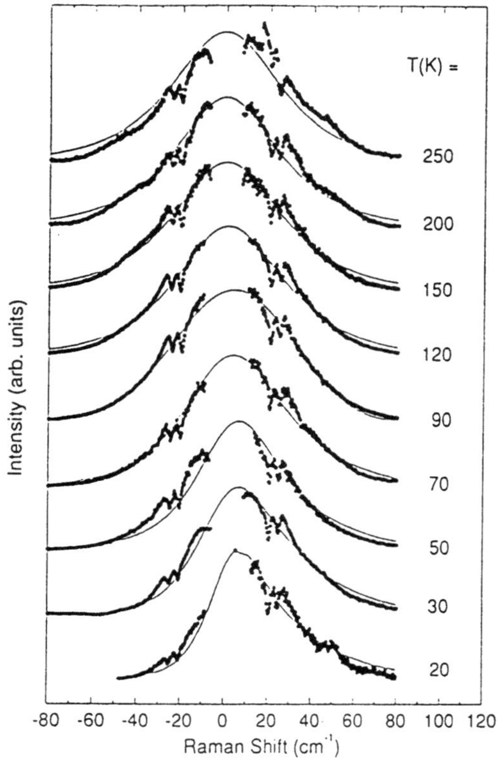

Fig. 3.29. Continuous emission spectra of a (6/6) ml GaAs/AlAs SL at different temperatures. The *solid lines* are fits to the data (*points*) by varying Γ_{hom}. All spectra have been normalized to the same intensity

Fig. 3.30 therefore cannot be explained by a lack of sensitivity in the fits to the data since the saturation appears far below threshold and the largest values determined for the (10/10) ml sample. We conjecture that these effects are caused by changes of the electronic structure with decreasing layer thickness, i. e., the increasing miniband width E_{bw} and the simultaneously increasing energy E_{ml} due to monolayer fluctuations. Estimating the influence of these two effects with a Kronig–Penney model we find that the ratio $E_{\text{ml}}/E_{\text{bw}}$ which amounts to 7 for the (10/10) ml sample is reduced to 0.9 in the (6/6) ml GaAs/AlAs SL. For the sample with 10 ml the electronic states remain confined to individual QWs even in the presence of disorder due to layer thickness fluctuations. A distribution of transition energies emerges which is described by the inhomogeneous broadening Γ_{inh}. This is the case for which the model of continuous emission as crystal-momentum nonconserving Raman scattering by acoustic phonons in single QWs was developed in Sect. 3.2 and which is the basis for the determination of Γ_{hom} in the fits to the experimental spectra. With decreasing SL period, however, the confinement of the electronic states decreases as well. The miniband width becomes comparable to monolayer fluctuations and the disorder is not sufficient to confine the wave functions in single QWs. In this case the present theory loses its applicability, and extended miniband states of the SL have to be invoked

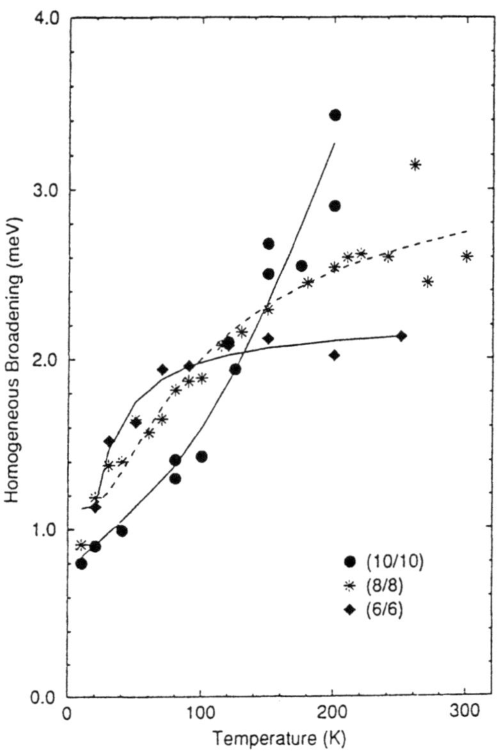

Fig. 3.30. $\Gamma_{\mathrm{hom}}(T)$ from continuous emissions spectra in short-period GaAs/AlAs SLs with 6, 8, and 10 ml thickness. The data for the (10/10) ml sample are from Fig. 3.28; the fit (*solid line*) corresponds to (3.16). The lines drawn for the other two samples only underline trends in the temperature behavior

for the description of the continuous emission. The smaller values of Γ_{hom} obtained from the fits for the 6 ml and 8 ml samples at higher temperatures and the observed saturation confirm these ideas. The increasing penetration of the wave functions in the barriers leads to an effective increase of the layer thickness and therefore to narrower continuous emission spectra (see (3.5)). Disregarding this effect in the fits one obtains, as in Fig. 3.20 (b), an apparent decrease of Γ_{hom} until, finally, the sensitivity limit is reached and saturation occurs. More detailed investigations of these effects require calculations of the continuous emission in ideal and disordered SLs.

3.7 Magneto-Raman Spectroscopy with the Continuous Emission

In this section we exploit the resonant character of the continuous emission to investigate the QW electronic structure. This property was already demonstrated in Sect. 3.1 where Fig. 3.3 shows magneto-oscillations of the crystal-momentum nonconserving scattering background at different frequency shifts of the detection energy from the exciting laser. The resonances of the continuous emission at interband transitions between Landau levels in QWs provide a new way to investigate their electronic properties which has several advantages compared to the measurement of magneto-Raman oscillations via LO phonons which is the topic of Chap. 4. LO phonon magneto-Raman profiles, i.e., the oscillations of the Raman scattering intensity in a magnetic field B for fixed excitation energy, exhibit a plethora of incoming or outgoing resonances which occur whenever the energies of the incident laser or the scattered photons coincide with magneto-optical interband transitions. Even for the strongest routinely accessible magnetic fields, however, typical LO phonon energies in III-V semiconductors are much larger than the electron or hole cyclotron energies. The interpretation of experimental LO phonon magneto-Raman fan plots is thus complicated by the multitude of resonances which are possible within mixed valence band states and, additionally, their outgoing or incoming character. Changing the detection energy in magneto-Raman profiles from optic to acoustic phonons, however, reduces the difference between the incoming and outgoing resonances at each transition from about 40 to 1 – 2 meV. Therefore the measurement of acoustic-phonon magneto-Raman profiles allows one to perform detailed investigations of individual interband transitions without perturbations from other resonances. This difference also facilitates theoretical calculations of the intensity profiles. Another advantage arises from the fact that Raman processes involving acoustic phonons can occur under conditions which are very close to double resonance (see (2.3)). Thus both, incoming and outgoing channels contribute to the signal enhancement. This is possible because the acoustic-phonon energies are of the same order or even smaller than typical homogeneous broadenings of the electronic structure.

Experiment. In the following we discuss a magneto-optical investigation of the electronic structure in a $GaAs/Al_{0.35}Ga_{0.65}As$ MQW using resonances of the continuous emission [1.26]. We determine the effective electron masses responsible for subband confinement along the growth direction and the free dispersion perpendicular to it. The influence of the band nonparabolicity on the observed mass anisotropy and the enhancement of the effective masses compared to the bulk are also discussed. Experiments were performed at 6 K on a (001)-oriented sample with 40 periods of (100/103) Å $GaAs/Al_{0.35}Ga_{0.65}As$. Well and barrier widths were determined by X-ray diffraction.

Figure 3.31 shows intensity profiles of the continuous emission vs. magnetic field for different excitation energies measured in the Faraday backscattering geometry (magnetic field parallel to the propagation direction of the light) with $\bar{z}(\sigma^+,\sigma^+)z$ (a) and $\bar{z}(\sigma^-,\sigma^-)z$ polarizations (b) [1.26]. The intensity profiles were recorded using conventional single-photon counting techniques and a double monochromator as a spectral bandpass with a width of $0.3\,\mathrm{cm}^{-1}$ set to a Stokes-Raman shift of $3\,\mathrm{cm}^{-1}$. The closest folded-phonon doublets in this sample are at 7.3 and $9.4\,\mathrm{cm}^{-1}$. At the chosen Raman shift one thus detects pure continuous emission. In the 10 meV range of excitation energies between 1581.4 and 1591.5 meV of Fig. 3.31 one observes a series of sharp intensity oscillations with a period approximately proportional to the

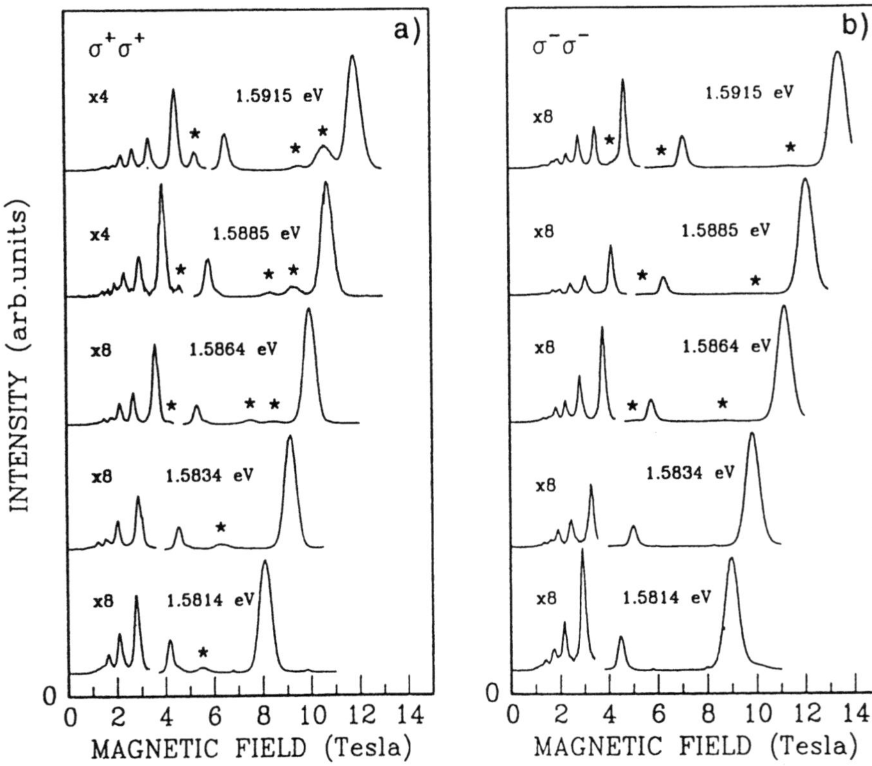

Fig. 3.31. Magneto-Raman intensity profiles of the acoustic-phonon continuous emission in a (100/103) Å $Al_{0.35}Ga_{0.65}As$ MQW for different laser energies. The difference between the detection and the excitation is $3\,cm^{-1}$. **(a)** $\bar{z}(\sigma^+,\sigma^+)z$, **(b)** $\bar{z}(\sigma^-,\sigma^-)z$ backscattering geometry. In the region of small magnetic fields the spectra were scaled by the factors indicated

inverse of the magnetic field, typical for magneto-optical effects. For larger excitation energies the resonances shift towards higher magnetic fields. A series of weaker resonances, marked by asterisks, is also observed. It shows the same behavior. For fixed excitation energy resonances in $\bar{z}(\sigma^+,\sigma^+)z$ geometry (Fig. 3.31 (a)) appear at smaller magnetic fields than their counterparts in $\bar{z}(\sigma^-,\sigma^-)z$ configuration (Fig. 3.31 (b)).

Figures 3.32 and 3.33 show fan plots (excitation energy vs. magnetic field) of the resonances for the two polarizations of Fig. 3.31 [1.26]. The measured data are marked by open symbols. For vanishing magnetic field the fan lines converge at the energy of 1563 meV, indicated by a horizontal bar on the energy scale. Some lines from the series of weaker resonances (open triangles) converge at a higher energy which, however, cannot be determined accurately. The two lowest-energy lines in Figs. 3.32 and 3.33 are from magneto-luminescence measurements. For $B = 0$ the luminescence maximum is at 1554 meV.

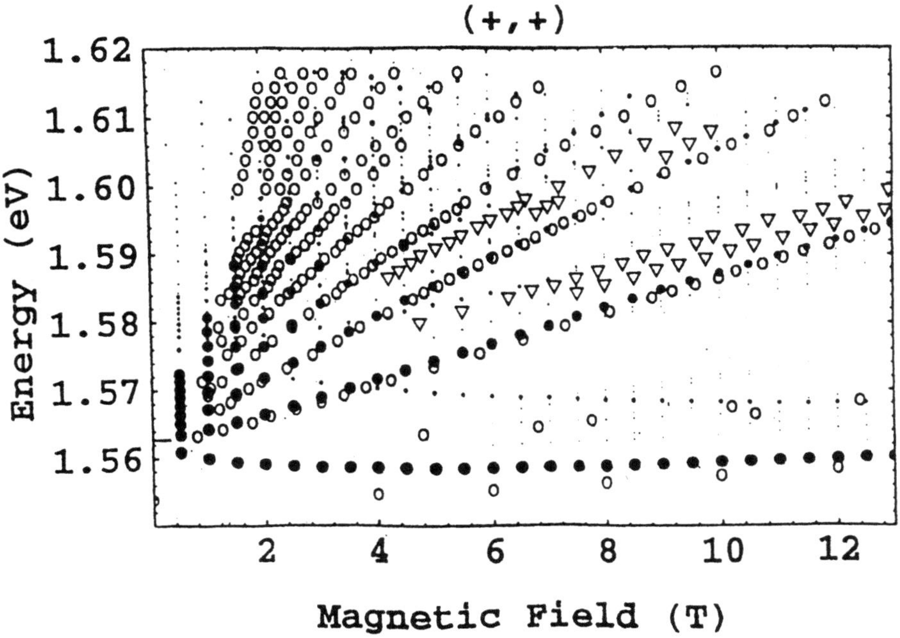

Fig. 3.32. Fan plot of the continuous emission magneto-Raman oscillations (*open symbols*) in a (100/103) Å GaAs/$Al_{0.35}Ga_{0.65}As$ MQW: $\bar{z}(\sigma^+, \sigma^+)z$ geometry. The *filled circles* correspond to calculated interband transition energies. See text for details

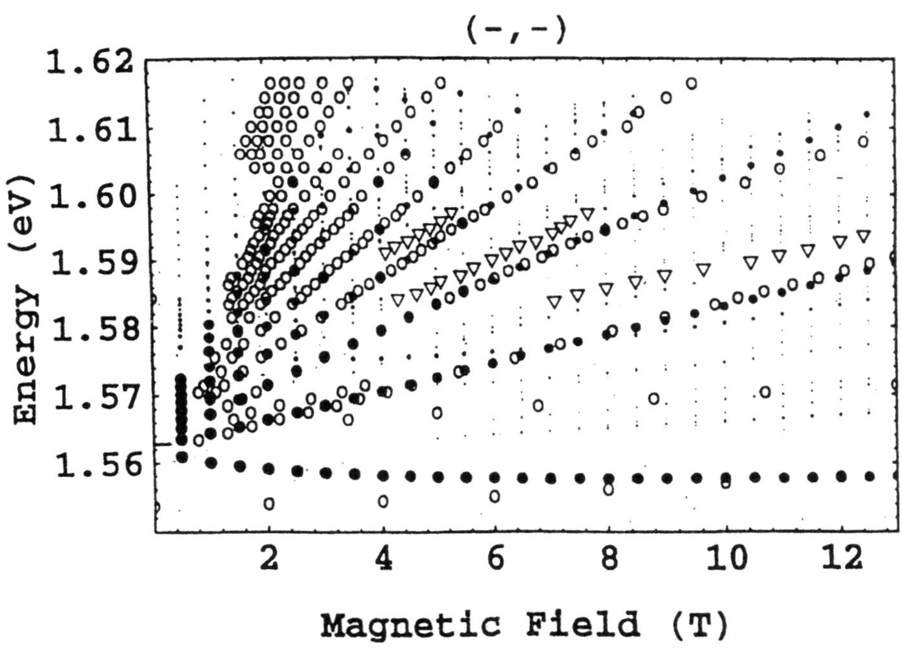

Fig. 3.33. As Fig. 3.32, $\bar{z}(\sigma^-, \sigma^-)z$ geometry

Theory. A quantitative analysis of the resonances in Figs. 3.32 and 3.33 requires the calculation of the magneto-optical transitions in QWs, including the conduction band nonparabolicity, the complicated, mixed valence band structure and the corrections due to excitonic effects. The transition energies as well as the intensities are influenced by these factors. The conduction band nonparabolicity in bulk GaAs contributes significantly to the confinement-induced part of the transition energy between hole and electron subbands which is dominated by the electron states [1.35]. The mixing of the valence bands, on the other hand, is responsible for the relaxation of the selection rules which hold for a simple parabolic model of the band structure [3.70]. Due to the increasing Coulomb interaction in low-dimensional systems even the exciton binding energy may yield contributions to the transition energies which cannot be neglected in comparison to the effects of confinement [3.70]. In an earlier investigation, both the nonparabolicity and the valence band mixing were taken into account in a calculation of the interband magneto-absorption in a six-band model of QW envelope functions [3.70]. The couplings between the upper Γ_8^v heavy-hole (HH) and light-hole (LH) bands and the lowest Γ_6^c conduction band were considered in the frame of $\boldsymbol{k} \cdot \boldsymbol{p}$ perturbation theory [3.70]. However, by comparison with experiment it was found that nonparabolicity effects are larger than predicted by this model and that a larger conduction band effective mass compared to bulk GaAs is necessary in order to describe the data [3.70]. For these reasons we chose a different approach [1.26]. Using different Hamiltonians for the valence and conduction bands we include a rather precise expression for the conduction band nonparabolicity, based on a (14×14)-$\boldsymbol{k} \cdot \boldsymbol{p}$-model [3.71]. The valence band mixing is treated separately. Neglecting a small contribution from the electron g factor, the bulk conduction band dispersion $E(\boldsymbol{k})$ is expanded up to fourth order in $\boldsymbol{k}$ and reads [3.71]

$$E(\boldsymbol{k}) = \frac{\hbar^2 k^2}{2m_\mathrm{e}} + \alpha_0 k^4 + \beta_0 (k_x^2 k_y^2 + k_y^2 k_z^2 + k_z^2 k_x^2) \; . \tag{3.17}$$

In this expression m_e is the effective mass at the conduction band edge of the bulk. The nonparabolicity and anisotropy of the conduction band are described by the parameters α_0 and β_0 whose dependence on the interband matrix elements and band gaps is given in [3.71]. We also neglect the spin-splitting due to the lack of inversion symmetry in zincblende III-V semiconductors (k^3-term). This is justified below.

In a QW the conduction band nonparabolicity leads to a renormalized effective mass along the growth direction, $m_\mathrm{e}^\perp$, which determines the confinement of states, and to a renormalized effective mass for the dispersion in the QW plane, $m_\mathrm{e}^\parallel$, which is reflected in the cyclotron motion of Landau levels due to a magnetic field perpendicular to the plane [3.72]. An important result of [3.72] is that these masses are generally different from each other. Taking the QW growth direction along $\hat{z}$ and the bulk dispersion of (3.17) one obtains implicit expressions for the energies due to confinement, ϵ, and the in-plane dispersion, $E(\boldsymbol{k}_\parallel)$, where $\boldsymbol{k}_\parallel = (k_x, k_y)$. For small ϵ the following effective masses can be defined:

$$m_{\mathrm{e}}^{\perp w}(\epsilon) = m_{\mathrm{e}}^{w}(1 + \alpha'_w \epsilon) \, , \tag{3.18a}$$

$$m_{\mathrm{e}}^{\perp b}(\epsilon) = m_{\mathrm{e}}^{b}(1 - \alpha'_b(V - \epsilon)) \, , \tag{3.18b}$$

$$m_{\mathrm{e}}^{\parallel w}(\epsilon) = m_{\mathrm{e}}^{w}(1 + (2\alpha'_w + \beta'_w)\epsilon) \, . \tag{3.18c}$$

In these equations one has $\alpha'_i = -(2m_{\mathrm{e}}^i/\hbar^2)^2\alpha_{i0}$ and $\beta'_i = -(2m_{\mathrm{e}}^i/\hbar^2)^2\beta_{i0}$, where m_{e}^i, α_{i0} and β_{i0} are the material parameters from (3.17) and $i = w, b$ denotes the well or barrier. While the increase of $m_{\mathrm{e}}^{\perp w}$ in comparison to the three-dimensional crystal ($\epsilon = 0$) is determined by α_0, the increase of $m_{\mathrm{e}}^{\parallel w}$ depends on the bulk parameters via a factor $2\alpha_0 + \beta_0$ which leads to a larger effective mass. The expressions (3.18a) and (3.18c) hold for infinite barriers [3.72]. The formula in (3.18b) is obtained by expanding the general expression, given by equation (15) in [3.72], to the lowest order in $V - \epsilon$, where V is the discontinuity of the conduction band minima for the well and barrier materials. These theoretical predictions were already compared with experiments in [3.72]. However, most of the data analyzed there are from cyclotron resonance measurements on modulation-doped samples for which this theory does not directly apply [3.72]. Even though the trends for the effective QW masses expected according to (3.18a) and (3.18c) are observed in the studies discussed in [3.72], further tests of the theory's validity seemed necessary, especially by magneto-optical interband experiments [3.72]. Another test of these considerations, using optically detected cyclotron resonance as an intraband method to determine the effective masses even in undoped QWs, can be found in [3.73].

The energy ϵ due to the electronic confinement is calculated by integrating an effective-mass equation [1.26]. We introduce the usual substitution $\frac{1}{m}\frac{\partial^2}{\partial z^2} \to \frac{\partial}{\partial z}\frac{1}{m(z)}\frac{\partial}{\partial z}$ where $m(z)$ stands for the well and barrier masses, $m_{\mathrm{e}}^{\perp w}$ and $m_{\mathrm{e}}^{\perp b}$, respectively. Boundary conditions require the continuity of the envelope function and its first derivative divided by the effective mass $m_{\mathrm{e}}^{\perp}(\epsilon)$ at the interfaces of the two materials. Note that in these calculations we use energy-dependent effective masses, corresponding to the boundary condition $N = 1$ in [3.72], and not the band edge values of the bulk constituents. The energies of Landau levels for small magnetic fields B can be derived from (3.17) by applying perturbation theory with respect to $k_\parallel$. One finds that [3.72]

$$\begin{aligned} E(B) = \epsilon_\lambda + \left(n + \frac{1}{2}\right)\frac{\hbar e B}{m_{\mathrm{e}}^{\parallel w}} + \\ - \frac{1}{8}\left[(8n^2 + 8n + 5)\alpha'_w + (n^2 + n + 1)\beta'_w\right]\left(\frac{\hbar e B}{m_{\mathrm{e}}^{w}}\right)^2 \, , \end{aligned} \tag{3.19}$$

where ϵ_λ is the energy of subband λ due to the electronic confinement. The first two terms in (3.19) correspond to an electron in a parabolic band whose effective mass is given by (3.18c). The third term originates from further contributions of the conduction band nonparabolicity and anisotropy to the dispersion in the QW plane which result from (3.17) [3.72]. While this term can be neglected for small B, its contribution to the Landau level energies in the range of fields considered here amounts to several percent. It is thus taken into account in the data analysis.

In the following calculation of the fan plots in Figs. 3.32 and 3.33 the valence band states are described by a (4×4)-Hamiltonian H_L in the basis of $|J = \frac{3}{2}, J_z\rangle$ quasi-angular-momentum states for the heavy and light holes, which goes back to Luttinger [3.74]. It reads

$$H_L = \begin{array}{c} \begin{matrix} +3/2 & -1/2 & +1/2 & -3/2 \end{matrix} \\ \begin{pmatrix} H_h & c & b & 0 \\ c^\dagger & H_l & 0 & -b \\ b^\dagger & 0 & H_l & c \\ 0 & -b^\dagger & c^\dagger & H_h \end{pmatrix} \end{array} , \tag{3.20}$$

where H_h and H_l as well as the terms b and c, responsible for the dispersion mixing in the QW plane, are given by

$$H_h = -\frac{\hbar^2(k_x^2 + k_y^2)}{2m_0}(\gamma_1 + \gamma_2) - \frac{\hbar^2 k_z^2}{2m_0}(\gamma_1 - 2\gamma_2) , \tag{3.21a}$$

$$H_l = -\frac{\hbar^2(k_x^2 + k_y^2)}{2m_0}(\gamma_1 - \gamma_2) - \frac{\hbar^2 k_z^2}{2m_0}(\gamma_1 + 2\gamma_2) , \tag{3.21b}$$

$$b = \sqrt{3}\frac{\hbar^2}{m_0}\gamma_3(k_x - ik_y)k_z , \tag{3.21c}$$

$$c = \sqrt{3}\frac{\hbar^2}{2m_0}[\gamma_2(k_x^2 - k_y^2) - 2i\gamma_3 k_x k_y] . \tag{3.21d}$$

The zero of energy is located at the top of the valence band. The dimensionless constants γ_1, γ_2 and γ_3 are the Luttinger parameters [2.153, 3.74]. In this model we assume that the spin–orbit split-off valence states do not influence the heavy- and light-hole dispersions. In (3.21a) and (3.21b) one recognizes an interesting effect which has a profound influence on the band structure. The effective mass of the HH (H_h) and LH (H_l) states along k_z is given by $1/(\gamma_1 \mp 2\gamma_2)$. This leads to a smaller confinement energy of the HH subbands than for the LH ones. However, in the perpendicular plane the effective masses of these bands are proportional to $1/(\gamma_1 \pm \gamma_2)$. Thus the HH dispersion is larger than that of the LH states. This mass reversal parallel and perpendicular to the growth direction leads to strong mixing and anisotropy effects of valence band states in QWs [3.70, 3.72] which are further discussed in [1.26].

For the comparison between theory and experiment the Coulomb interaction was considered by a correction to the calculated transition energies. For the exciton ground state of quasi-two-dimensional systems in a strong external magnetic field the following expression for the field- and Landau-index-dependent binding energy has been obtained [3.75]

$$E_{\mathrm{ex}}(n, B) = 3R^* D \sqrt{\frac{eB}{2(2n+1)\mu R^*}}\,, \tag{3.22}$$

where n is the Landau quantum number of the electron level involved, μ is the reduced effective mass of the exciton and $R^* = \mu e^4/2\epsilon^2$ is an effective Rydberg constant with the dielectric constant ϵ of the QW. D is a dimensionality parameter with values between 1/4 for three dimensions and 1 for a strictly two-dimensional system.

Effective-Mass Anisotropy. The filled circles in Figs. 3.32 and 3.33 are the energies of the magneto-optical resonances calculated with the model described. The diameter of these points is proportional to the oscillator strength of the transitions. The dominating fan lines correspond to resonances between Landau levels of the first HH and electron subbands. Several weaker structures can be assigned to transitions from Landau levels of the first LH subband to electron states. Some parameters in the calculation were adjusted in order to obtain optimum agreement with the experiment. At first the valence band Landau levels were calculated from known constants which are summarized in Table 3.3. The electron Landau levels were calculated with (3.19). From fits to the fan plots we obtain $m_{\mathrm{e}}^{\|} = (0.073 \pm 0.001)\, m_0$. The exciton effects were taken into account by subtracting the ground state binding energy according to (3.22) from the calculated transitions using the parameters given in Table 3.3. The fact that (3.22) only holds for high magnetic fields

Table 3.3. Material parameters for the calculation of the magneto-optical interband transitions which were fitted to the fan plots in Figs. 3.32 and 3.33

	GaAs	$Al_{0.35}Ga_{0.65}As$
E_g (eV)	1.519 [2.153]	2.023 [3.76]
E_c (eV)		0.335 [3.77]
α_0 (eVÅ^4) / α' (eV^{-1}) [3.71,3.72]	−2107 / 0.642	−1071 / 0.753
β_0 (eVÅ^4) / β' (eV^{-1}) [3.71,3.72]	−2288 / 0.697	−1508 / 1.060
m_{e} (m_0) [2.153]	0.0665	0.101
$m_{\mathrm{e}}^{\perp}$ (m_0)	0.068[(a)]	0.078
$m_{\mathrm{e}}^{\|}$ (m_0)	0.073[(a)]	
D	0.8	
E_{ex}^{hh} (meV)	9 [(a)]	
E_v (eV)		0.168 [3.77]
γ_1 [2.153]	6.85	5.66
γ_2 [2.153]	2.1	1.6
γ_3 [2.153]	2.9	2.34
κ [2.153]	1.2	

(a) experimental value

and large Landau indices n leads to a systematic deviation of the measured fan lines from the calculation for $B < 3\,\mathrm{T}$. This range was therefore not considered in the determination of $m_e^{\parallel}$. From the difference of the convergence points of fan lines for $B \to 0$ and the luminescence data in Figs. 3.32 and 3.33 we estimate the zero-field exciton binding energy to be about $E_{ex}^{hh} = 9\,\mathrm{meV}$. This value is in good agreement with an exciton energy of 10 meV calculated in [3.78] for a 100 Å GaAs QW surrounded by $Al_{0.4}Ga_{0.6}As$ barriers. With the parameters for the well and barrier materials in Table 3.3 and the theory outlined above one obtains energies of 34 meV and 7.4 meV for the first electron and heavy-hole subbands. With this value of the hole subband and the measured transition energies at $B = 0$ one obtains an experimental electron subband energy of $\epsilon = 36.5\,\mathrm{meV}$. According to (3.18a) this corresponds to an effective mass of $m_e^{\perp} = 0.068\,m_0$. The difference between the measured and calculated value of ϵ can partially be attributed to uncertainties in the determination of the QW layer thickness. Note, however, that different boundary conditions in the envelope function approximation can also lead to differences in ϵ of this order [3.72]. It has also been pointed out that modifications of the usual boundary conditions are necessary in order to maintain the conservation of the probability current for the case of nonparabolic bands [3.72, 3.79]. For a 100 Å GaAs QW this causes an increase of ϵ by 2 meV [3.72]. Both effects were phenomenologically considered in the calculations for Figs. 3.32 and 3.33 by setting the well width to 96 Å. The experimentally determined masses $m_e^{\parallel}$ and $m_e^{\perp}$ are slightly smaller than the values of $m_e^{\parallel} = 0.074\,m_0$ and $m_e^{\perp} = 0.069\,m_0$ calculated from (3.18c) and (3.18a) assuming infinitely high barriers. These differences could be due to the penetration of the subband wave functions into the barrier material.

Further Effects. This section should not be concluded without some remarks on issues which have not been discussed in the data analysis so far. In the expansion of $E(\boldsymbol{k})$ up to fourth order in $\boldsymbol{k}$ one obtains in (3.17) also a cubic contribution which is described by a constant γ_0 [3.71]. It is responsible for the conduction band spin splitting even at zero field and arises from the lack of inversion symmetry in III-V semiconductors [3.80]. In bulk GaAs this effect is so small that it can be neglected [3.80]. In a QW of width d, however, it can become large since the replacement of k_z by the wavevector of size quantization π/d leads to terms which are effectively linear in $k_{\parallel}$ [3.81]. We have thus estimated the increase of $m_e^{\parallel}$ by the γ_0-term averaged over the QW plane. For the sample considered we find a contribution which is about one-third of the change due to the anisotropic β_0-term of fourth order in $\boldsymbol{k}$. On the other hand the β_0-term itself contributes another 50% to the increase of $m_e^{\parallel}$ which is mainly given by $\alpha_0 k^4$ (see (3.18c)). Thus the change of $m_e^{\parallel}$ due to the cubic term can be neglected within the experimental accuracy.

Strictly speaking, the confinement energies of the electron and hole states as well as the energy-dependent effective masses should be calculated in a self-consistent way. In the above considerations this procedure was approximated by using subband energies and effective masses obtained from a simple Kronig–Penney calculation. With these values we then solved the more complicated effective-mass equation. This procedure is equivalent to the first step of a self-consistency algorithm and yields subband energies which are precise enough for the description of the experimental findings. It should also be noted that we neglected the influence of the hole quantization on the electron mass and changes of the valence band masses (Luttinger parameters) due to effectively larger band gaps, since these effects are small in comparison with those in the conduction band.

Another important point which has not been considered in connection with the electronic structure of QWs so far is the electron mass anisotropy due to different subband energies of the heavy and light hole states. In measurements of bulk semiconductors under uniaxial stress X it has been shown that the valence band splittings and the resulting different interband transition energies cause an anisotropy of the effective electron mass parallel and perpendicular to X [3.82–3.84]. In a QW whose symmetry is lowered in analogy to a three-dimensional crystal under uniaxial stress one should thus use different effective masses $m_{\mathrm{e},0}^{\perp w}$ and $m_{\mathrm{e},0}^{\| w}$ in (3.18a) and (3.18c) instead of the bulk band edge mass m_{e}^{w}. A quantitative calculation of this mass anisotropy for uniaxial stress is not possible within a two-band model since the contributions of higher bands are important [3.82, 3.83]. In order to estimate the effect for QWs we have calculated the mass anisotropy $\Delta m/m = |(m_{\mathrm{e},0}^{\|} - m_{\mathrm{e},0}^{\perp})/((m_{\mathrm{e},0}^{\|} + m_{\mathrm{e},0}^{\perp})/2)|$ in bulk GaAs for values of the uniaxial stress which lead to the same splitting between the heavy- and light-hole states as the electronic confinement. For this purpose we have used data from infrared measurements of the free-carrier plasma frequency under uniaxial stress given in [3.84]. For $X = -0.3\,\mathrm{GPa}$, corresponding to a well width of 100 Å, we find $\Delta m/m = 0.7\%$. This value is an order of magnitude smaller than the experimental mass anisotropy due to the nonparabolicity and anisotropy of the conduction band (7.1 %) and can thus be neglected within the experimental accuracy. For a (44/44) Å GaAs/AlAs MQW, which has been investigated by optic-phonon magneto-Raman spectroscopy in [1.39] (see Sect. 4.3), one obtains $\Delta m/m = 2.0\%$ for $X = -1.0\,\mathrm{GPa}$. The experimental mass anisotropy is 20.9% [1.39].

Even though these points have been neglected in the data analysis so far, a better theoretical treatment of the QW electronic structure in a multiband model is necessary in order to determine their respective contributions to the mass anisotropy. More precise measurements are required for their experimental verification.

4. Optic-Phonon Magneto-Raman Scattering

In this Chapter we discuss the basic features of magneto-Raman scattering by optic phonons and its applications to the investigation of the electronic structure and the electron–phonon interaction in bulk semiconductors (GaAs, InP, $Cd_{0.95}Mn_{0.05}Te$) and GaAs/AlAs QWs.

4.1 Introduction to Magneto-Raman Scattering

It has been mentioned already in Chap. 1 that the periodic modulation of the semiconductor electronic structure by a phonon allows one to consider Raman scattering as a form of modulation spectroscopy which enhances features around critical points of the band structure [1.12]. Before the advent of continuous emission, which has been observed so far only in quantum wells and superlattices due to its origin in crystal-momentum-nonconserving scattering processes, resonant magneto-Raman scattering by optic phonons, mostly with longitudinal (LO) character, has been established as a unique technique for the investigation of the electronic structure and the electron–phonon interaction in bulk semiconductors in high magnetic fields [1.28–1.38]. Since a phonon-Raman process involves steps mediated by electron–photon as well as electron–phonon coupling (see (2.3)), magneto-Raman scattering offers additional degrees of freedom compared with the conventional modulated magneto-optical spectroscopies [4.1]. The multitude of scattering processes possible within a real semiconductor band structure and the different mechanisms of the electron–phonon interaction lead to rather distinctive selection rules which make detailed investigations possible. It is beyond the scope of this chapter to present a comprehensive review of the magneto-Raman studies which have been performed in recent years. Instead, this section focuses on the general features and emphasizes the basic concepts. In the following two sections we give specific examples of LO-phonon magneto-Raman scattering from bulk semiconductors (Sect. 4.2) and quantum wells (Sect. 4.3).

4.1.1 Resonant Raman Scattering at Landau Levels

As discussed before, the energy denominators of the expression for the Raman intensity in (2.3) lead to resonances in the incoming or outgoing channel when the energies of the incident or scattered photons coincide with electronic interband transitions. In some cases both denominators are simultaneously small, and double resonance occurs. This enhances the signal even more. A strong magnetic field, where the electrons or holes are quantized in Landau levels and perform a cyclotron motion perpendicular to the field direction, offers the possibility to easily tune electronic resonances over a wide range of energies. Furthermore, the characteristic change in the density of states which accompanies Landau quantization sharpens the resonances and thus leads to stronger signals and more structured spectra.

The different possibilities for Raman resonances between Landau levels are schematically shown in Fig. 4.1. For a specific combination of laser energy $E_L = E^{DR}$ and magnetic field $H = H^{DR}$ one obtains a double resonance (b). At this point the separation between the hole states (h and l) is equal to the LO phonon energy. Away from double resonance, incoming ("in") and outgoing ("out") resonances for the same laser energy occur at different magnetic fields (a,c).

Magneto-Raman profiles measure the LO phonon scattering intensity vs. magnetic field for a fixed excitation energy. One expects resonances for each interband transition between Landau levels. The intensity profiles should thus exhibit oscillations with a characteristic periodicity proportional to the inverse of the magnetic field. These peaks are due to incoming and outgoing resonances whose magnetic field positions change with the excitation energy. This behavior is modeled in Fig. 4.2 using parameters typical for GaAs [1.32].

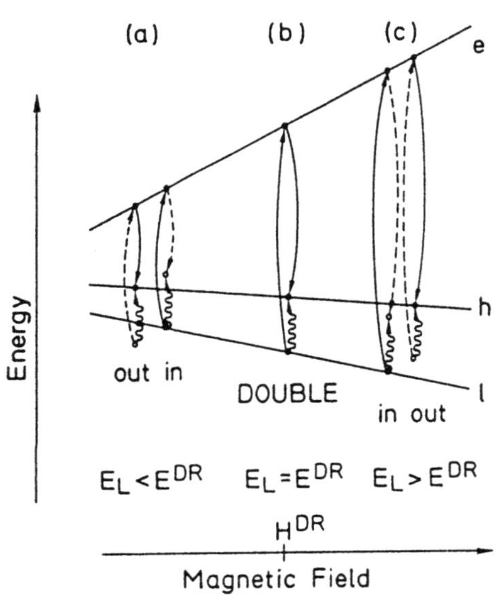

Fig. 4.1. Schematic illustration of resonant magneto-Raman scattering processes in a semiconductor. The states e, h, and l present Landau levels of electrons, heavy, and light holes and depend linearly on the field. The *solid/dashed arrows* denote real/virtual optical transitions; the *wavy lines* repesent the emitted optical (LO) phonon. See text for details

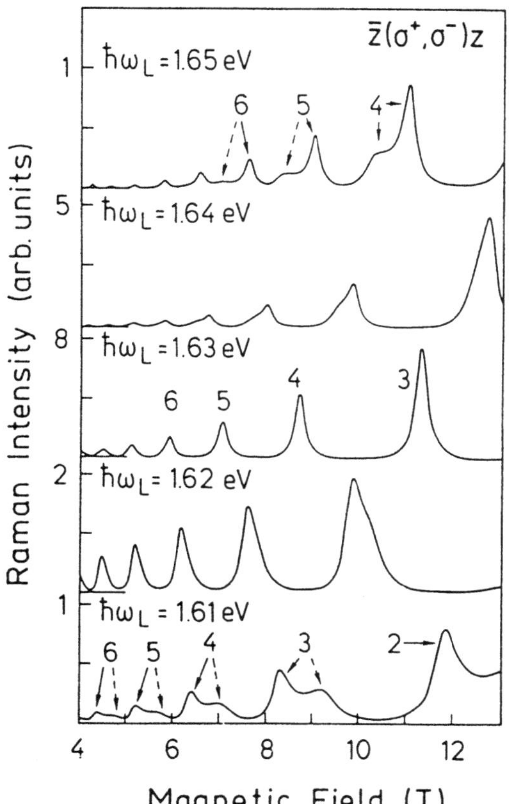

Fig. 4.2. Theoretical LO phonon magneto-Raman profiles for the simple model of Landau levels in a semiconductor given in Fig. 4.1. Double resonance occurs at the laser energy $\hbar\omega_L = 1.63$ eV. Away from this energy the spectra show separate incoming and outgoing resonances, marked by *dashed* and *solid arrows*

At the lowest and highest laser energy the profiles exhibit two well-separated peaks for each Landau level index (assumed to be the same for electrons and holes). They correspond to incoming and outgoing resonances with the participation of light and heavy holes, respectively. Between these energies the two types of resonances approach each other. At $\hbar\omega_L = 1.63$ eV the transition energies only differ by the LO phonon energy. A series of double resonances occurs (see Fig. 4.1 (b)), and the signal is strongly enhanced.

4.1.2 Electronic Structure in a Magnetic Field

Idealized case. The resonance behavior of the magneto-Raman intensity just discussed is based on a rather simple model for the semiconductor electronic structure. The linear magnetic-field dependence of the Landau levels reflects constant effective masses m^* and a parabolic bulk dispersion. Comparing the kinetic energy of a carrier without field with its cyclotron energy one obtains the following relation between the crystal momentum $k_{\parallel}$ perpendicular to the field B and the index n of a Landau orbit [4.2]

$$\frac{\hbar^2 k_{\parallel}^2}{2m^*} = \frac{\hbar e B}{m^*}\left(n + \frac{1}{2}\right) \tag{4.1}$$

which holds in the quasiclassical limit of large n and small B where the Landau levels are closely spaced. Distinguishing between heavy- and light-hole states one assumes that these are pure and do not interact with each other, i. e., they are regarded as $|J_z = \pm 3/2\rangle$ and $|J_z = \pm 1/2\rangle$ components of a $|J = 3/2, J_z\rangle$ quasi-angular-momentum manifold, respectively.

Real Semiconductors. For real semiconductors, like the direct gap III-V and II-VI compounds of interest here, these assumptions have to be modified in several ways. The bands are nonparabolic and can be described by constant effective masses only in the close vicinity of the gap. Away from the band extrema one has to choose more accurate approximations, such as expansions to higher orders in the wavevector (see Sect. 3.7). In the framework of $\boldsymbol{k} \cdot \boldsymbol{p}$-theory this leads to complicated expressions for the dispersion, like the one given by (3.17), and additional parameters which involve matrix elements to higher bands appear. In this case the Landau levels no longer depend linearly on n and B (see (3.19)). A common way to take such deviations from a parabolic behavior into account is the introduction of energy-dependent effective masses.

At the top of the valence band the heavy- and light-hole states are fourfold degenerate and for arbitrary directions in $\boldsymbol{k}$-space their wave functions contain admixtures of different $|J, J_z\rangle$ components. The dispersion is also affected by interactions with higher bands, and for a precise description one usually resorts to an (8×8)-$\boldsymbol{k} \cdot \boldsymbol{p}$-matrix which includes the twofold degenerate Γ^{6c} conduction band, the Γ^{8v} heavy- and light-hole states and the Γ^{7v} spin–orbit split-off band (twofold degenerate) [3.74, 4.3–4.8]. In a magnetic field couplings between these states also modifiy the valence band Landau levels. From numerical solutions of such rather complicated Hamiltonians one finds that the wave functions sometimes contain admixtures of up to four different bands with almost equal weights. This has profound consequences for interband magneto-optical transitions and Raman processes.

Even though many quantitative comparisons of magneto-Raman fan plots with theory have been performed using (8×8)-Hamiltonians, one gains more of a physical insight into the relevant mechanisms from smaller models. An approximate analytical solution for valence band Landau levels has been given by Luttinger who showed that the (4×4)-matrix for the Γ^{8v} manifold can be decoupled into two (2×2)-blocks [3.74]. This is possible for $k_z = 0$, $q = 0$, and $\gamma_2 = \gamma_3$, i. e., for zero crystal momentum along the field direction, neglecting the anisotropic contribution to the valence band g factor, and disregarding the valence band warping, proportional to the difference in the Luttinger parameters $\gamma_2 - \gamma_3$, respectively. In this case one obtains the following eigenvalue problems for the hole Landau levels [3.74]:

$$\begin{pmatrix} (\gamma_1+\gamma')\left(n-\frac{3}{2}\right)+\frac{3}{2}\kappa-\epsilon & -\sqrt{3}\gamma''\sqrt{n(n-1)} \\ -\sqrt{3}\gamma''\sqrt{n(n-1)} & (\gamma_1-\gamma')\left(n+\frac{1}{2}\right)-\frac{1}{2}\kappa-\epsilon \end{pmatrix} \Psi_a = 0 \ , \tag{4.2a}$$

$$\begin{pmatrix} (\gamma_1-\gamma')\left(n-\frac{3}{2}\right)+\frac{1}{2}\kappa-\epsilon & -\sqrt{3}\gamma''\sqrt{n(n-1)} \\ -\sqrt{3}\gamma''\sqrt{n(n-1)} & (\gamma_1+\gamma')\left(n+\frac{1}{2}\right)-\frac{3}{2}\kappa-\epsilon \end{pmatrix} \Psi_b = 0 \ . \tag{4.2b}$$

The eigenvectors are [3.74]

$$\Psi_a(n) = \begin{pmatrix} a_1(n)\ |n-2, J_z=+\frac{3}{2}\rangle \\ a_2(n)\ |n, J_z=-\frac{1}{2}\rangle \end{pmatrix} , \tag{4.3a}$$

$$\Psi_b(n) = \begin{pmatrix} b_1(n)\ |n-2, J_z=+\frac{1}{2}\rangle \\ b_2(n)\ |n, J_z=-\frac{3}{2}\rangle \end{pmatrix} . \tag{4.3b}$$

They are mixtures of products between the $|J = 3/2, J_z\rangle$ quasi-angular-momentum wave functions and Landau states $|n\rangle$ $(n \geq 0)$. The Landau oscillator quantum numbers of the two components in each wave function differ by two. The coupled J_z components are $|J_z = \pm 3/2\rangle$ and $|J_z = \mp 1/2\rangle$, respectively. Admixing coefficients for oscillators with negative indices are zero. The matrices (4.2a) and (4.2b) were set up for arbitrary directions of the magnetic field in the (110) plane. The anisotropy of the valence band is partly considered by the parameters γ' and γ'' which, using $c = \cos\theta$ and $s = \sin\theta$, are defined as

$$\begin{aligned} \gamma' &= \frac{1}{4}\left[(3c^2-1)^2\ \gamma_2 + 3s^2\ (3c^2+1)\ \gamma_3\right] , \\ \gamma'' &= \frac{1}{8}\left[(3-2c^2+3c^4)\ \gamma_2 + (5+2c^2-3c^4)\ \gamma_3\right] . \end{aligned} \tag{4.4}$$

where θ is the angle between the field $\boldsymbol{B}$ and the z direction. For $\boldsymbol{B} \parallel \hat{z}$ we have $\theta = 0°$ and

$$\gamma' = \gamma_2 \ , \qquad \gamma'' = \frac{1}{2}\ (\gamma_2+\gamma_3) \ . \tag{4.5}$$

For $\boldsymbol{B} \parallel [110]$ $(c = 0, s = 1)$ these parameters are

$$\gamma' = \frac{1}{4}\ (\gamma_2+3\gamma_3) \ , \qquad \gamma'' = \frac{1}{8}\ (3\gamma_2+5\gamma_3) \ . \tag{4.6}$$

If necessary the valence band warping proportional to $\gamma_2 - \gamma_3$, which couples the two (2×2)-blocks ((4.2a) and (4.2b)) of the $\Gamma^{8\mathrm{v}}$ matrix, can be treated by perturbation theory. Expressions are given in [3.74]. For the hole Landau level energies one finds the following expressions (in units of $\hbar eB/m_0$):

Table 4.1. Valence band Landau levels $E(n, 1\pm)$ and $E(n, 2\pm)$ and wave function mixing coefficients for GaAs calculated from (4.2a) and (4.2b). Energies are given in units of $\hbar eB/m_0 = 0.1158\,\mathrm{meV/T}\cdot B$

			Heavy-mass levels				
		$E(n,1-)$				$E(n,2-)$	
		$\lvert n-2,+\frac{3}{2}\rangle$	$\lvert n,-\frac{1}{2}\rangle$			$\lvert n-2,+\frac{1}{2}\rangle$	$\lvert n,-\frac{3}{2}\rangle$
n	E	a_1^-	a_2^-	n	E	b_1^-	b_2^-
2	−2.2	0.830	0.558	2	−1.1	0.954	0.299
3	−5.0	0.720	0.694	3	−3.4	0.927	0.376
4	−7.4	0.666	0.746	4	−5.6	0.910	0.415
5	−9.6	0.635	0.772	5	−7.8	0.899	0.439
6	−11.7	0.616	0.788	6	−9.9	0.891	0.454
7	−13.8	0.602	0.798	7	−12.0	0.885	0.466
			Light-mass levels				
		$E(n,1+)$				$E(n,2+)$	
		$\lvert n-2,+\frac{3}{2}\rangle$	$\lvert n,-\frac{1}{2}\rangle$			$\lvert n-2,+\frac{1}{2}\rangle$	$\lvert n,-\frac{3}{2}\rangle$
n	E	a_1^+	a_2^+	n	E	b_1^+	b_2^+
0	−1.8	0.000	−1.000	0	−2.7	0.000	−1.000
1	−6.5	0.000	−1.000	1	−11.6	0.000	−1.000
2	−15.4	0.558	−0.830	2	−22.5	0.299	−0.954
3	−26.2	0.694	−0.720	3	−33.8	0.376	−0.927
4	−37.6	0.746	−0.666	4	−45.3	0.415	−0.910
5	−49.1	0.772	−0.635	5	−56.9	0.439	−0.899

$$E(n,1\pm) = \gamma_1 n - (\frac{1}{2}\gamma_1 + \gamma' - \frac{1}{2}\kappa) + \pm\sqrt{\left(\gamma' n - (\frac{1}{2}\gamma' + \gamma_1 - \kappa)\right)^2 + 3\gamma''^{\,2}\, n\,(n-1)} \tag{4.7a}$$

$$E(n,2\pm) = \gamma_1 n - (\frac{1}{2}\gamma_1 - \gamma' + \frac{1}{2}\kappa) + \pm\sqrt{\left(\gamma' n + (-\frac{1}{2}\gamma' + \gamma_1 - \kappa)\right)^2 + 3\gamma''^{\,2}\, n\,(n-1)} \tag{4.7b}$$

where $n \geq 0$ for the "+" ladders (light-mass levels) and $n \geq 2$ for the "−" states (heavy masses). As an example, calculated energies and admixture coefficients are given in Table 4.1 using parameters for GaAs ($\gamma_1 = 6.85$, $\gamma_2 = 2.10$, $\gamma_3 = 2.90$, $\kappa = 1.20$ [2.153]). Note the irregular spacing of the levels for small n. For large n the Landau levels become equidistant and the wave function mixing coefficients approach $\pm 1/2$ or $\pm\sqrt{3}/2$. The dominating admixture of $|\pm\sqrt{3}/2| = 0.866$ occurs for the heavy-mass holes at the component with $J_z = \pm 1/2$. For the light-mass holes it appears at the $(J_z = \pm 3/2)$ admixtures.

4.1.3 Magneto-Raman Processes and Selection Rules

The effects of band nonparabolicity and valence band mixing just mentioned have to be taken into account in order to understand magneto-Raman fan plots of "real" semiconductors. One of the reasons why magneto-Raman scattering is such a sensitive probe of the electronic structure and its modifications due to elementary excitations is the fact that Raman processes involve steps mediated by both the electron–photon and the electron–phonon interactions. This gives rise to unique and stringent selection rules and allows one to subdivide the large number of possible resonances into small sets which can then be compared with carefully chosen approximations of theories with otherwise overwhelming complexity.

We first illustrate the possible Raman processes for an idealized semiconductor with three parabolic bands, characterized by effective masses m^*, g factors and pure $|S = 1/2, S_z\rangle$ (electrons), $|J = 3/2, J_z\rangle$ (heavy and light holes), and $|J = 1/2, J_z\rangle$ (spin–orbit split-off holes) wave functions. This situation is schematically shown in Fig. 4.3 (a) [1.32]. The material parameters can be obtained from known band and Luttinger parameters using expressions derived from $\boldsymbol{k} \cdot \boldsymbol{p}$-theory [4.9]. We assume that the wave functions do not mix; their Landau oscillator components are $|n\rangle$.

Deformation-Potential Coupling. For the deformation-potential electron–phonon interaction (see (2.4)) one obtains the Raman processes compiled in Table 4.2 in the configurations $\bar{z}(\sigma^{\pm}, \sigma^{\mp})z$, i.e., for backscattering geometries with complementary circular polarizations ($e^{\pm} = \sigma^{\pm} =$

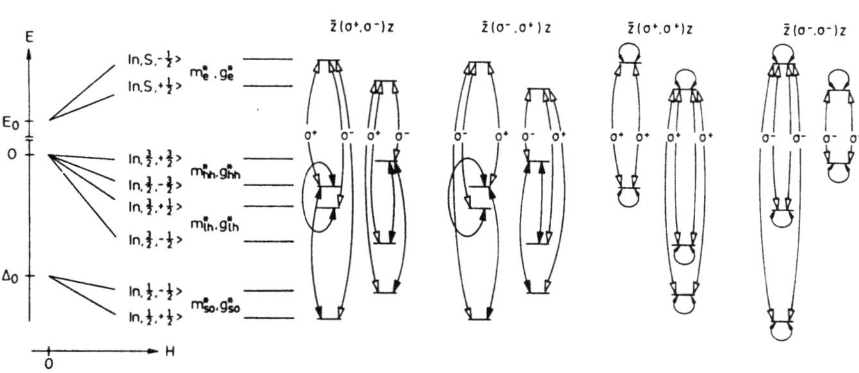

Fig. 4.3. Raman processes between Landau levels in a parabolic three-band model. **(a)** Schematic presentation of parameters and wave functions; **(b)** Raman processes for deformation-potential interaction ($\bar{z}(\sigma^{\pm}, \sigma^{\mp})z$); **(c)** Raman processes for Fröhlich interaction ($\bar{z}(\sigma^{\pm}, \sigma^{\pm})z$). *Filled (open) arrows* indicate transitions due to electron–phonon (electron–photon) coupling. See text for details

Table 4.2. Raman processes with deformation-potential scattering in the $\bar{z}(\sigma^{\pm},\sigma^{\mp})z$ geometries. The "transitions" denote the step of phonon emission where a hole is scattered between different valence levels v_i and v_j ($|J,J_z\rangle$). The interaction matrix elements (see (2.3)) and the spin of the electron involved are indicated for each case. From this information the magneto-optical interband transitions can be reconstructed. P is the interband matrix element between the Γ^{8v} valence and Γ^{6c} conduction bands; d_0 is the optic deformation-potential constant

$\langle c\,|e_{\pm}^{*}\cdot\boldsymbol{p}\,|\,v_j\rangle\, D_{v_i}^{v_j}\,\langle v_i\,|e_{\mp}\cdot\boldsymbol{p}\,|\,c\rangle$			
Transition	$\bar{z}(\sigma^{-},\sigma^{+})z$	Transition	$\bar{z}(\sigma^{+},\sigma^{-})z$
$|\frac{3}{2},+\frac{3}{2}\rangle \to |\frac{3}{2},-\frac{1}{2}\rangle$	$i\frac{|P|^2}{3}d_0$ (↑)	$|\frac{3}{2},-\frac{3}{2}\rangle \to |\frac{3}{2},+\frac{1}{2}\rangle$	$-i\frac{|P|^2}{3}d_0$ (↓)
$|\frac{3}{2},+\frac{1}{2}\rangle \to |\frac{3}{2},-\frac{3}{2}\rangle$	$i\frac{|P|^2}{3}d_0$ (↓)	$|\frac{3}{2},-\frac{1}{2}\rangle \to |\frac{3}{2},+\frac{3}{2}\rangle$	$-i\frac{|P|^2}{3}d_0$ (↑)
$|\frac{1}{2},+\frac{1}{2}\rangle \to |\frac{3}{2},-\frac{3}{2}\rangle$	$i\frac{2|P|^2}{3}d_0$ (↓)	$|\frac{1}{2},-\frac{1}{2}\rangle \to |\frac{3}{2},+\frac{3}{2}\rangle$	$-i\frac{2|P|^2}{3}d_0$ (↑)
$|\frac{3}{2},+\frac{3}{2}\rangle \to |\frac{1}{2},-\frac{1}{2}\rangle$	$i\frac{2|P|^2}{3}d_0$ (↑)	$|\frac{3}{2},-\frac{3}{2}\rangle \to |\frac{1}{2},+\frac{1}{2}\rangle$	$-i\frac{2|P|^2}{3}d_0$ (↓)

$(e_x \pm ie_y)/\sqrt{2})$. The Raman processes in configurations where deformation-potential scattering is allowed thus consist of the following steps: absorption of a photon and creation of an electron with spin ↑ (↓) and a hole in state v_i, (Stokes) phonon emission and hole scattering from $v_i \to v_j$, and recombination of the electron and the hole in v_j. The following selection rules hold [1.32]:

$$\begin{aligned} \Delta J &= 0,\ \pm 1 \\ \Delta S_z &= S_{z,e} - S_{z,h} = 0 \\ \Delta n &= 0 \\ \Delta J_z &= J_{zi} - J_{zj} = \begin{cases} 2 & \text{for } \bar{z}(\sigma^{-},\sigma^{+})z \\ -2 & \text{for } \bar{z}(\sigma^{+},\sigma^{-})z\,. \end{cases} \end{aligned} \tag{4.8}$$

Dipole-allowed magneto-optical interband transitions are governed by

$$\begin{aligned} \Delta n &= 0 \\ \Delta J_z &= \pm 1 \end{aligned} \tag{4.9}$$

in $\sigma^{\pm}$ polarization, respectively. The processes allowed for both scattering geometries are given schematically in Fig. 4.3 (b). The Raman intensity is obtained from a coherent summation over all these possibilities. Tuning the Landau levels in a magnetic field should lead to double resonances for two of the processes in each geometry when their energy separation coincides with the LO phonon. In GaAs or InP, however, where the spin–orbit coupling Δ_0 is much larger than the phonon energy, only one of them remains as a serious candidate. In $\bar{z}(\sigma^{+},\sigma^{-})z$ a double resonance could occur for hole scattering between $|3/2, -1/2\rangle \to |3/2, +3/2\rangle$; in $\bar{z}(\sigma^-, \sigma^+)z$ the hole needs to go from $|3/2, +1/2\rangle \to |3/2, -3/2\rangle$.

Fröhlich interaction. Since the Fröhlich electron–phonon interaction is diagonal with respect to the electronic states (see (2.5)), the selection rules and possible Raman processes are determined by the magneto-optical transitions, i. e., the electron–photon interaction. Table 4.3 summarizes the Raman processes possible for hole scattering involving the different valence states in the $\bar{z}(\sigma^\pm, \sigma^\pm)z$ geometries. The selection rules are

$$\begin{aligned} \Delta J &= 0 \\ \Delta S_z &= 0 \\ \Delta n &= 0 \\ \Delta J_z &= 0 \end{aligned} \tag{4.10}$$

for electron–phonon scattering in both configurations. Application of the dipole selection rules (see (4.9)) explains why Raman processes in this case are only allowed for incident and scattered photons with the same circular polarizations. Figure 4.3 (c) schematically shows the processes of Table 4.3. In addition to scattering by holes, analogous terms where phonons are emitted by electrons have to be taken into account. No double resonances are expected for Fröhlich coupling.

Valence Band Mixing Effects. Comparison with the experimental magneto-Raman fan plots, to be discussed in the following section, shows that the model for the electronic structure used in Fig. 4.3 is oversimplified. The strongly mixed $\Gamma^{8\mathrm{v}}$ valence levels have to be considered at least in the (2×2)-approximation due to Luttinger given above [3.74]. The results just derived can be easily extended to take the valence band mixing into account when the hole Landau levels are described by the $E(n, 1\pm)$ and $E(n, 2\pm)$ ladders of (4.7a) and (4.7b) and the corresponding wave functions given by (4.3a) and (4.3b) (see Table 4.1). For deformation-potential coupling one then obtains

Table 4.3. Raman processes involving the Fröhlich electron–phonon interaction. For details see the caption of Table 4.2

$\langle c\,\vert \boldsymbol{e}_\pm^* \cdot \boldsymbol{p}\,\vert v_i\rangle \langle v_i\,\vert \boldsymbol{e}_\pm \cdot \boldsymbol{p}\,\vert c\rangle$			
Transition	$\bar{z}(\sigma^-, \sigma^-)z$	Transition	$\bar{z}(\sigma^+, \sigma^+)z$
$\vert\frac{3}{2}, +\frac{3}{2}\rangle \to \vert\frac{3}{2}, +\frac{3}{2}\rangle$	$\vert P\vert^2$ (↑)	$\vert\frac{3}{2}, -\frac{3}{2}\rangle \to \vert\frac{3}{2}, -\frac{3}{2}\rangle$	$\vert P\vert^2$ (↓)
$\vert\frac{3}{2}, +\frac{1}{2}\rangle \to \vert\frac{3}{2}, +\frac{1}{2}\rangle$	$\frac{\vert P\vert^2}{3}$ (↓)	$\vert\frac{3}{2}, -\frac{1}{2}\rangle \to \vert\frac{3}{2}, -\frac{1}{2}\rangle$	$\frac{\vert P\vert^2}{3}$ (↑)
$\vert\frac{1}{2}, +\frac{1}{2}\rangle \to \vert\frac{1}{2}, +\frac{1}{2}\rangle$	$\frac{2\vert P\vert^2}{3}$ (↓)	$\vert\frac{1}{2}, -\frac{1}{2}\rangle \to \vert\frac{1}{2}, -\frac{1}{2}\rangle$	$\frac{2\vert P\vert^2}{3}$ (↑)

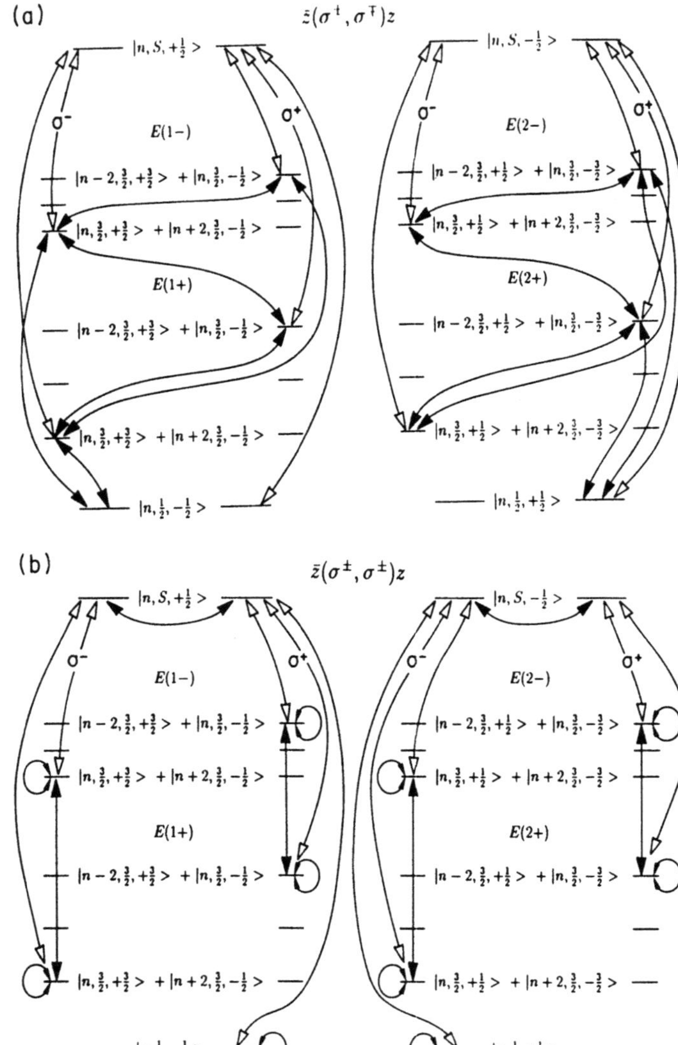

Fig. 4.4. Possible Raman processes in the **(a)** $\bar{z}(\sigma^{\pm},\sigma^{\mp})z$ (deformation potential) and **(b)** $\bar{z}(\sigma^{\pm},\sigma^{\pm})z$ (Fröhlich mechanism) scattering geometries when mixed heavy- and light-mass valence bands are considered in the (2×2)-model of (4.2a) and (4.2b). See text for details

the Raman processes shown in Fig. 4.4 (a) [1.32]. Compared with Fig. 4.3 (b) there are additional terms, and one expects new resonances in the spectra. Note also that there are now two likely candidates for double resonances as compared to just one in Fig. 4.3 (b). For $\bar{z}(\sigma^{+},\sigma^{-})z$ these involve scattering between the $|n,3/2,-1/2\rangle \rightarrow |n,3/2,+3/2\rangle$ and $|n,3/2,-3/2\rangle \rightarrow |n,3/2,+1/2\rangle$ admixtures of light- and heavy-mass holes in the $E(1\pm)$ and $E(2\pm)$ ladders, respectively. In $\bar{z}(\sigma^{-},\sigma^{+})z$ geometry these processes contain steps of hole scattering between the $|n,3/2,+3/2\rangle \rightarrow |n,3/2,-1/2\rangle$ and

$|n, 3/2, +1/2\rangle \rightarrow |n, 3/2, -3/2\rangle$ admixtures in the same ladders. Due to g-factor-induced splittings which are rather important for the valence bands the two double resonances for each geometry should occur at different values of laser energy and magnetic field. They can thus be distinguished in an experiment.

Raman processes for the geometries with Fröhlich interaction are shown in Fig. 4.4 (b). Their number has increased significantly as compared to the model of Fig. 4.3 (c). Contrary to the case discussed above, where no double resonance is expected for $\bar{z}(\sigma^\pm, \sigma^\pm)z$, there are now two processes for each geometry where it could occur. For $\bar{z}(\sigma^+, \sigma^+)z$ these involve scattering a hole from $|n, 3/2, -1/2\rangle$ in $E(1+)$ to the same admixture in $E(1-)$ or from $|n, 3/2, -3/2\rangle$ in $E(2+)$ to the same state in $E(2-)$. In $\bar{z}(\sigma^-, \sigma^-)z$ geometry the coupling occurs between the $|n, 3/2, +3/2\rangle$ admixtures in $E(1\pm)$ and the $|n, 3/2, +1/2\rangle$ states in $E(2\pm)$. If one considers, however, that the whole wave functions between which these couplings should take place are orthogonal, the occurrence of double resonances seems unlikely. This puzzle is solved by taking the small but nonzero crystal-momentum transfer for the dipole-forbidden Fröhlich interaction and the dispersion of the Landau bands along the magnetic-field direction into account [4.10]. This leads to signal intensities which are comparable to those of deformation-potential-mediated Raman processes [4.10].

4.2 Magneto-Raman Scattering in Bulk Semiconductors

4.2.1 Magneto-Raman Scattering in GaAs

Fan Plots. Figures 4.5 and 4.6 show the LO phonon magneto-Raman resonances for GaAs as fan plots of laser energy vs. magnetic field measured in the polarization geometries $\bar{z}(\sigma^\pm, \sigma^\mp)z$ and $\bar{z}(\sigma^\pm, \sigma^\pm)z$, respectively [1.33]. These fan plots show a rich and complex structure which cannot be understood in the simple model for the electronic structure used at the beginning of Sect. 4.1. However, considerable insight can be gained using the slightly more complicated (2×2)-models for the hole Landau levels of (4.2a) and (4.2b) and the selection rules derived above for Raman processes in a magnetic field.

When the magnetic field tends to zero, most of the fan lines in Figs. 4.5 and 4.6 approach the energy $E_0 + \hbar\omega_0$, i.e., one phonon ($\hbar\omega_0 = 37\,\text{meV}$ in GaAs [2.153]) above the direct band gap ($E_0 = 1.519\,\text{eV}$ [2.153]). This is characteristic of outgoing resonances (see Fig. 4.1). The close spacing of these fan lines is due to the participation of heavy-mass hole states in the magneto-Raman resonances. Calculations of magneto-Raman profiles from the possible scattering processes of Fig. 4.4 show that for each geometry there are two terms which dominate [1.31–1.33]. They are those which may also lead to double resonances (see Sect. 4.1.3 and Tables 4.2 and 4.3). Away from these "singular" combinations of laser energy and magnetic field, however,

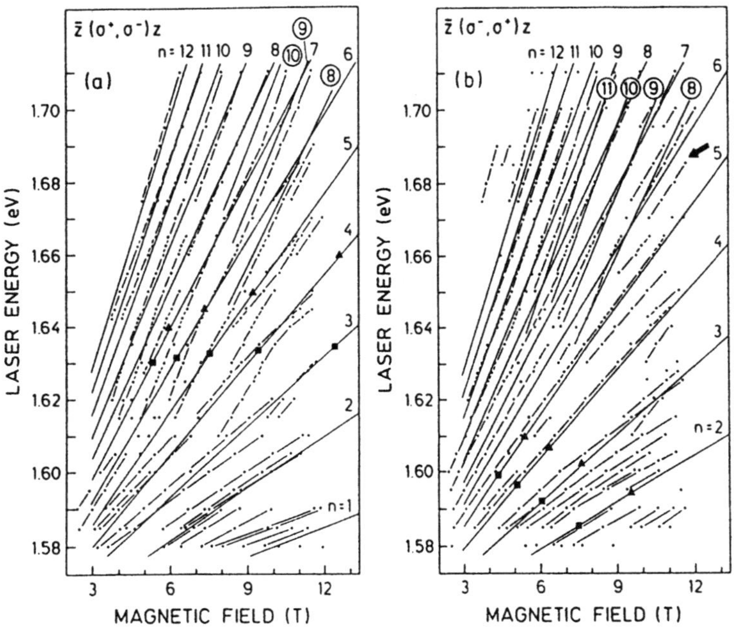

Fig. 4.5. Fan plots of LO phonon magneto-Raman resonances for (001)-oriented GaAs in the $\bar{z}(\sigma^{\pm},\sigma^{\mp})z$ configurations (T = 10 K). The experimental resonances (*points*) are connected by *lines*. The *solid lines* are theoretical interband transitions involving heavy mass valence levels in outgoing (Landau indices n indicated by bare numbers) or incoming resonance (Landau indices encircled). The *triangles* and *squares* denote theoretical double resonances for different scattering terms. See text for further explanations

they only contribute to outgoing resonances. In these processes the hole is scattered between light- and heavy-mass Landau levels, thereby emitting an LO phonon.

The question arises why these strong Raman processes do not lead to pronounced incoming resonances with light-mass valence levels as well. Fan lines from incoming resonances can be easily distinguished in fan plots by their different convergence behavior, approaching the gap energy for vanishing field. In fact, such resonances need to occur, otherwise double resonances would not be possible. However, calculations show that they are much weaker than their outgoing counterparts due to the larger broadening of the light-mass Landau levels which has also been observed in other experiments [4.11]. Therefore incoming resonances with the light-mass levels are generally hard to observe, except for situations close to double resonance discussed below. Note that this effect was taken into account in the schematic illustration of magneto-Raman profiles in Fig. 4.2.

The outgoing experimental fan lines in Figs. 4.5 and 4.6 are in good agreement with calculations (solid lines) based on an (8×8)-$\boldsymbol{k} \cdot \boldsymbol{p}$-Hamiltonian [1.31–1.33]. Before plotting, the transition energies were corrected for exciton

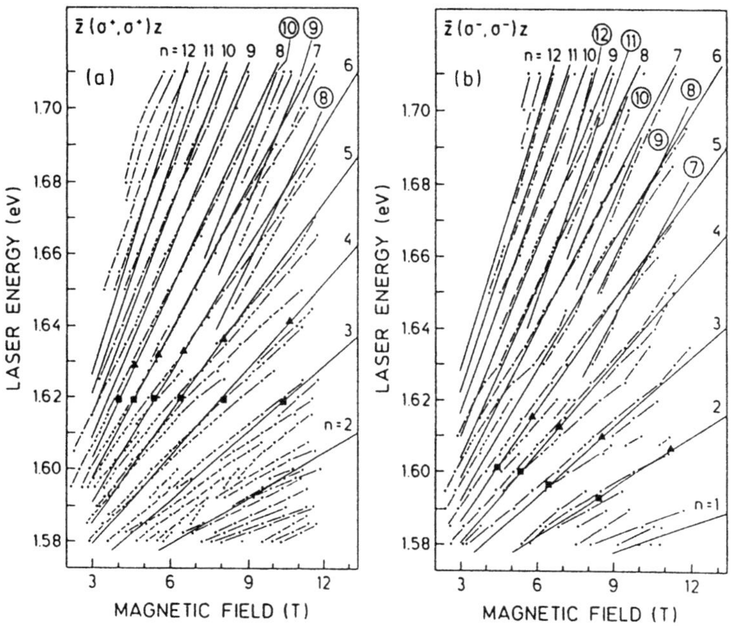

Fig. 4.6. Fan plots of LO phonon magneto-Raman resonances for (001)-oriented GaAs in the $\bar{z}(\sigma^{\pm},\sigma^{\pm})z$ configurations. See caption of Fig. 4.5 and text for further explanations

effects by subtracting a magnetic-field and Landau-index dependent binding energy [1.30]. The theoretical heavy-mass level to electron transitions shown belong to the two possible double resonances in each geometry (see Fig. 4.4, and Sect. 4.1.3). Since these resonances are rather close in energy only one of them is plotted [1.33]. The relevant admixtures of the wave functions obtained from the larger model were determined by comparison with those of the (2×2)-model given in (4.3a) and (4.3b) (see Table 4.1). As an example, in $\bar{z}(\sigma^+,\sigma^-)z$ geometry the two transitions assigned to outgoing resonances in σ^- polarization with participation of heavy-mass Landau levels are $|J=3/2, J_z=+1/2\rangle \to |\downarrow\rangle$ and $|J=3/2, J_z=+3/2\rangle \to |\uparrow\rangle$, involving admixtures of the $E(n,2-)$ and $E(n,1-)$ ladders, respectively. As mentioned in Sect. 4.1.3, the two double resonances for each geometry occur at different points in the fan plot. In Figs. 4.5 and 4.6 they are marked by squares and triangles. A compilation of the processes leading to double resonance is given in Table 4.4.

Double Resonances. Figure 4.7 shows a comparison between experimental (left) and theoretical (right) magneto-Raman profiles for GaAs in $\bar{z}(\sigma^+,\sigma^-)z$ configuration (deformation-potential scattering) [1.31,1.33]. The spectra were calculated including all terms in Fig. 4.4 [1.32]. Broadenings of 1, 3, and 10 meV were chosen for transitions involving heavy-mass, light-mass, and spin–orbit split-off levels. They are similar to values used in other resonance Raman experiments [4.11] and yield reasonable agreement with the measured

Table 4.4. Raman processes which can become doubly resonant due to hole scattering between light- and heavy-mass Landau levels. The *squares* and *triangles* correspond to those of Figs. 4.5 and 4.6. Only the J_z components of the $|n, J = 3/2, J_z\rangle$ admixtures for the different ladders (see Sect. 4.1.2) are given. The conduction electron levels $|n, S = 1/2, S_z = \pm 1/2\rangle$ are abbreviated by the spins ↑ and ↓

Scattering configuration	Symbol	Light-mass level (J_z, ladder)	Heavy-mass level (J_z, ladder)	Conduction level
$\bar{z}(\sigma^+, \sigma^-)z$	■	$-\frac{3}{2}$ E(2+)	$+\frac{1}{2}$ E(2−)	↓
	▲	$-\frac{1}{2}$ E(1+)	$+\frac{3}{2}$ E(1−)	↑
$\bar{z}(\sigma^-, \sigma^+)z$	■	$+\frac{1}{2}$ E(2+)	$-\frac{3}{2}$ E(2−)	↓
	▲	$+\frac{3}{2}$ E(1+)	$-\frac{1}{2}$ E(1−)	↑
$\bar{z}(\sigma^+, \sigma^+)z$	■	$-\frac{3}{2}$ E(1+)	$-\frac{3}{2}$ E(1−)	↓
	▲	$-\frac{1}{2}$ E(2+)	$-\frac{1}{2}$ E(2−)	↑
$\bar{z}(\sigma^-, \sigma^-)z$	■	$+\frac{1}{2}$ E(2+)	$+\frac{1}{2}$ E(2−)	↓
	▲	$+\frac{3}{2}$ E(1+)	$+\frac{3}{2}$ E(1−)	↑

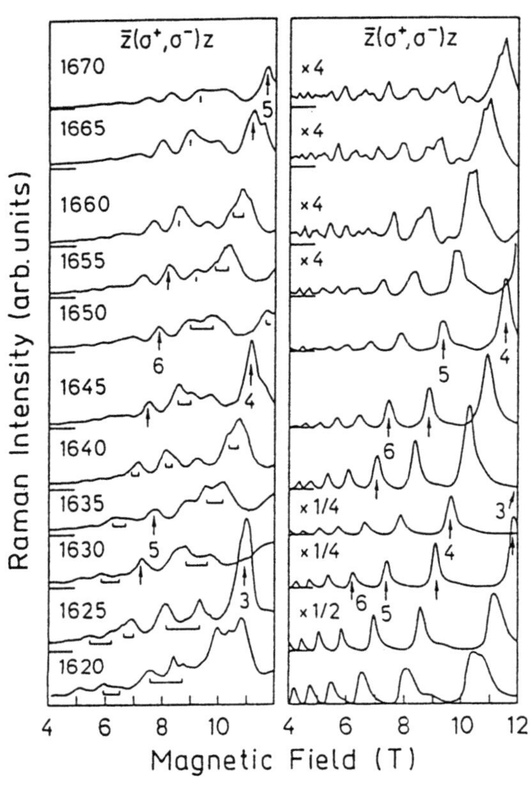

Fig. 4.7. Experimental (left) and theoretical (right) magneto-Raman profiles for the $\bar{z}(\sigma^+, \sigma^-)z$ geometry. The *arrows* denote double resonances for Landau indices n. The *brackets* in the experimental data connect incoming and outgoing resonances. The laser energy at each curve is in meV. Before plotting, some spectra were multiplied by the factors indicated

profiles. Calculated double resonances, corresponding to the squares and triangles of the $\bar{z}(\sigma^+,\sigma^-)z$ fan plot in Fig. 4.5, are marked by arrows and the respective Landau quantum numbers (right graph). Experimental features attributed to double resonances are also indicated by arrows (left graph). Incoming and outgoing partners around experimental double resonances are connected by brackets. More details of the double resonance region in the fan plot are given in Fig. 4.8 [1.33]. In addition to the outgoing resonances of Fig. 4.5 the incoming ones are also shown for each of the two processes. Double resonances occur at the crossing points of the calculated fan lines.

Figure 4.9 shows experimental (left) and theoretical (right) magneto-Raman fan plots for the $\bar{z}(\sigma^-,\sigma^-)z$ geometry (Fröhlich interaction) [1.31, 1.33]. In this case no incoming resonances are observed. However, the peaks at outgoing resonances are strongly enhanced when double resonance occurs. Figure 4.10 (a) shows a detailed plot of the fan lines and double resonances, in analogy to Fig. 4.8 [1.31]. Figure 4.10 (b) shows the Raman intensity along different fan lines for Landau indices $n = 2-4$. Theoretical double resonances, marked by arrows, are in good agreement with the experiment. Note the decreasing signal intensity with increasing Landau index.

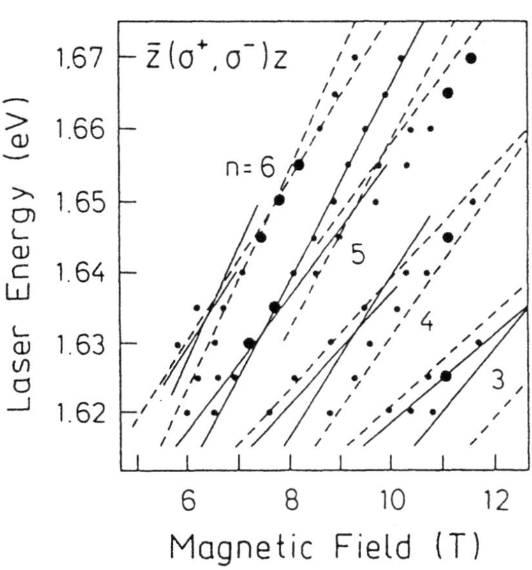

Fig. 4.8. Experimental resonances (*points*) from Fig. 4.7 compared to a calculated fan plot. The *solid* and *dashed* lines denote incoming (steeper slope, light-mass levels) and outgoing (shallower slope, heavy-mass levels) resonances for each of the two double resonance terms. Double resonances are expected for each n where the respective incoming and outgoing resonance fan lines cross each other. These crossing points correspond to the squares and triangles in Fig. 4.5 and Table 4.4. The *small points* denote incoming or outgoing resonances, the *large points* mark double resonances

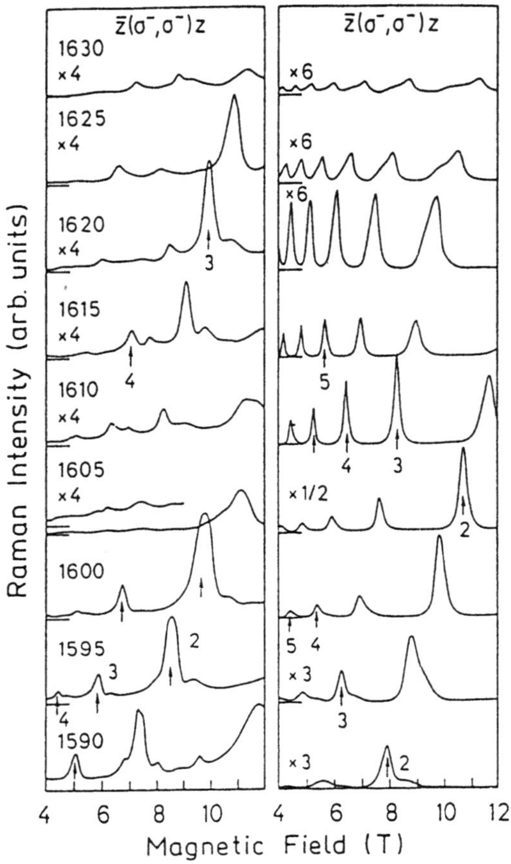

Fig. 4.9. Experimental (left) and theoretical (right) magneto-Raman profiles for GaAs in $\bar{z}(\sigma^-, \sigma^-)z$ configuration. The notation is as in Fig. 4.7

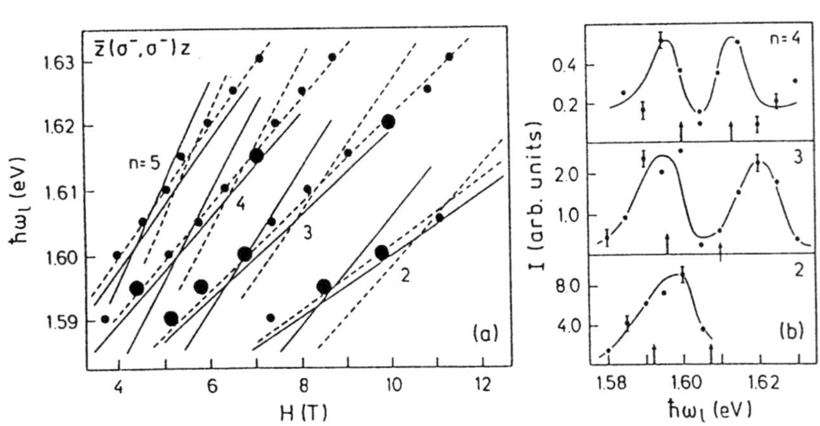

Fig. 4.10. **(a)** Experimental resonances from Fig. 4.9 compared to a theoretical fan plot. **(b)** Raman intensity along different fan lines. Theoretical double resonances are marked by *arrows*. See text, Fig. 4.8, and Table 4.4 for details

Conduction Band Nonparabolicity. As has been mentioned in Sects. 3.7 and 4.1.2 the electron dispersion in direct gap semiconductors strongly deviates from a parabolic behavior away from the Brillouin zone center. Since magneto-Raman profiles show pronounced oscillations also for energies far above the direct gap, one of the first applications of the technique was to determine the nonparabolicity of the Γ^{6c} conduction band in GaAs [1.28]. The interpretation was based on the following simple expression for the dispersion [1.28, 1.35], derived from a five-band $\boldsymbol{k}\cdot\boldsymbol{p}$-model containing the Γ^{7v}, Γ^{8v}, Γ^{6c}, Γ^{7c}, and Γ^{8c} bands:

$$E(k) = -\frac{E_0^*}{2} + \sqrt{\frac{E_0^{*2}}{4} + \frac{\hbar^2 k^2}{2m_0} P^2} + \frac{\hbar^2 k^2}{2m_0}(1 + C^*) \,. \tag{4.11}$$

Here $E_0^* = E_0 + \Delta_0/3$ is an average gap which takes the spin–orbit splitting of the valence band into account. The constant C^* represents the contribution of the higher conduction bands to the nonparabolicity and can be written as [4.12]

$$C^* = -\frac{P'^2}{3}\left(\frac{2}{E_0' + \Delta_0' - E_0} + \frac{1}{E_0' - E_0}\right) + C \,. \tag{4.12}$$

The matrix element P' connects Γ^{6c} with the Γ^{7c} and Γ^{8c} bands with gaps E_0' and $E_0' + \Delta_0'$, respectively [4.12, 4.13]. The small constant C takes the influence of remote p-like bands into account. A detailed discussion of other $\boldsymbol{k}\cdot\boldsymbol{p}$-terms which contribute to the nonparabolicity is given in [1.35]. The advantage of (4.11) is that it only contains one parameter, C^*, to be determined experimentally. Introducing a magnetic field as in (4.1) one obtains the following expression which can be fitted to magneto-Raman fan plots:

$$E(n,H) = -\frac{E_0^*}{2} + \sqrt{\frac{E_0^{*2}}{4} + E_0^*\left(\frac{m_0}{m_{e0}^*} - 1 - C^*\right)\frac{\hbar e H}{m_0 c}\,(n + \frac{1}{2})} + \frac{\hbar e H}{m_0 c}(1 + C^*)\,(n + \frac{1}{2}) \,. \tag{4.13}$$

Since deviations from the parabolic behavior are strongest far away from the direct gap, it is necessary to obtain data in this range. Figure 4.11 shows magneto-Raman fan plots for laser energies above 1.7 eV, thus extending the range of Figs. 4.5 and 4.6 [1.35]. For small laser energies and magnetic fields the resonances have mostly outgoing character. For larger laser energies and magnetic fields incoming resonances dominate. In the $\bar{z}(\sigma^-, \sigma^+)z$ configuration only incoming resonances are observed. The solid lines are fits to the data using (4.13) for the electron Landau levels. Although the contribution of heavy-mass holes to the transition energies is small, they have to be treated as accurately as possible if one attempts to extract information on the electron dispersion. This is especially important for the large oscillator indices $10 \lesssim n \lesssim 30$ of the resonances in this region. A comparison of calculations with the simple (2×2)-model (see (4.7a) and (4.7b)) and larger Hamiltonians

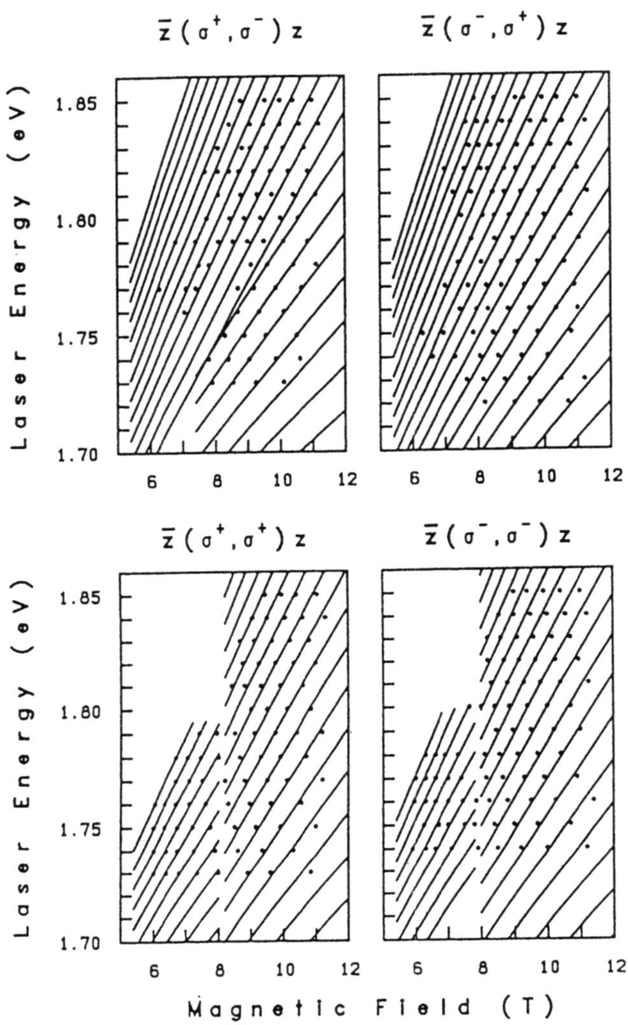

Fig. 4.11. LO phonon magneto-Raman fan plots for GaAs in the range where the conduction band nonparabolicity is important. The *solid lines* are fan plots of incoming and outgoing resonances fitted to the data (*points*) with (4.13) and a model of the valence Landau levels discussed in the text. Fit parameters and Landau quantum numbers of the fan lines are compiled in Table 4.5

shows significant deviations [1.35]. The hole contributions to the transitions in Fig. 4.11 were thus calculated numerically from the coupled (8×8)-matrices for 30 oscillator indices [4.14]. We only considered the $|n, J = 3/2, J_z = \pm 1/2\rangle$ admixtures of the heavy-mass levels since they are the strongest components of the wave functions for large n (see Table 4.1). They take part in $\sigma^{\mp}$ transitions to $|n, \downarrow (\uparrow)\rangle$, respectively. Table 4.5 shows the ranges of Landau levels, the types of resonances, and the values of C^* determined from the fits in Fig. 4.11. In all calculations a value of $m_{e0}^* = 0.0665\, m_0$ [2.153] was taken for the band edge electron mass in (4.13). The effective mass is related to P

Table 4.5. Experimental values of the nonparabolicity constant C^* determined from fits to the one-LO-phonon magneto-Raman profiles in Fig. 4.11. The types of resonances and the range of Landau levels for the different scattering geometries are also given

Scattering configuration	Type of resonance	C^*	Range of Landau indices in Fig. 4.11
$\bar{z}(\sigma^+,\sigma^-)z$	incoming	−1.90	$16 \leq n \leq 28$
	outgoing	−2.85	$7 \leq n \leq 13$
$\bar{z}(\sigma^-,\sigma^+)z$	incoming	−2.40	$9 \leq n \leq 28$
$\bar{z}(\sigma^+,\sigma^+)z$	incoming	−2.00	$9 \leq n \leq 24$
	outgoing	−2.65	$10 \leq n \leq 19$
$\bar{z}(\sigma^-,\sigma^-)z$	incoming	−2.00	$9 \leq n \leq 25$
	outgoing	−2.15	$12 \leq n \leq 20$

and C^* by ${m_{e0}^*}^{-1} = 1 + P^2/E_0^* + C^*$ [1.35]. The average of all measurements, including two-phonon fan plots not discussed here [1.35], is

$$C^* = -2.3 \pm 0.3 \ . \tag{4.14}$$

Figure 4.12 shows the conduction band dispersion of GaAs (solid line) obtained from magneto-Raman scattering in comparison with other models [1.35]. High above the gap, taken as zero of the energy, strong deviations from the parabolic behavior (dashed curve) occur. A comparison with other calculations, plotted in the inset, shows that the simple approximation used here gives an accurate description of the GaAs dispersion up to about 600 meV above the gap. A fit of the dispersion in (3.17) with $\beta_0 = 0$ to the magneto-Raman results gives $\alpha_0 = (-2370 \pm 100)\,\mathrm{eV\AA^4}$ [1.35]. This is in good agreement with a theoretical estimate of $\alpha_0 = -1969\,\mathrm{eV\AA^4}$ obtained by applying perturbation theory to a (14×14)-$\boldsymbol{k}\cdot\boldsymbol{p}$-Hamiltonian up to fourth order in $\boldsymbol{k}$ [3.71]. See Sect. 5.1.2 for another estimate of α_0 [1.35].

As mentioned in Sects. 3.7 and 4.1.2 it is often convenient to express the band nonparabolicity in terms of an energy-dependent effective mass which can be defined by

$$\frac{1}{m_e^*} = \frac{1}{\hbar^2 k}\frac{\mathrm{d}E(k)}{\mathrm{d}k} \ . \tag{4.15}$$

This is a dynamical effective mass which enters the semiclassical equation of motion for an electron under the action of an external force [1.35]. We find

$$m_e^*(E) = m_{e0}^*\,(1 + 1.447\,\mathrm{eV}^{-1}\,E + 0.245\,\mathrm{eV}^{-2}\,E^2) \tag{4.16}$$

which is shown by the solid line with error bars in Fig. 4.13. The results are in good agreement with optical and transport measurements (infrared reflectivity, Faraday rotation, Shubnikov-de-Haas oscillations) on n-doped GaAs [4.16].

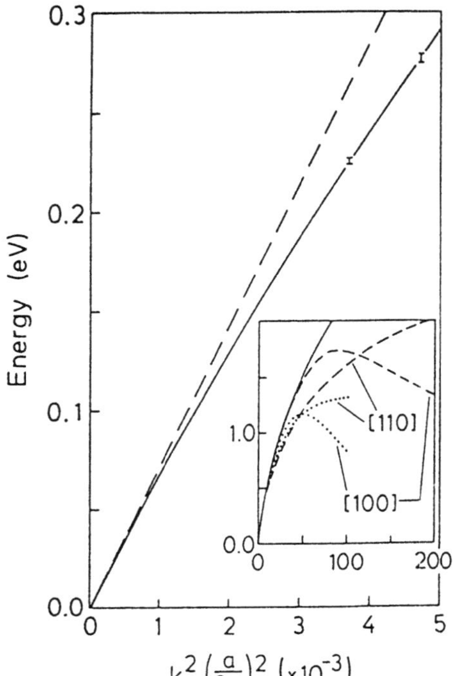

Fig. 4.12. Dispersion of the Γ^{6c} conduction band in GaAs (001). The *solid line* was calculated using (4.11) with $C^* = -2.3 \pm 0.3$ and $m^*_{e0} = 0.0665\, m_0$. The *dashed line* represents the parabolic dispersion for this value of m^*_{e0}. In the inset the magneto-Raman result (*solid line*) is compared to dispersions from a (16×16)-$\boldsymbol{k} \cdot \boldsymbol{p}$-Hamiltonian (*dotted curves*) [4.13] and from a local empirical pseudopotential calculation (*dashed curves*) [4.15] in the [100] and [110] directions

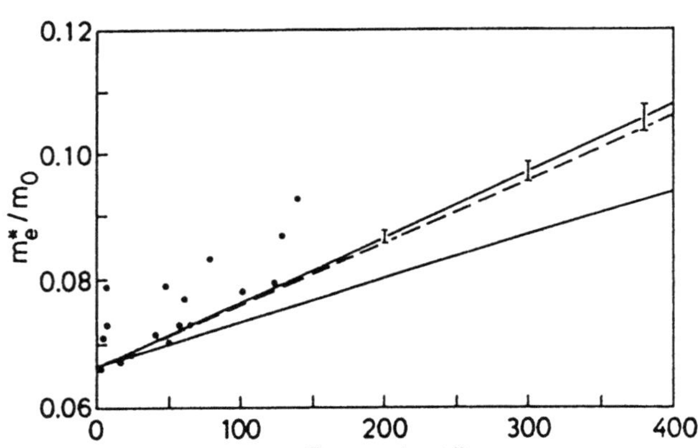

Fig. 4.13. Conduction band effective mass of GaAs (001) vs. energy above the bottom of the band. The *solid line with error bars* is calculated from (4.16) and (4.11) with $m^*_{e0} = 0.0665\, m_0$ and $C^* = -2.3 \pm 0.3$. The *dashed line* is from the fit of a (16×16)-$\boldsymbol{k} \cdot \boldsymbol{p}$-Hamiltonian to the experimental dispersion. The lower solid line was calculated for $C^* = 0$. The *data points* are from measurements on n-doped GaAs [4.16]

4.2.2 Resonant Magneto-Polarons in InP

Two-Level Systems. It has been noticed early that magneto-Raman scattering allows one to investigate the renormalization of interband transitions between electron and hole Landau levels due to the electron–phonon interaction [1.29, 1.30]. Resonant magneto-polarons occur when the energy between two Landau levels at the cyclotron frequency $\omega_c = eB/m^*$ corresponds to that of the LO phonon $\hbar\Omega_{LO}$ [4.17]. In this case electrons of a higher level $|n\rangle$ can be scattered into lower states $|m\rangle$, accompanied by the emission of an LO phonon. The total wave function of the system is given by the product of Landau levels $|n\rangle$ with oscillator index n and a vibrational part $|LO\rangle$, which reflects the phonon population. The coupling is usually mediated by the Fröhlich electron–phonon interaction [4.18] and leads to a splitting of the degenerate states $|n, 0\,LO\rangle$ and $|m, 1\,LO\rangle$ $(n > m)$ at magnetic fields which fulfill

$$(n - m)\hbar\omega_c = \hbar\Omega_{LO} \; . \tag{4.17}$$

This holds for parabolic bands whose magnetic-field dependence is given by

$$|n, 0\,LO\rangle : \quad E_n(B) = \hbar\omega_c\,(n + \frac{1}{2}) \tag{4.18a}$$

$$|m, 1\,LO\rangle : \quad E_m(B) = \hbar\omega_c\,(m + \frac{1}{2}) + \hbar\Omega_{LO} \tag{4.18b}$$

and causes anticrossings of the Landau level transitions which otherwise vary linearly with the field. Around the resonance magnetic field, where (4.17) is exactly fulfilled, the excitations have mixed electron–phonon character.

Magneto-polaron effects may in principle occur for both, electrons and holes. However, they have never been observed for the latter, presumably due to the more complicated valence band structure. In the following we therefore only consider magneto-polarons related to electron Landau levels. According to (4.17) magneto-polaron resonances are possible between levels with arbitrary Landau indices. The most pronounced effects, however, occur at the so-called polaron threshold which is given by $|n = 0, 1\,LO\rangle$, i. e., the lowest Landau level plus the energy of an LO phonon (see (4.18b)). This state is renormalized by the levels $|n, 0\,LO\rangle$ (see (4.18a)) at the respective crossing points and one obtains a series of split lines. Magneto-polaron effects reflect the underlying electron–phonon interaction. Therefore they have been investigated intensively in bulk semiconductors as well as low-dimensional systems by magneto-optical inter- and intra-band experiments [1.29, 1.30, 1.36, 1.37, 4.17, 4.19–4.22], far-infrared spectroscopy of shallow defects [4.23], or by transport measurements [4.24–4.27]. Theories of the effect have been published in [4.17, 4.28–4.34]. A theory for coupled magneto-polaron resonances based on Green's functions has been given recently in [1.37].

In the simplest approximation the coupling of Landau levels by optic phonons and the anticrossings at a polaron resonance can be described by a

(2×2)-model [4.34]. The character of a level $|n, 0\,\mathrm{LO}\rangle$ below the threshold changes to $|n = 0, 1\,\mathrm{LO}\rangle$ and follows that state for larger magnetic fields. Above the threshold and at larger fields the character of the $|n = 0, 1\,\mathrm{LO}\rangle$ state, which dominates for small B changes continuously to become the $|n, 0\,\mathrm{LO}\rangle$ state. In magneto-optical interband transitions with the electron component $|n, 0\,\mathrm{LO}\rangle$ this effect causes strongly nonlinear fan lines and splittings at the polaron threshold. In addition, the signal intensities are modified. In three dimensions the free dispersion of the Landau states along the the magnetic field and the influence of other levels has to be taken into account as well. This causes further couplings and makes the problem rather complicated, especially for the upper polaron branch [4.17, 4.19, 4.20, 4.28–4.34].

In second-order perturbation theory the correction ΔE_n of a Landau level $|n\rangle$ (4.18a) by magneto-polaron effects is [4.33]

$$\Delta E_n = -\sum_{m=0}^{\infty} \sum_{q} \frac{|M_{nm}(\boldsymbol{q})|^2}{D_{nm}} \tag{4.19}$$

where D_{nm} is the energy denominator

$$D_{nm} = \hbar\Omega_{\mathrm{LO}} - \Delta_n + \frac{\hbar^2 q_z^2}{2m^*} + \hbar\omega_{\mathrm{c}}(m - n) \tag{4.20}$$

and $M_{nm}(\boldsymbol{q})$ the matrix element of the electron–phonon interaction between the states $|n, 0\,\mathrm{LO}\rangle$ and $|m, k_z, 1\,\mathrm{LO}, \boldsymbol{q}\rangle$ with electron crystal momentum k_z and phonon wavevector $\boldsymbol{q}$. Via the energy shift Δ_n in (4.20) the change of the Landau level considered can be taken into account self-consistently [4.33]. In [4.33] it is shown that the shift of $|n, 0\,\mathrm{LO}\rangle$ below the polaron threshold, assuming weak electron–phonon interaction ($\alpha \ll 1$), is given by

$$|\Delta E_n| = \left(\frac{\alpha}{2n}\right)^{2/3} \hbar\Omega_{\mathrm{LO}} \,. \tag{4.21}$$

The total splitting is approximately twice this value. The coupling constant α is given by [4.18, 4.29, 4.34]

$$\alpha = e^2 \left(\frac{1}{\epsilon_\infty} - \frac{1}{\epsilon_0}\right) \sqrt{\frac{m^*}{2\hbar^3 \Omega_{\mathrm{LO}}}} \,. \tag{4.22}$$

The consideration of electron–phonon interaction contributions to the real and imaginary part of the Landau level self-energy in a Green's function formalism allows one to calculate the upper and lower polaron branches at a crossing point [4.34]. Magneto-polaron resonances obtained this way for the conduction band of GaAs are shown in Fig. 4.14. The dashed lines indicate the unperturbed states $|n, 0\,\mathrm{LO}\rangle$ and the threshold $|n = 0, 1\,\mathrm{LO}\rangle$. The solid curves are calculated lower and upper polaron branches at each resonance, in analogy to interacting two-level systems. The energy splitting depends on n as predicted by the analytical expression in (4.21) [4.34]. Magneto-optical interband experiments in GaAs confirm these results, however, a comparison of theory and experiment was not attempted for whole fan plots [4.34, 4.35].

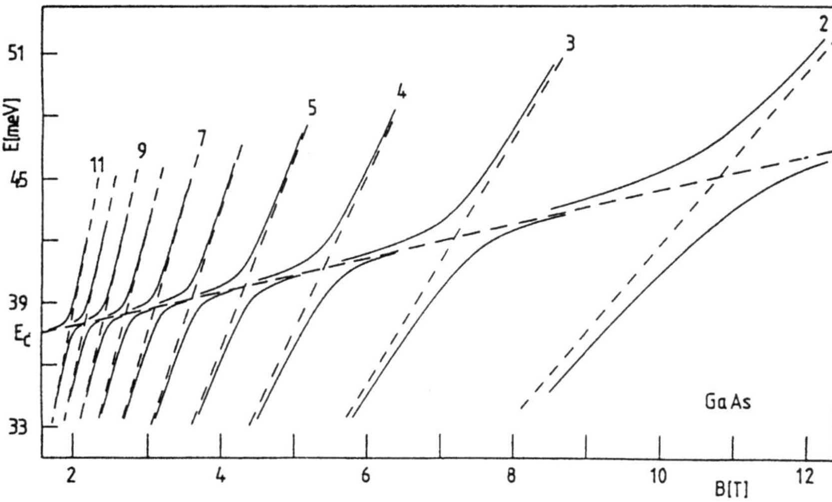

Fig. 4.14. Resonant magneto-polarons in GaAs. The *dashed lines* denote the unperturbed electron Landau levels $|n, 0\,\mathrm{LO}\rangle$ (indices n are indicated) and the polaron threshold $|n = 0, 1\,\mathrm{LO}\rangle$. The *solid curves* were calculated using Green's functions. Note the reduced energy splitting for increasing n (from [4.34])

Magneto-Raman Data. Magneto-Raman profiles and the related fan plots are particularly suitable for the investigation of magneto-polaron effects [1.29, 1.30, 1.36, 1.37]. The character of Raman scattering as a form of internal modulation spectroscopy [1.12] allows one to observe structures above the polaron threshold. Even resonances at higher thresholds, which are hard to measure, e. g., by absorption, can be clearly observed [1.30]. Figure 4.15 shows a series of LO phonon magneto-Raman profiles for a (001)-oriented surface of InP [1.36]. The spectra were measured at 10 K in the $\bar{z}(\sigma^-, \sigma^+)z$ Faraday backscattering geometry. For excitaton below the threshold one observes sharp and pronounced intensity oscillations (peaks a, b, d, f). In earlier magneto-Raman investigations these structures were attributed to magneto-optical interband transitions between light-mass and electron Landau levels in outgoing resonance [1.29, 1.30]. With increasing energy their intensity decreases and new resonances (peaks c, e, g) appear between the originally strong lines. The intensity profiles at the largest energies shown are dominated by these structures which are, however, significantly broader than those below threshold. The resonances at each excitation energy are plotted in Fig. 4.16 vs. the magnetic field [1.36]. By comparison with the calculated interband transitions from [1.30] (solid lines) it is found that the measured fan lines are in good agreement with the theory for the lowest and largest laser energies. In the intermediate range around 1.545 eV characteristic deviations due to magneto-polaron effects occur. However, the fan lines do not immediately bend over and vanish as expected from Fig. 4.14 and the changes

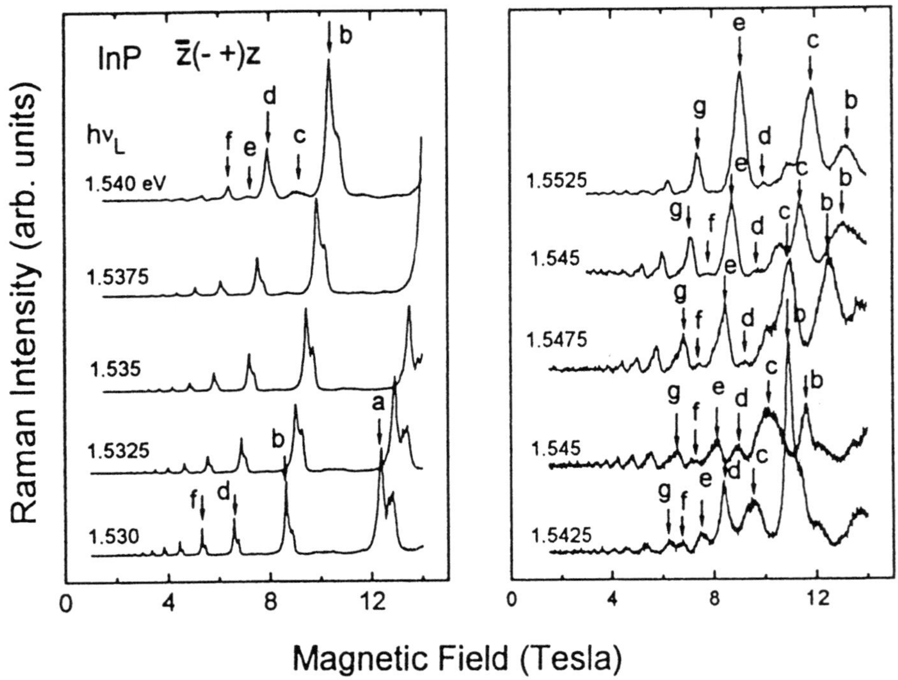

Fig. 4.15. LO phonon magneto-Raman profiles for InP measured with different excitation energies around the polaron threshold ($\bar{z}(\sigma^-, \sigma^+)z$ geometry, $T = 10\,\mathrm{K}$). See text for details

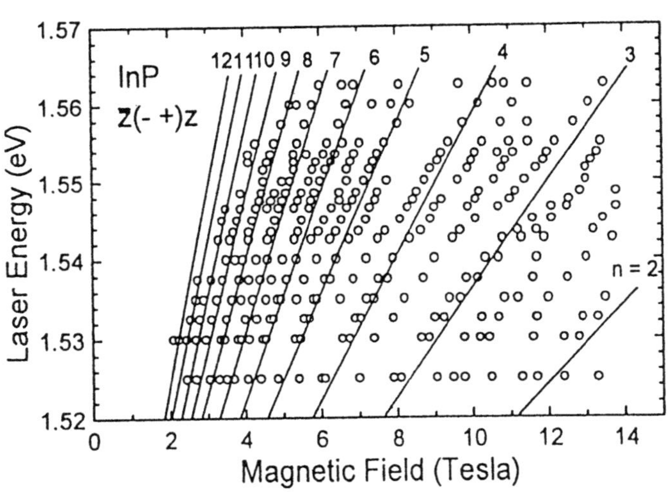

Fig. 4.16. Fan plot of magneto-Raman resonances (*open circles*) for InP in the $\bar{z}(\sigma^-, \sigma^+)z$ geometry. The *solid lines* represent calculated interband transitions between light-hole and electron Landau levels in outgoing resonance. Their indices n are given next to the fan lines

in their wave functions mentioned above. Instead, they can be observed far beyond resonance and are only slightly shifted compared to their original direction.

Coupled Magneto-Polarons. This behavior can only be understood if further couplings between the Landau levels which are contained in (4.19), but were so far not considered for the sake of simplicity, are taken into account. Neglecting dispersion effects along the magnetic field due to the smaller density of states, the main candidates for such interactions are the $|n \pm 1\rangle$ states neighboring each level $|n\rangle$. Electron–phonon interactions between these states are mediated by their coupling to the polaron threshold. An extension of the simple (2×2)-model is thus possible for coupled $|n, 0\,\mathrm{LO}\rangle$, $|n \pm 1, 0\,\mathrm{LO}\rangle$ and $|n = 0, 1\,\mathrm{LO}\rangle$ states. The influence of these additional couplings is shown qualitatively in Fig. 4.17 [1.37]. In this example we consider the magneto-polaron resonance between the threshold and a fan line with Landau index $n = 5$ using effective masses of $m_{\mathrm{e}}^* = 0.077\,m_0$ and $m_{\mathrm{lh}}^* = 0.12\,m_0$ for electrons and light holes in InP [2.153]. This results in slopes $\hbar e/m^*$ of 1.5 and 1.0 meV/T, respectively. The phonon energy is 43 meV [2.153]. For simplicity, g factors are neglected. The interband transitions in Fig. 4.17 are denoted by E_n^m where the lower (upper) number is the hole (electron) Landau index, respectively. The pure electronic levels for $n = 5$ and the (lowest) polaron threshold are denoted by E^5 and $E^{0,\mathrm{LO}}$, respectively. In interband magneto-optical transitions the polaron threshold is observed via the renormalization

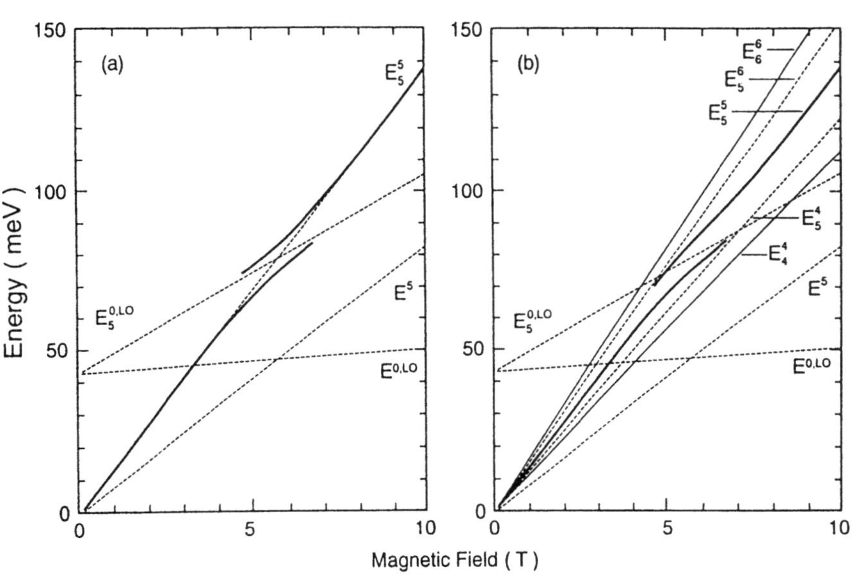

Fig. 4.17. Magneto-polaron effects on interband transitions (*solid lines*) for **(a)** a two-level system and **(b)** coupling of electron states with neighboring Landau levels. The *dashed lines* represent the unrenormalized transitions. See text for details

of the electron part of the E_n^m fan lines. The relevant asymptotes are thus $E_n^{0,\mathrm{LO}}$.

The solid lines in Fig. 4.17 (a) correspond to resonances expected in a two-level model if one assumes that mixing with the threshold $E_5^{0,\mathrm{LO}}$, which is approached asymptotically by the renormalized states, reduces its wave function such that it can only be observed within 1 T of the crossing point. The unrenormalized states are indicated by dashed lines. Let us now consider the coupling of E_5^5 with neighboring transitions shown in Fig. 4.17 (b). Approaching the threshold from lower (higher) energies, it first bends over towards $E_5^{0,\mathrm{LO}}$. However, before reaching the transition with the next lower (higher) index its curvature changes again and it follows fan lines for which the electron Landau index has been lowered (raised) by one unit, i. e., E_5^4 (E_5^6). The hole index does not change at all. This behavior is indicated by the solid lines in Fig. 4.17 (b) where we have assumed again that the resonances can be experimentally observed only within 1 T away from the anticrossing point. In contrast to Fig. 4.17 (a) the renormalized fan lines approach significantly different asymptotes away from the resonance region. This case corresponds to the experimental observations in Fig. 4.16 [1.30, 1.36]. Note that the comparison of the asymptotes for the renormalized transitions $E_n^{n\pm1}$ with the main lines E_n^n should allow one to determine the electron and hole masses separately. Usually, only the reduced effective mass can be determined from fan lines of interband magneto-optical experiments.

After subtraction of the gap energy, the valence band contributions, which are well known from earlier investigations, and corrections due to exciton effects [1.30] from the transition fan lines in Fig. 4.16 one obtains the data points in Fig. 4.18, i. e., the evolution of the pure electron Landau levels vs. magnetic field [1.36]. The renormalizations due to magneto-polaron effects and the level coupling are obvious. Due to the small electron g factor the levels in both configurations, $\bar{z}(\sigma^-,\sigma^+)z$ and $\bar{z}(\sigma^+,\sigma^-)z$, nearly coincide. Note also that the two separate fan lines $E_{n+1}^{n+1} \to E_{n+1}^n$ and $E_n^{n+1} \to E_n^n$, which appear between two transitions E_{n+1}^{n+1} and E_n^n for increasing magnetic field in Fig. 4.17, coalesce into a single curve, corresponding to the $E^{n+1} \to E^n$ transition between electron levels after the subtraction of their respectively different valence band contributions.

For the theoretical description of the magneto-polaron effects due to coupled Landau levels (solid lines in Fig. 4.18) we use a simplified ansatz. The diagonal basis contains the states $|n, 0\,\mathrm{LO}\rangle$ and the polaron threshold $|n = 0, 1\,\mathrm{LO}\rangle$ (assuming $k_z = 0$). Their energies are given by (4.18a) and (4.18b), respectively. For the interactions between the Landau levels and the threshold we use coupling constants corresponding to the splitting of (4.21) which depends on the Landau index n. For α we use the literature value of 0.11 [2.153]. Numerical diagonalization of the larger matrix set up this way yields a fan plot of coupled Landau levels, shown by the solid lines in Fig. 4.18. These results confirm the predicted decrease of the splittings with

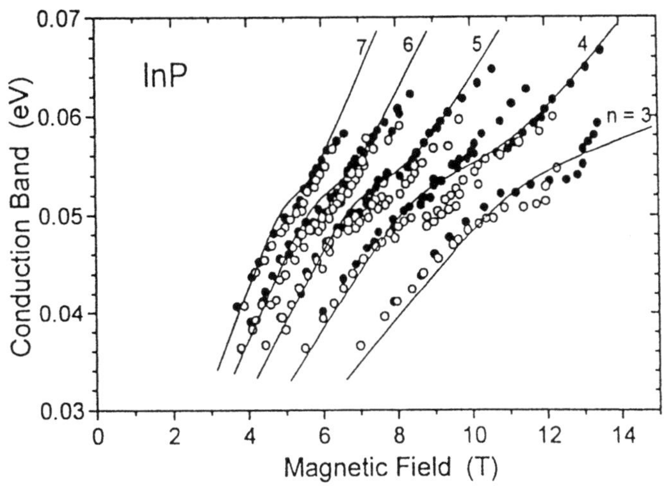

Fig. 4.18. Electron contributions to interband transitions between Landau levels (index n) in InP. The *open (filled) circles* are for the $\bar{z}(\sigma^-,\sigma^+)z$ ($\bar{z}(\sigma^+,\sigma^-)z$) configuration. Calculated Landau levels, renormalized by magneto-polarons are shown as *solid lines*

increasing Landau index according to (4.21) and highlight the importance of couplings between neighboring Landau levels for magneto-polaron effects. The data cannot be explained with a constant splitting of all resonances by $\alpha\hbar\Omega_{\mathrm{LO}}$. A fan plot from a more complete self-energy calculation based on Green's functions and results for anticrossings at higher thresholds are given in [1.37]. The comparison with that work shows that the simple extension of the two-level model presented here contains the essential features of coupled magneto-polarons.

4.2.3 Double Resonances in $Cd_{0.95}Mn_{0.05}Te$

We have already discussed the fact that strong double resonances can occur for Raman processes where semiconductor holes are scattered between light- and heavy-mass bands (see Fig. 4.1). Since the energy separation between these states is generally different from the phonon energy one uses external parameters in order to tune a system into double resonance. With this approach double resonances have been observed in bulk GaAs, e.g., by applying uniaxial stress [3.6] or high magnetic fields as discussed in Sect. 4.2.1 [1.31–1.33]. In quantum wells the separation between the heavy- and light-hole subbands depends on the layer thickness, and double resonances have been demonstrated in a series of samples with different well widths [3.9–3.11]. The energy levels in QW's can also be tuned by an electric field. Double resonances between Stark states of valence or conduction levels have been obtained in this way [3.3].

Semimagnetic Semiconductors. In the following we discuss Raman measurements in the semimagnetic semiconductor $Cd_{0.95}Mn_{0.05}Te$ [1.38]. The exchange coupling of paramagnetic manganese ions (Mn^{2+}, S=5/2) with electrons and holes causes a strong splitting of excitonic states in a magnetic field [4.36] which can be as large as 100 meV and thus easily exceeds typical LO phonon energies. This offers another possibility to obtain double resonances.

The coupling of the Mn spins $\boldsymbol{S}_{\text{ion}}$ with the electron spin $\boldsymbol{\sigma}_{\text{el}}$ is described by the exchange interaction Hamiltonian [4.37]

$$H_{\text{ex}} = -\sum_{\boldsymbol{R}_{\text{ion}}} J(\boldsymbol{r} - \boldsymbol{R}_{\text{ion}})\, \boldsymbol{S}_{\text{ion}}\, \boldsymbol{\sigma}_{\text{el}} \; . \tag{4.23}$$

To restore the translational invariance one makes the virtual-crystal approximation and replaces the magnetic moments located at individual lattice sites by an average spin at each cation which depends on the Mn concentration x. In the molecular-field approximation this average spin is given by $\langle S_z \rangle$. The perturbation operator then reads

$$H_{\text{ex}} = -\sum_{\boldsymbol{R}_{\text{cation}}} J(\boldsymbol{r} - \boldsymbol{R}_{\text{cation}})\, x\, \langle S_z \rangle\, \sigma_z \; . \tag{4.24}$$

This operator is formally equivalent to a magnetic field which acts on the electronic states and causes a Zeeman effect. For the interaction with s- or p-like bands one introduces the constants [4.38]

$$A = \frac{1}{6} x N_0 \alpha \langle S_z \rangle_{|B,T} \tag{4.25a}$$

$$B = \frac{1}{6} x N_0 \beta \langle S_z \rangle_{|B,T} \tag{4.25b}$$

where N_0 is the number of unit cells per unit volume; $\alpha = \langle S|J|S \rangle$ and $\beta = \langle X|J|X \rangle$ are the exchange integrals for the interaction of the Mn spins with electrons ($s-d$ exchange) and holes ($p-d$ exchange), respectively. For $Cd_{1-x}Mn_xTe$ their values are $N_0\alpha = 0.22$ eV and $N_0\beta = -0.88$ eV [4.37]. The average spin $\langle S_z \rangle$ depends on the temperature T and the magnetic field B. From thermodynamical considerations one finds [4.37]

$$\langle S_z \rangle = S_0 \cdot B_S \left(\frac{S g \mu_B B}{k_B (T + T_0)} \right) , \tag{4.26}$$

with the modified Brillouin function $B_S(y)$ for spin $S = 5/2$ given by [2.4]

$$B_S(y) = \frac{2S+1}{2S} \coth\left(\frac{2S+1}{2S} y \right) - \frac{1}{2S} \coth\left(\frac{y}{2S} \right) . \tag{4.27}$$

The parameters S_0 and T_0 take the clustering of the Mn ions and the resulting antiferromagnetic order into account. For a Mn content of 5 % this causes a reduction of the saturation value for $\langle S_z \rangle$ from the expected 5/2 to $S_0 = 1.54$ for large B/T and a Neel temperature of $T_0 = 2.3$ K [4.37].

The influence of the exchange interaction on excitons in $Cd_{1-x}Mn_xTe$ has been thoroughly investigated [4.36–4.41]. Neglecting the much smaller

Zeeman contributions the Γ_6^c electron states $E^c(\sigma_z)$ with spin 1/2 and the Γ_8^v valence band $E^v(J_z)$ with quasi-angular-momentum $J = 3/2$ are split in the following way [4.38]:

$$\begin{aligned} E^c(\pm 1/2) &= E_0 \mp 3\,A \\ E^v(\pm 3/2) &= \pm 3\,B \\ E^v(\pm 1/2) &= \pm B\,. \end{aligned} \tag{4.28}$$

In (4.28) the quantization axis has been chosen according to the exchange field, i. e., the spin polarization direction of the Mn system $\langle S_z \rangle$.

Due to the ferromagnetic character of the $s-d$ exchange ($\alpha > 0$) the electron ground state has a spin component $\sigma_z = +1/2$, while $\sigma_z = -1/2$ has a higher energy. The $p-d$ exchange is antiferromagnetic ($\alpha < 0$). The lowest energy state of the ($J = 3/2$) multiplet is $J_z = -3/2$. In decreasing order the allowed optical interband transitions in Faraday geometry have energies of

$$\begin{aligned} E_1(-3/2 \to -1/2) &= E_0 - 3B + 3A - E_R \quad (\sigma^+) \\ E_2(-1/2 \to +1/2) &= E_0 - \;\; B - 3A - E_R \quad (\sigma^+) \\ E_3(+1/2 \to -1/2) &= E_0 + \;\; B + 3A - E_R \quad (\sigma^-) \\ E_4(+3/2 \to +1/2) &= E_0 + 3B - 3A - E_R \quad (\sigma^-)\,. \end{aligned} \tag{4.29}$$

They occur with different circular polarizations $\sigma^\pm$. The exciton binding energy E_R due to Coulomb interaction is also indicated.

Double Resonances between Exchange-Split Excitons. Figure 4.19 shows the energies of these transitions vs. magnetic field for $Cd_{0.95}Mn_{0.05}Te$ as determined from the reflectivity measurements [1.38]. The maximum splitting of these transitions at 10 T is about 80 meV, much larger than the energy $E(LO_1) = 21$ meV of the CdTe-like LO phonon in this compound which will be considered in the following. With these numbers the expected double resonances can be determined. From the energy level scheme in the inset of Fig. 4.19 one finds two possible candidates. One is the process indicated by the dashed lines, where an electron is excited from $J_z = -3/2$ to the $(-1/2)$-conduction state and recombines to the ($J_z = +1/2$)-level into which the hole is scattered by deformation-potential interaction with a phonon. This process should take place in $\bar{z}(\sigma^+, \sigma^-)z$ geometry, and double resonance occurs when the separation between these holes equals $E(LO_1)$. This situation is indicated by the arrows in Fig. 4.19. The double resonance is expected at a field $H_{DR} \simeq 2.5$ T and a laser energy of $\hbar\omega_l = 1.697$ eV.

Figure 4.20 shows magneto-Raman profiles for different excitation frequencies (a) and the dependence of the LO_1 phonon intensity on the laser energy (b) [1.38]. According to the transitions of Fig. 4.19 one observes a resonance which shifts towards larger magnetic fields with increasing excitation energy (a). The resonance is superimposed on a luminescence background from the lowest transition which decreases for higher fields. The dependence of the LO_1 intensity on the excitation energy is shown in (b). The width of

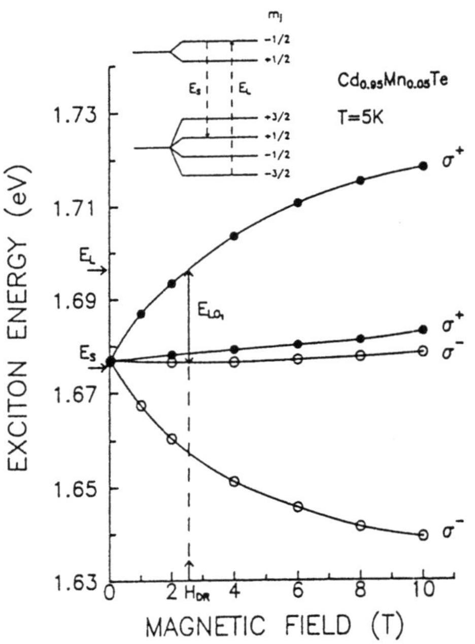

Fig. 4.19. Magnetic-field dependence of exciton energies in $Cd_{0.95}Mn_{0.05}Te$ for different circular polarizations. The *solid lines* underline the observed (*open and filled circles*) trend of large splittings due to the exchange interaction. The inset shows the energy level scheme for electrons and holes. See text for details

the observed resonance is the same as in the reflectivity measurements. The comparison of Figs. 4.20 (a) and (b) highlights the singular character of double resonances which occur only when maxima are found simultaneously for both, measurements vs. magnetic field and vs. excitation energy. The double resonance parameters are $\hbar\omega_l = E_{\mathrm{DR}} = 1.6965\,\mathrm{eV}$ and $H_{\mathrm{DR}} = 2.6\,\mathrm{T}$.

In double resonance the energies of the two hole states involved differ by the LO_1 phonon energy. This can be used to determine the $p-d$ exchange integral $N_0\beta$ from

$$\Delta E = E^v(+1/2) - E^v(-3/2) = \frac{2}{3}\, x\, N_0\, |\beta|\, \langle S_z \rangle|_{H_{DR},T} \equiv \hbar\Omega_{\mathrm{LO}_1}\,. \quad (4.30)$$

With the parameters given above and $T = 5\,\mathrm{K}$ we find $N_0\beta = -0.83\,\mathrm{eV}$, in good agreement with the literature of $-0.88(4)\,\mathrm{eV}$ [4.37]. The sign of $N_0\beta$ is determined by the scattering geometry for which the double resonance occurs.

The profile at $\hbar\omega_l = 1.6965\,\mathrm{eV}$ in Fig. 4.20 (a) measured in $\bar{z}(\sigma^+, \sigma^+)z$ configuration (dashed line) only shows the luminescence background. From the selection rules one expects in this geometry only single resonances due to Fröhlich interaction (see (4.11)). The intensity profiles in Fig. 4.20 (a) demonstrate the dramatic enhancement of the double resonance signal compared to the single resonance which for $\bar{z}(\sigma^+, \sigma^+)z$ cannot be observed due to the luminescence background.

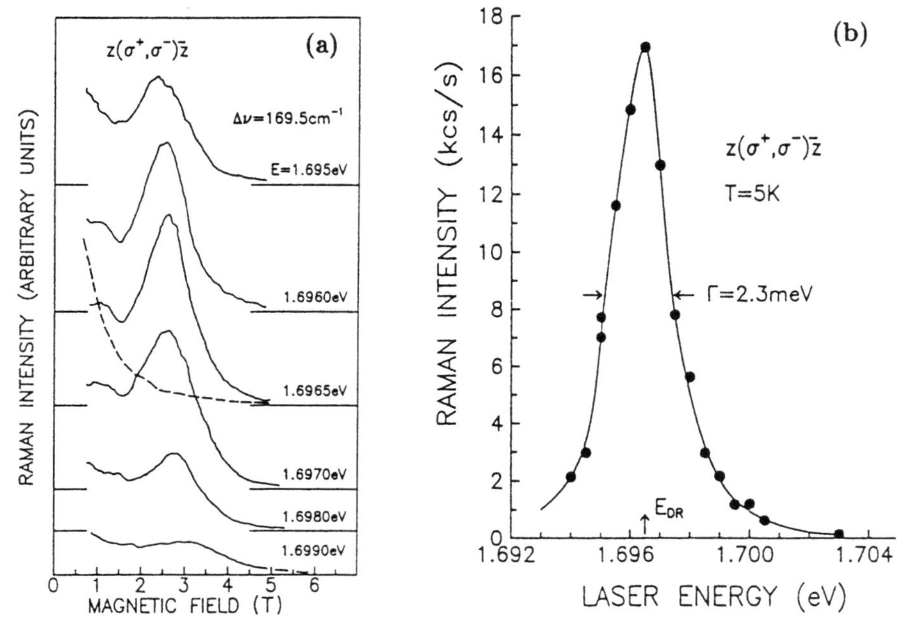

Fig. 4.20. Double resonance in $Cd_{0.95}Mn_{0.05}Te$: **(a)** Magneto-Raman profiles of the LO_1 phonon for different excitation energies in $\bar{z}(\sigma^+,\sigma^-)z$ configuration. For comparison, a spectrum in $\bar{z}(\sigma^+,\sigma^+)z$ geometry (no resonance) is also shown (*dashed line*). **(b)** Resonance behavior of the LO_1 intensity determined from the maxima of the magneto-Raman profiles

For the energy levels shown in the inset of Fig. 4.19 double resonances are only possible in $\bar{z}(\sigma^+,\sigma^-)z$ geometry. Another double resonance should occur for a process in which the hole is scattered between the ($J_z = -1/2$)- and (+3/2)-states. In this case, however, the outgoing resonance occurs at the lowest-energy exciton transition and the observation of the double resonance is not possible due to the much stronger luminescence.

4.3 Magneto-Raman Scattering in Quantum Wells

Continuous emission resonances involving acoustic phonons were used in Sect. 3.7 for a magneto-optical investigation of the QW electronic structure. In this section we discuss the application of optic-phonon magneto-Raman scattering for this purpose [1.39].

In contrast to bulk semiconductors (see Sects. 4.1 and 4.2) few experimental magneto-Raman results are available for QWs [4.42–4.46]. Investigations of optic-phonon resonances in GaAs/Al_xGa_{1-x}As samples in a magnetic field showed a much stronger increase of the interface phonon intensity compared to that of confined phonons [4.42–4.44]. In temperature-dependent measurements it was shown that this effect is related to the increasing localization of the electronic intermediate states with increasing field [4.43, 4.44]. Similar to the case of acoustic phonons discussed in Chap. 3, this favors scattering of modes with crystal-momentum components in the QW plane and is therefore especially important for interface modes [4.43, 4.44]. In [4.46] outgoing resonances of the LO phonon Raman signal at heavy- and light-mass exciton transitions of a 100 Å GaAs/AlAs MQW were followed over a wide energy region. Some of the observed intensity oscillations were attributed to double resonances. In theoretical investigations the difficulties of obtaining an unambiguous assignment of double resonances in QWs which arise from wave function mixing and excitonic effects as well as strongly broadened resonances were pointed out [4.47, 4.48].

In the following we present a systematic investigation of optic-phonon-related magneto-Raman resonances in GaAs/AlAs MQWs from which we obtain information about the electronic structure [1.39]. A theory of magneto-Raman scattering in QWs [4.47, 4.48] is applied to assign the observed fan lines, to determine selection rules and to identify the relevant scattering mechanisms. We use [001]-oriented samples grown by MBE with nominal layer thicknesses of (44/44) Å and (99/99) Å. They consist of 80 and 100 periods, respectively.

Theory. The theory used to interpret the data is based on the following model for the electronic structure [1.39, 4.47, 4.48]. The degenerate Γ_8^{v} valence band is described by a (4×4)-Luttinger Hamiltonian [3.74] for the $(J = 3/2)$-manifold. In this approximation the most important valence band mixing effects are taken into account. The QW subband energies are calculated self-consistently using an energy-dependent effective mass which is obtained from a $\boldsymbol{k} \cdot \boldsymbol{p}$-model [3.79, 4.49]. In the description of the fan plots the $|J = 3/2, J_z = \pm 3/2\rangle$ states are called heavy holes ($\mathrm{hh}^{\pm}$), and the $|J = 3/2, J_z = 1/2\rangle$ ones are called light holes ($\mathrm{lh}^{\pm}$). This assignment refers to the effective hole masses which determine the subband energies and are responsible for the confinement along the growth direction. For the in-plane dispersion which is quantized into Landau levels by a perpendicular magnetic field a mass reversal occurs. Consequently the $\mathrm{hh}^{\pm}$ states have a smaller effective mass than the $\mathrm{lh}^{\pm}$ bands. This effect causes a band crossing which, however, is lifted by further couplings in the (4×4)-model considered. It results in a strong nonparabolicity which is also reflected in magneto-optical experiments, e. g., by intensity changes of fan lines. We neglect the Γ_7^{v} spin–orbit split-off band. The effective conduction band masses for the free dispersion in the QW plane (perpendicular to the growth direction) $m_{\mathrm{e}}^{\parallel}$ and for the subband quantization $m_{\mathrm{e}}^{\perp}$ are determined from fits to the experiment. When comparing magneto-optical interband transitions with measurements they are denoted by the dominating component of the mixed hole state and the electron Landau level with index n including its spin ($\uparrow$ or $\downarrow$). As an example, the resonance involving the $|3/2, +3/2\rangle$ hole state and the $\uparrow$-electron with $n = 3$ is abbreviated by $e \uparrow \rightarrow \mathrm{hh}^{+}\ n = 3$. We only consider transitions between Landau levels from the first electron and hole subbands.

The calculation of magneto-Raman profiles was described in [4.47] and [4.48]. In the $\bar{z}(\sigma^{\pm},\sigma^{\pm})z$ geometries Fröhlich electron–phonon interaction has to be considered and scattering is allowed by confined phonons with even index even in the dipole approximation (note that Fröhlich scattering in the bulk is forbidden in the dipole approximation). For the $\bar{z}(\sigma^{\pm},\sigma^{\mp})z$ geometries the deformation-potential mechanism contributes and phonons with odd index are important. The influence of resonances with different n on the various confined modes was investigated theoretically in [4.47, 4.48]. Here we only consider the strongest modes, LO_1 and LO_2, whose frequencies are quite close.

As mentioned in Sect. 3.7, the theoretical transitions have to be corrected for exciton effects before they can be compared to the experiment. For this we use the expression given in (3.22) for the exciton ground state which depends on the Landau index and the magnetic field [3.70, 3.75].

Fan Plots. In Figures 4.21 and 4.22 we show fan plots of measured magneto-Raman oscillations for the two samples investigated in comparison with the theory [1.39]. The theoretical fan plots were calculated with the parameters given in Table 4.6. Parameters which were determined from fits to the data are discussed below. The fan plots can be separated into groups of transitions involving heavy (hh) or light holes (lh). Additionally, each transition may lead to an incoming or outgoing resonance when its energy is the same as that of the laser or the scattered photons, respectively. The corresponding in- or outgoing resonances only differ by the LO phonon energy. The given Landau indices n refer to the e-hh transitions in outgoing resonance (solid lines in Figs. 4.21 and 4.22).

For small excitation energies the magneto-Raman resonances in Fig. 4.21 are mostly due to e-hh transitions in incoming resonance (dotted fan lines). They correspond to Landau-index conserving transitions with σ^- (σ^+) polarization between $|3/2, \pm 3/2\rangle$ hole states and electrons with spin $\uparrow$ ($\downarrow$). For laser energies above 1760 meV several e-hh transitions in outgoing resonance are observed (solid lines). The first lh and hh subbands in this sample are separated by about 50 meV, more than the LO phonon energy. Due to the larger effective mass of the lh states in the QW plane their separation from the hh levels with the same Landau index decreases with increasing magnetic field. Depending on the polarization geometry incoming e-lh resonances (dashed-dotted fan lines) and e-hh transitions in outgoing resonance (solid lines) with the same n can be quite close. In some cases they are almost parallel or cross each other under a small angle. Consequently the separation between the hole states of the transitions is close to the phonon energy and a strong signal enhancement due to double resonances, similar to bulk GaAs, is expected to occur [1.31–1.33, 4.46–4.48]. In Fig. 4.21, however, the double resonance only seems to favor the appearance of e-hh transitions in outgoing resonance (solid lines) which have a strong intensity over a wide range. This behavior is attributed to the larger broadening of the e-lh transitions and to the fact

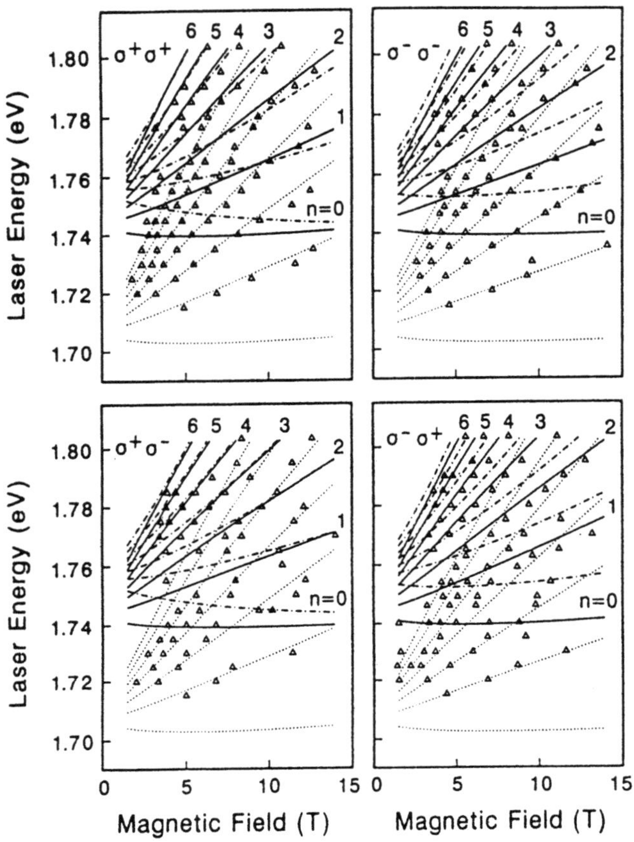

Fig. 4.21. Optic-phonon magneto-Raman fan plots of a (44/44) Å GaAs/AlAs MQW for different polarizations. The *triangles* are measured resonances. The *solid (dotted) fan lines* denote calculated transitions for outgoing (incoming) resonances between hh valence levels and electrons. The *dashed-dotted lines* are calculated incoming resonances between lh and electron states

that the lines are almost parallel to each other. Under these conditions the e-hh fan lines in outgoing resonance above 1760 meV are stronger than the e-hh (dotted lines) and the e-lh transitions (dashed-dotted lines) in incoming resonance. Only in a few cases, e. g., in $\bar{z}(\sigma^-, \sigma^-)z$ configuration in Fig. 4.21 some of the fan lines agree better with e-lh incoming resonances. Another effect which contributes to the change in character of the e-hh fan lines from incoming to outgoing resonance in Fig. 4.21 is the increasing broadening of Landau levels with larger indices. In the area where the e-hh transitions in in- and outgoing resonance cross each other, the Landau quantum numbers of the outgoing resonances are smaller by about three units than those of the incoming ones. The larger broadening of levels with larger n further reduces the Raman intensity compared to the outgoing e-hh resonances with smaller indices.

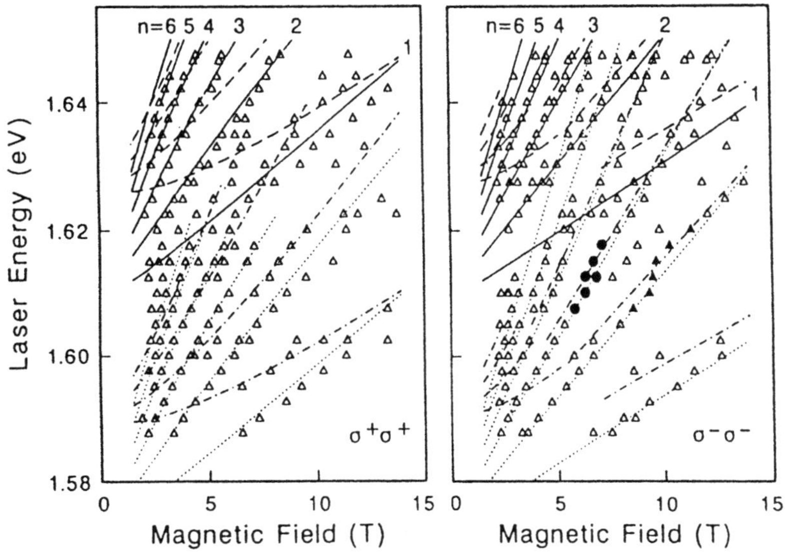

Fig. 4.22. Optic-phonon magneto-Raman fan plots of a (99/99) Å GaAs/AlAs MQW for the $\bar{z}(\sigma^{\pm}, \sigma^{\pm})z$ configurations. The *triangles* are measured resonances. The *solid (dotted) fan lines* denote calculated transitions for outgoing (incoming) resonances between hh valence levels and electrons. The *dashed (dashed-dotted) lines* are calculated outgoing (incoming) resonances involving lh states

Table 4.6. Material parameters used for the calculation of magneto-Raman fan plots and intensity profiles discussed in this section [2.153]

	GaAs	AlAs
E_0 (eV)	1.52	3.13
E_c (eV)		1.13
Δ_0 (eV)	0.341	0.275
m_{hh} (m_0)	0.34	0.4
m_{lh} (m_0)	0.092 [(a)]	0.194 [(a)]
m_e (m_0)	0.066 [(a)]	0.133 [(a)]
$(2m_0/\hbar^2)P^2$ (eV)	24.8	24.8
γ_1	7.1 [(b)]	3.45
γ_2	2.1 [(b)]	0.68
γ_3	2.9 [(b)]	1.29
κ	1.2	

[(a)] $\boldsymbol{k} \cdot \boldsymbol{p}$ calculation [3.79]

[(b)] [4.50]

The fan plots of the (99/99) Å sample in Fig. 4.22 are more complex than those in Fig. 4.21. The observed fan lines can be attributed to all four kinds of resonances. The energy difference between the lowest hh and lh subbands is only 16 meV. This proximity of the subbands enhances the valence band nonparabolicity and causes a strong mixing of the wave functions. The smaller broadening of the electronic resonances in this sample also allows one to observe weaker features and transitions with larger indices. Due to the different transition probabilities resonances with lh levels are expected to be ten times weaker than e-hh transitions. However, the strong mixing of the hole states increases the number of possible resonances, and e-lh transitions can also be observed via an hh component in the wave function. As in Fig. 4.21 the outgoing resonances are stronger than incoming ones for the same laser energy.

Electronic Structure. Band mixing effects are seen most clearly in the $\bar{z}(\sigma^-,\sigma^-)z$ geometry of Fig. 4.22. They manifest themselves in resonances and fan line discontinuities which do not appear in a simple parabolic model of the band structure. An example is indicated by the filled triangles. Between 9 and 10 T, the ($e\uparrow\rightarrow \mathrm{hh}^+\ n=2$) transition (dotted fan line) suddenly shifts towards higher energies but hardly changes its slope. This discontinuity arises from two different valence band states which change their admixtures with the magnetic field. Another example in the same fan plot, also due to changes in the transition oscillator strength, is marked by filled circles. These effects are further discussed in [1.39].

Table 4.7 gives the parameters which were used to fit the theory to the fan plots in Figs. 4.21 and 4.22. Some calculated values are included for comparison. The dependence of the broadening parameter Γ on the Landau index which is used to calculate magneto-Raman intensity profiles is also given. The effective electron masses $m_\mathrm{e}^{\|}$ were determined from the slope of the fan lines after subtraction of the calculated valence level and exciton energies. The electron masses $m_\mathrm{e}^{\perp}$ are obtained by a precise measurement of the subband energies ϵ from which they can be calculated using (3.18a) [3.72]. For this purpose the exciton binding energy E_B without field was determined from the convergence points of the fan plots for $B\rightarrow 0$ and the luminescence maximum of the lowest e-hh transition. The values obtained this way are given in Table 4.7. They are in good agreement with exciton energies calculated from [3.70, 3.75]

$$E_B = 4DR^* \tag{4.31}$$

using $D\simeq 0.7$ (see (3.22)). From the calculated hole subbands one finds an energy of $\epsilon = 150.6$ meV for the first electron subband in the 44 Å sample and a value of $\epsilon = 43.9$ meV for the one with 99 Å QW thickness. These numbers are used with (3.18a) to obtain the values of $m_\mathrm{e}^{\perp}$ given in Table 4.7. The dependence of the effective masses $m_\mathrm{e}^{\|}$ and $m_\mathrm{e}^{\perp}$ on the QW thickness determined from optic and acoustic-phonon magneto-Raman scat-

Table 4.7. Parameters obtained by fitting theoretical expressions to the magneto-Raman fan plots in Figs. 4.21 and 4.22. For comparison, effective masses calculated from (3.18a) and (3.18c) and the exciton binding energy from (4.31) are also given

	(44/44) Å GaAs/AlAs	(99/99) Å GaAs/AlAs
GaAs layer thickness (Å)	43.6	97
$m_e^{\parallel}$ (m_0)	0.090	0.076
$m_e^{\parallel\,(a)}$ (m_0)	0.086	0.072
$m_e^{\perp}$ (m_0)	0.073	0.069
$m_e^{\perp\,(a)}$ (m_0)	0.073	0.068
D	0.75	0.7
E_B hh (meV)	11.6	10.3
E_B hh$^{(b)}$ (meV)	13.5	11.2
Γ (meV)	$(4.5 + 0.3\,\mathrm{n})$	$(0.5 + 0.3\,\mathrm{n})$

$^{(a)}$ (3.18a) and (3.18c)

$^{(b)}$ (4.31)

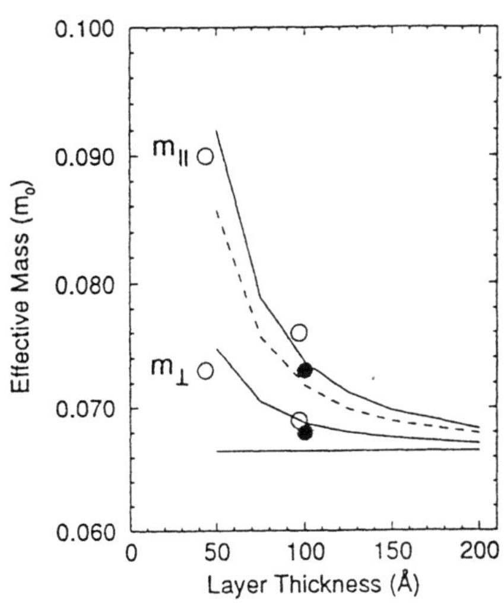

Fig. 4.23. Dependence of the effective masses $m_e^{\parallel}$ and $m_e^{\perp}$ on the layer thickness of GaAs QWs. *Open (filled) circles* are from magneto-Raman measurements with optic (acoustic) phonons. The *solid lines* are calculated with (3.18a) and (3.18c). The bulk band edge mass *(horizontal line)* is $m_e^* = 0.0665\,m_0$ [2.153]. The *dashed curve* was obtained from magneto-Raman measurements of the conduction band nonparabolicity in bulk GaAs (see (4.16)) [1.35]

tering (Sect. 3.7) is shown in Fig. 4.23 in comparison with calculations using (3.18a) and (3.18c) [4.51]. As mentioned in Sect. 3.7 the electron effective masses in QWs are anisotropic and larger than the band edge value due to the nonparabolicity and anisotropy of the bulk conduction band dispersion [3.72]. The experimental differences between $m_e^{\parallel}$ and $m_e^{\perp}$ in Table 4.7 are in good agreement with calculations from (3.18c) and (3.18a). This confirms the predictions of [3.72]. The lower effective masses at about 100 Å which

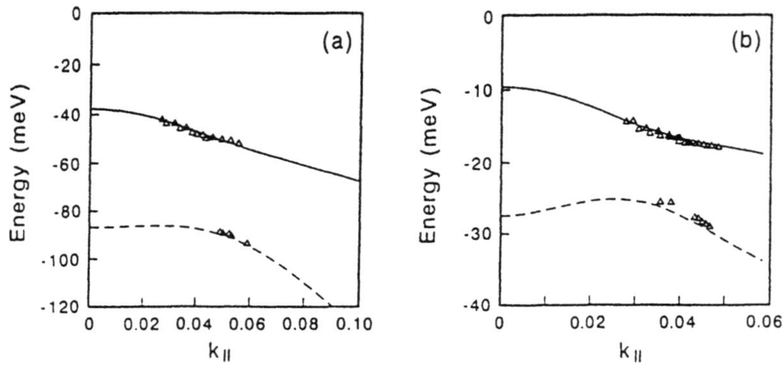

Fig. 4.24. In-plane dispersions for GaAs/AlAs QWs with a layer thickness of 44 **(a)** and 99 Å **(b)**. The *triangles* were determined from the measured magneto-Raman fan lines in Figs. 4.21 and 4.22. The *solid and dashed lines* are zero-field dispersion curves for the first hh and lh subbands. Values of $k_{\|}$ are given in units of π/a with the conventional lattice constant $a = 5.65$ Å of GaAs

were obtained in Sect. 3.7 for a sample with $Al_{0.35}Ga_{0.65}As$ barriers (filled circles in Fig. 4.23) compared to the values determined here for QWs with AlAs barriers (open circles) can be partly attributed to the different penetration depth of the wave functions into the barriers [3.78]. The dashed line in Fig. 4.23 represents the nonparabolic conduction band dispersion of bulk GaAs [1.35] (see (4.16)) which was transferred to the QW case by assigning an effective wavevector corresponding to the layer thickness [3.72]. As expected, the effective masses measured in bulk material are close to the $m_e^{\|}$ curve for QWs since they were derived from energy spacings between Landau levels and thus also reflect the free dispersion perpendicular to the magnetic field.

The consistency of these results can be checked by comparing the hole contributions which can be derived from the fan plots after subtraction of the electron Landau level energies with the zero-field in-plane dispersion curves. For the assignment of wavevectors $k_{\|}$ to the hole cyclotron energies we use (4.1). The data points (open triangles) in Fig. 4.24 were obtained this way [1.39]. They are in good agreement with the calculated in-plane dispersions (solid lines). Note that the fan lines analyzed here originate from regions with strong nonparabolicity, especially for the 99 Å sample. For the lh bands the strong mixing with hh states even causes an electron-like dispersion.

Intensity Profiles. Figure 4.25 shows measured magneto-Raman profiles for the (44/44) Å GaAs/AlAs sample (left column) compared to calculations (right column) [1.39]. The calculations were performed in the theoretical model outlined above [4.48] using the parameters given in Tables 4.6 and 4.7. Results are only given above a certain minimum value of the magnetic field since the number of Landau levels which have to be considered for $B \to 0$ and thus the size of the calculation increases very rapidly. Also, the correction

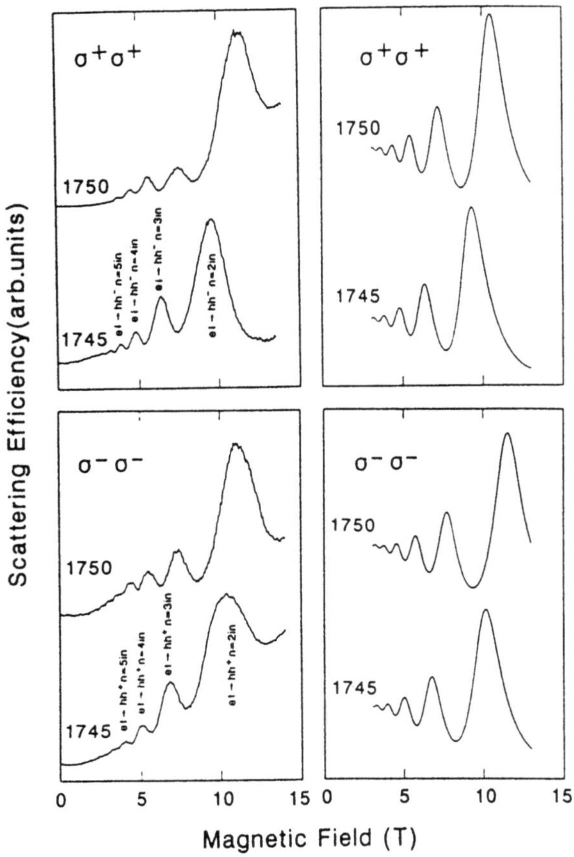

Fig. 4.25. Experimental *(left column)* and theoretical *(right column)* magneto-Raman profiles for a (44/44) Å GaAs/AlAs MQW in the $\bar{z}(\sigma^+, \sigma^+)z$ and $\bar{z}(\sigma^-, \sigma^-)z$ configurations. The excitation energies are indicated (in meV). The peaks are assigned to a series of incoming resonances ("in") with e-hh transitions

used for excitonic effects does not hold anymore at small fields [3.75]. The comparison in Fig. 4.25 shows that the observed resonances as well as their relative intensities are well reproduced by the theory. All fan lines are due to e-hh transitions in outgoing resonance. The differences in the magneto-Raman profiles for different geometries are mainly due to the hole g factors. Calculations for the $\bar{z}(\sigma^+, \sigma^-)z$ and $\bar{z}(\sigma^-, \sigma^+)z$ configurations, not shown here, are also in good agreement with the experiment.

Figure 4.26 shows measured (left column) and calculated (right column) magneto-Raman profiles for the (99/99) Å GaAs/AlAs MQW [1.39]. In contrast to the 44 Å sample of Fig. 4.25 the spectra vary strongly with the excitation energy and the magnetic field. The complexity of the profiles arises from the mixed hh and lh subbands. It increases for laser energies above 1620 meV due to the presence of both incoming and outgoing resonances. Despite these difficulties the agreement between experiment and theory is

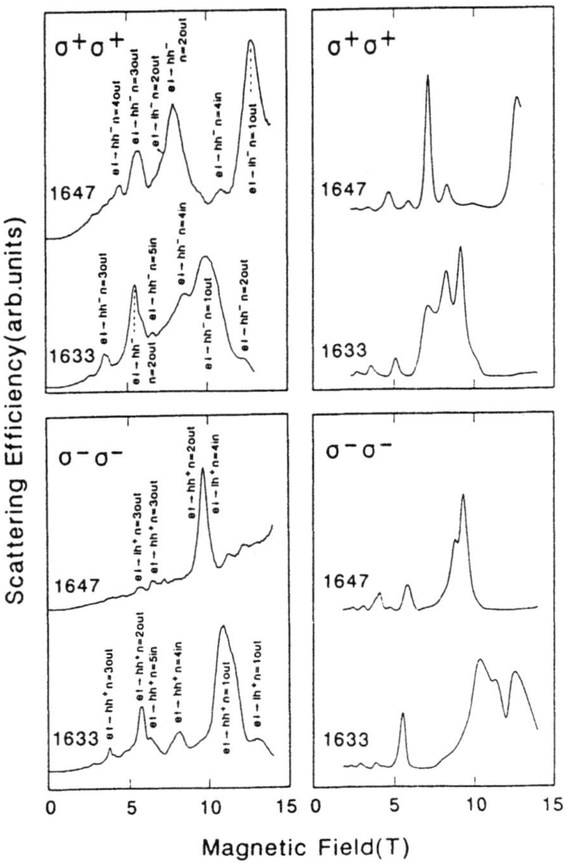

Fig. 4.26. Experimental *(left colum)* and theoretical *(right column)* LO phonon magneto-Raman profiles of a (99/99) Å GaAs/AlAs MQW in the $\bar{z}(\sigma^+, \sigma^+)z$ and $\bar{z}(\sigma^-, \sigma^-)z$ geometries. The excitation energies are indicated (in meV). The peaks are due to incoming ("in") and outgoing ("out") resonances involving e-hh and e-lh transitions

satisfactory. Resonances involving hh states are stronger than those where lh levels participate, because of the different broadenings of these transitions. Outgoing resonances have larger intensities than incoming ones.

For a quantitative comparison between theoretical and experimental magneto-Raman profiles in QWs and the bulk it is necessary to consider the broadening of the Landau levels. In the calculations of Figs. 4.25 and 4.26 it was assumed empirically that the linewidth of the resonances becomes larger with increasing Landau index. Good agreement is only obtained when a linear dependence of the broadening parameter on n (see Table 4.7) is taken into account.

5. Resonant Magneto-Luminescence

The optical excitation of bulk GaAs and GaAs/$Al_xGa_{1-x}As$ QWs with electron or hole excess energies up to an LO phonon leads to strongly resonant luminescence from magneto-optical interband transitions which is partially nonthermal. Fan plots obtained from this effect can be used for detailed investigations of the electronic structure and its renormalization by the electron–phonon coupling. The underlying scattering mechanisms are based on the relaxation of carriers between Landau levels and their interaction with acoustic phonons and interface roughness, similar to the Raman processes discussed in Sect. 3.

5.1 Bulk Semiconductors

5.1.1 Landau Level Fine Structure in GaAs

Investigations of GaAs Raman spectra in high magnetic fields lead to the discovery of other peaks which appear in addition to the LO phonon line [1.40, 1.41]. On an absolute energy scale the spectral positions of these features depend only on the magnetic field, and they do not change with the excitation wavelength. This is typical for hot luminescence recombination between electron and hole Landau levels. The relation of these resonances to the electronic structure is exploited in this section to investigate effects which arise from valence band mixing [1.40, 1.41].

Figure 5.1 shows typical Raman and luminescence spectra of a [001]-oriented undoped GaAs sample for different magnetic fields and laser energies $\hbar\omega_l$. The spectra were excited at low densities ($\ll 100\,\mathrm{W/cm^2}$) [1.41]. In addition to the LO phonon, many lines occur which move with the magnetic field. On an absolute energy scale, however, they stay fixed when the laser wavelength is varied for constant field. In Fig. 5.1 (a) the additional lines are closer to the excitation than the LO phonon (a); in Fig. 5.1 (b) they are further away. The relative intensities of the lines depend on the magnetic field and the laser energy. Similar structures are also observed for geometries with equal polarizations of incident and scattered photons [1.40].

Figure 5.2 shows fan plots of the hot magneto-luminescence peaks for four polarization geometries [1.41]. Data from spectra which were measured

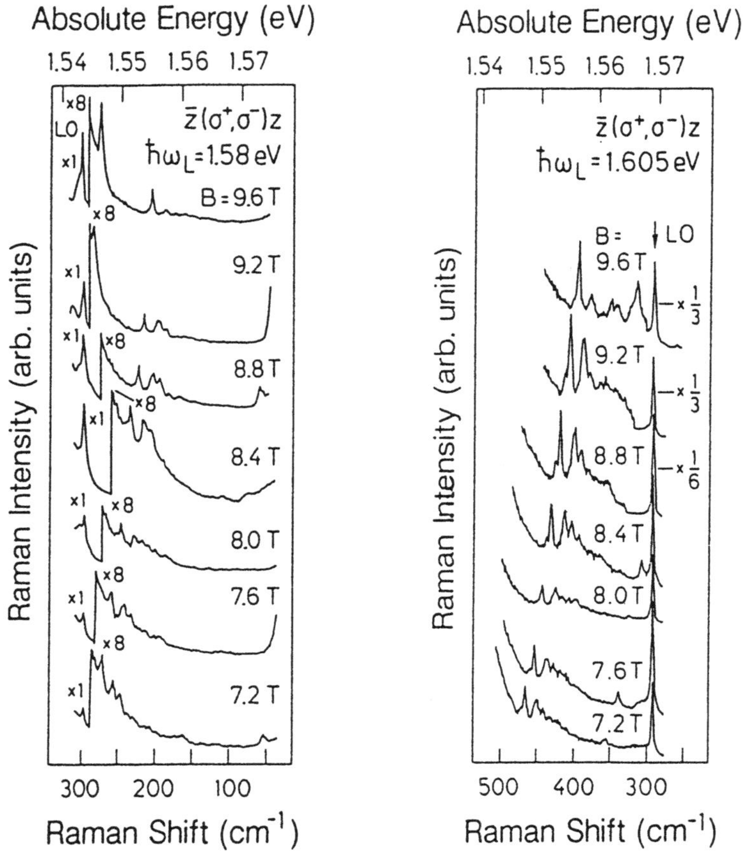

Fig. 5.1. Raman and luminescence spectra of GaAs for different magnetic fields and two laser energies $\hbar\omega_l$ (T=10 K). Parts of the spectra were multiplied with the factors indicated

with $\hbar\omega_l$ = 1.58 eV (filled circles) and 1.605 eV (crosses), respectively, coincide on an absolute energy scale. For $B \to 0$ all fan lines converge at (1.520±0.002) eV, the gap energy of GaAs at 10 K [2.153]. The intensity of the magneto-luminescence peaks depends linearly on the excitation power. This is a strong indication for the recombination of excitons which are associated with each Landau level transition and not of free electrons and holes [5.1].

Hot luminescence has been intensively investigated without a magnetic field in GaAs and other III–V semiconductors. Many studies focused on the behavior of carriers which are excited into the conduction band with a certain excess energy. During their relaxation they can recombine with acceptor levels. This causes hot luminescence which contains information about the electron–phonon interaction, inter-valley scattering times and band structure effects [5.1–5.7]. For bulk semiconductors and QWs these phenomena were also investigated in high magnetic fields where the polarization behavior of the hot luminescence allows one to determine scattering times [5.8, 5.9].

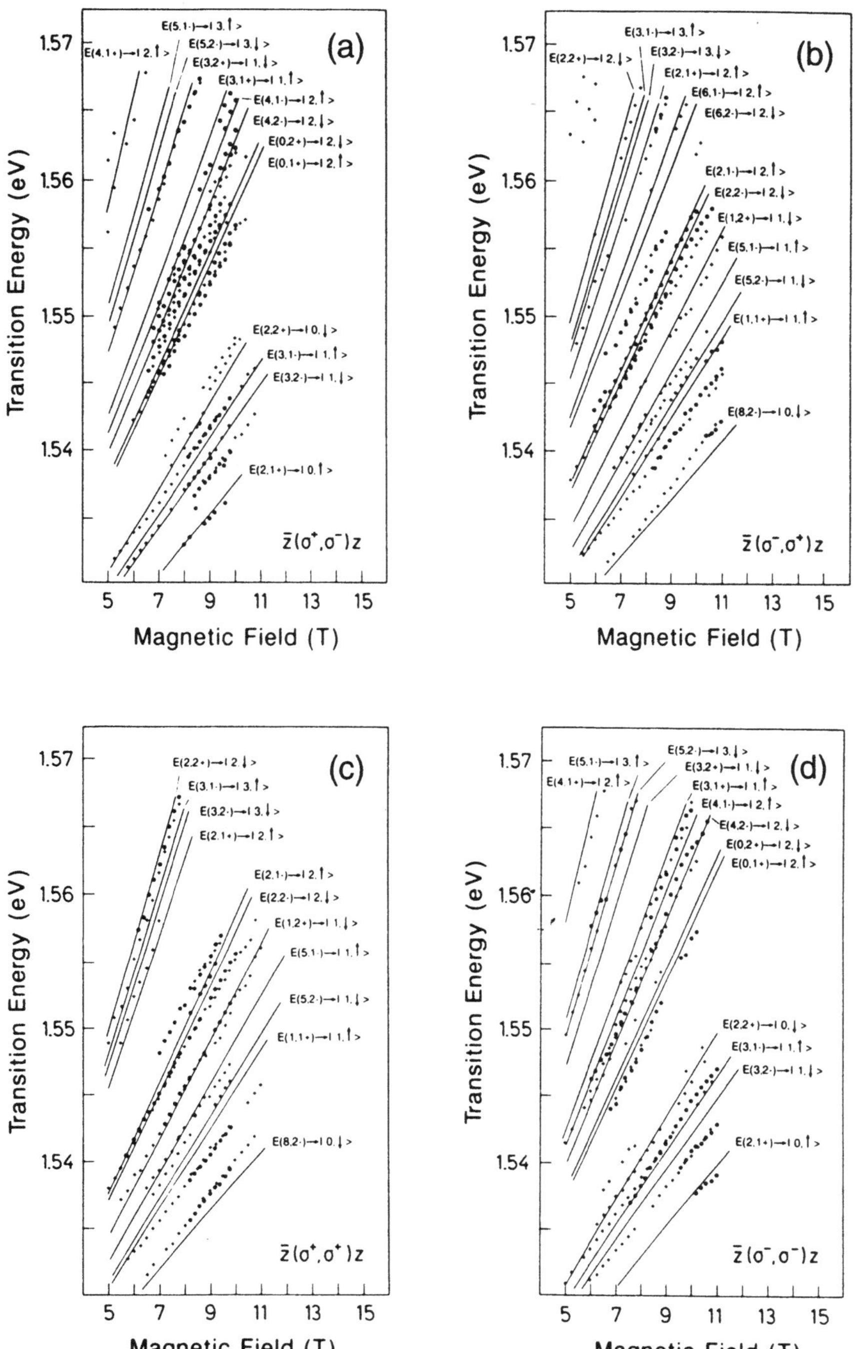

Fig. 5.2. Fan plots of hot magneto-luminescence peaks in GaAs for different polarizations. On an absolute energy scale the data measured with laser energies of 1.580 eV *(filled circles)* and 1.605 eV *(crosses)* coincide. The *solid lines* are calculated transitions considering valence band mixing effects. See the text for a description of the notation

In two-dimensional GaAs/$Al_xGa_{1-x}As$ heterostructures hot luminescence has been observed in modulation-doped samples as well as under high-excitation conditions [5.10, 5.11]. The recombination to acceptor states or between electron and hole Landau levels was used to investigate effects of the electron–electron interaction vs. carrier density. The filling and saturation of higher Landau level transitions has been demonstrated under these conditions [5.11]. Recently, the spin relaxation of carriers [5.12], Auger recombination involving Landau levels [5.13], and shake-up processes [5.14] in the electron gas of heterostructures in high magnetic fields have been observed by luminescence techniques. The characteristic feature of the results presented here, however, is that the hot magneto-luminescence between electron and hole Landau levels occurs in nominally undoped samples at quite small excitation densities while in most of the other studies large carrier densities were obtained by doping or high excitation.

The question arises as to whether the hot luminescence fan lines in Fig. 5.2 can be attributed to resonances of the electronic structure. An empirical approach is the comparison with LO-phonon magneto-Raman profiles for which detailed studies of the electronic structure and assignments to theoretical interband transitions have been performed [1.31–1.33]. An analysis of the fan plots of Fig. 5.2 shows that the hot magneto-luminescence and magneto-Raman resonances indeed have the same origin [1.41]. It should thus be possible to explain the fan lines in Fig. 5.2 with the model outlined in Sect. 4.1.2 and applied to GaAs in Sect. 4.2.1. The strongest outgoing resonances of the magneto-Raman profiles in GaAs (see Figs. 4.5 and 4.6) are due to magneto-optical interband transitions between heavy-mass valence and electron Landau levels [1.33]. Wave functions of the $E(n,1-)$ and $E(n,2-)$ ladders (see (4.3a), (4.3b), (4.7a), and (4.7b)) from a (2×2)-model were used to identify the strongly mixed valence Landau levels obtained from more precise calculations in a larger (8×8)-model and to classify the observed resonances. One should thus expect that the theoretical fan plots of Figs. 4.5 and 4.6 [1.33] are sufficient to explain the resonances in Fig. 5.2. However, considerably more transitions are experimentally observed. On the other hand, the magneto-Raman fan plots (Figs. 4.5 and 4.6) also exhibit a fine structure at small energies which was not considered in Sect. 4.2.1. The resonances which were used to assign the fan lines in Figs. 4.5 and 4.6 are denoted in Fig. 5.2 by $E(n+2,2-) \rightarrow |n,\downarrow\rangle$ for scattering geometries where σ^- photons are analyzed and by $E(n,1-) \rightarrow |n,\uparrow\rangle$ where σ^+ photons are detected. In all these (dominating) transitions $|n, J_z = \pm 1/2\rangle$ admixtures of heavy-mass valence levels and $|n,\downarrow(\uparrow)\rangle$ electrons are involved. The fine structure of transitions in Fig. 5.2, however, can only be explained by a more detailed analysis of all possible transitions which also considers the weaker resonances.

In a first step we therefore take into account that the observed fan lines are doublets, split by a g factor corresponding to the sum of the electron and hole values. This effect, which is dominated by the larger heavy hole g

factor, was neglected in Sect. 4.2.1 since it is smaller than the experimental accuracy [1.33]. The more precise data in Fig. 5.2, however, allow us to distinguish between these different contributions and they are thus indicated together with their respective partners. Doublets of $E(n+2,2-) \rightarrow |n,\downarrow\rangle$ and $E(n+2,1-) \rightarrow |n,\uparrow\rangle$ in Figs. 5.2 (a) and 5.2 (d), responsible for transitions in σ^- polarization, as well as $E(n,1-) \rightarrow |n,\uparrow\rangle$ and $E(n,2-) \rightarrow |n,\downarrow\rangle$ in Figs. 5.2 (b) and 5.2 (c), which lead to resonances in σ^+ polarization, can be identified. Transitions with the same polarization occur between suitable $\pm 3/2$ and $\pm 1/2$ admixtures of the hole states (see (4.3a), (4.3b) and Table 4.1) and electron levels with spins $\uparrow$ ($\downarrow$). The oscillator strengths of transitions calculated in a (8×8)-model [4.7] are smaller for resonances between light-mass valence and electron levels. These transitions can also be assigned to fan lines in Fig. 5.2. They are denoted by $E(n,1+) \rightarrow |n,\uparrow\rangle$ and $E(n,2+) \rightarrow |n,\downarrow\rangle$ for σ^+ detection (Fig. 5.2 (b) and (c)) and by $E(n+2,1+) \rightarrow |n,\uparrow\rangle$ and $E(n+2,2+) \rightarrow |n,\downarrow\rangle$ when σ^- photons are analyzed (Fig. 5.2 (a) and (d)). The irregular spacing of these transitions for small n is caused by "quantum effects" [3.74,4.7]. In addition to these resonances, which can be derived in a systematic way, further fan lines are plotted in Fig. 5.2. The calculations show that they have comparable oscillator strengths. The admixture responsible for the coupling, however, only appears at the third or even fourth position in the wave function which was nevertheless labeled according to the strongest component. If these transitions are also taken into account, most of the fan lines observed by hot luminescence or magneto-Raman scattering can be identified [1.41].

Since the Landau level calculations were performed in a single-particle picture [4.7], exciton effects have to be taken into account separately. This was done using the variational ansatz described in [1.30] which yields the ground state of magneto-excitons with reduced effective mass μ, depending on the magnetic field and the Landau index. For mixed valence states the assignment of an effective mass is no longer unique, and values of μ were determined directly from the slopes of the fan lines. The Landau quantum number was taken from the electron state involved [1.40,1.41]. This procedure yields exciton energies which have to be subtracted from the calculated transitions before comparing them with the experiment. The fan lines in Fig. 5.2 have only small Landau indices ($n \leq 3$). In this range the exciton binding energies exceed 4 meV and thus represent a correction to the transition energies which cannot be neglected, especially when considering a level fine structure. The irregularly spaced low-index transitions are not simply shifted uniformly when the Coulomb correction is applied but their order may also change. Without the consideration of exciton effects a comparison with the experiment is therefore hardly possible.

The mechanism responsible for the occurence of hot luminescence between Landau levels can be considered as an intraband relaxation of carriers accompanied by the emission of phonons. Electrons and holes reach a dynamic equilibrium at the minima of the Landau bands from where recombination takes place. A characteristic feature of this effect is its appearance for laser energies below the polaron threshold where optic phonons cannot be emitted. Therefore carriers can only relax via acoustic phonons and get rid of their energy much more slowly [5.15]. This causes a population of higher Landau levels and hot luminescence occurs. When the emission of optic phonons is possible the cariers relax rapidly to the lowest Landau level [5.15], and the recombination from higher-energy states is suppressed. A detailed understanding of these processes requires theoretical investigations of the carrier relaxation between Landau levels.

5.1.2 Magneto-Luminescence in the Quasi-Classical Limit

It was mentioned in Sect. 5.1.1 that the intensities of hot luminescence lines vary with the excitation energy while their spectral position does not change. It was further pointed out that the observed lines are only strong as long as the excited carriers cannot emit optic phonons and rapidly decay to the lowest Landau state. In the following we discuss these effects for excitation energies which are very close to interband magneto-optical transitions [1.42, 1.43]. In this case the signal is strongly enhanced and very sharp resonances of transitions with large Landau indices can be observed already at small magnetic fields. This makes the analysis of fan plots in the quasiclassical limit possible which allows one to determine effective masses and g factors.

Figure 5.3 shows luminescence spectra of GaAs for different magnetic fields, measured in $\bar{z}(\sigma^+,\sigma^+)z$ configuration at $T = 5\,\mathrm{K}$ with a laser energy of $\hbar\omega_l = 1.583\,\mathrm{eV}$ [1.43]. The laser energy is close to a magneto-optical interband transition where hot luminescence is resonantly excited. One observes a sharp line which shifts with the magnetic field and rapidly loses its intensity when the separation from the excitation increases. With a double monochromator as spectral band pass at an energy $\hbar\omega_s$, which is shifted by $\Delta = \hbar\omega_l - \hbar\omega_s$ with respect to the excitation, one obtains the intensity profiles shown in Fig. 5.4 [1.43]. Pronounced magneto-oscillations occur already below 1 T. The well-defined differences between the resonance positions in the $\bar{z}(\sigma^+,\sigma^+)z$ and $\bar{z}(\sigma^-,\sigma^-)z$ configurations indicate that light-mass valence levels which have a larger g factor than the heavy-mass states participate [1.33, 2.153]. The electron g factor contribution is much smaller and can be neglected [2.153]. Intensity profiles in the $\bar{z}(\sigma^\pm,\sigma^\mp)z$ geometries are about 20 times weaker. These resonances depend on the polarization of the analyzer. In measurements with different excitation energies and fixed detection the large resonances in Fig. 5.4 remain at the same positions. However, their intensity rapidly decreases with increasing detuning Δ [1.43]. These oscillations thus have the character of "outgoing resonances" which occur when their energy coincides with $\hbar\omega_s$. The weaker resonances in Fig. 5.4 move with Δ. They appear when the magneto-optical transitions coincide with $\hbar\omega_l$. At

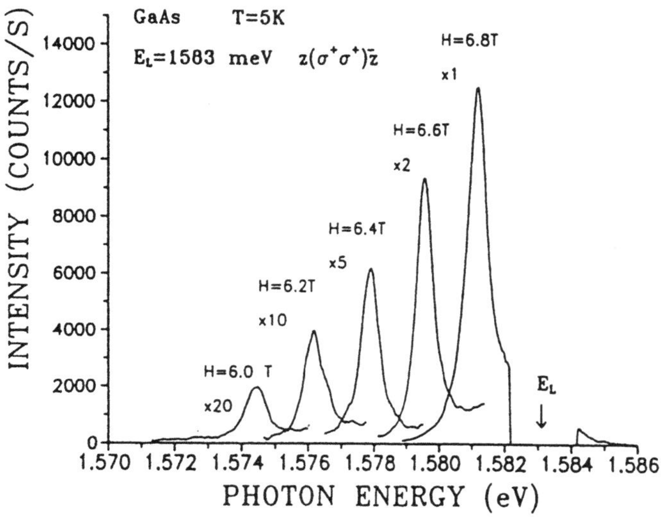

Fig. 5.3. Luminescence spectra of GaAs for different magnetic fields. For excitation close to an interband magneto-optical resonance hot luminescence occurs. This leads to a sharp emission line which depends on the magnetic field. The spectra were multiplied with the factors indicated

these points $\hbar\omega_s$ is smaller than the resonance energy, and the signal enhancement is due to the luminescence tail from the lowest Landau level which can be considered as an "incoming" resonance. This effect appears at quite low excitation densities ($\ll$ 10 W/cm^2). It allows one to perform very precise measurements of the fan lines over a certain range, limited towards the lower energies ($\hbar\omega_s < 1.525$ eV) by the much stronger fundamental exciton luminescence. In the direction of increasing energies the resonances rapidly broaden and become weaker when the excitation energy exceeds the polaron threshold only by a few meV [1.43].

Figure 5.5 shows fan plots determined from resonant magneto-luminescence intensity profiles as in Fig. 5.4 [1.42, 1.43]. The energy scale refers to the detection energy $\hbar\omega_s$. The spectra were measured with $\Delta = 2$ meV. These rather precise magneto-luminescence measurements can be used to determine material parameters of GaAs. For data at small magnetic fields and large Landau indices, as is the case in Fig. 5.5, the quasiclassical limit of Landau quantization is realized, which allows the most precise mapping of level energies on the bulk dispersion and the application of simple formulae. This is not only true for expressions concerning g factors and effective masses but also for more complex phenomena such as the corrections due to Coulomb interaction. Exciton effects were considered in the fit of a theoretical model to the data in Fig. 5.5 by using an approximation for large Landau quantum numbers which gives analytical expressions for the magneto-exciton ground state vs. n and B [1.43, 5.16]. In addition to this correction, which is small

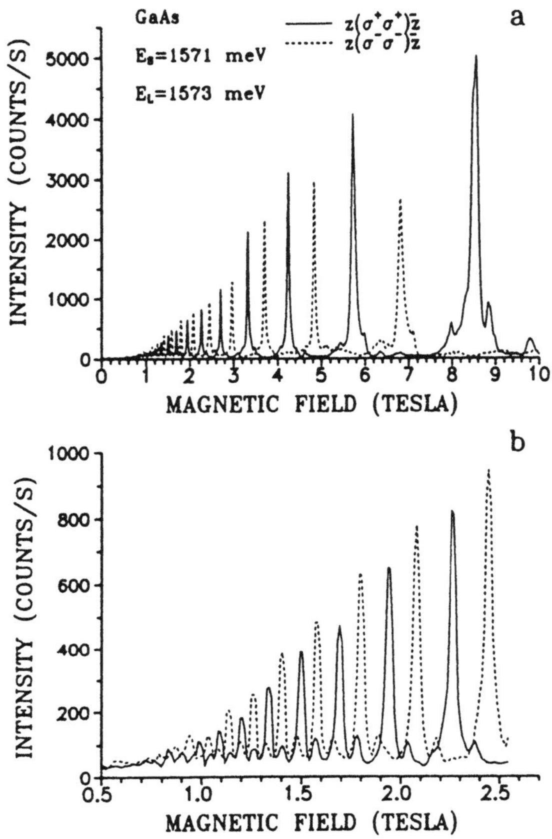

Fig. 5.4. (a) Hot luminescence intensity profiles due to magneto-optical interband transitions vs. magnetic field. Measurements for $\bar{z}(\sigma^+,\sigma^+)z$ ($\bar{z}(\sigma^-,\sigma^-)z$) configuration are shown by *solid (dashed) lines*, respectively. The range of small fields is enlarged **(b)**

in the range investigated, the theoretical transitions in Fig. 5.5 were fitted to the following empirical expression quadratic in B [1.43]:

$$E(n) = E_0 + \epsilon_1(n)\,B - \epsilon_2(n)\,B^2 \ . \tag{5.1}$$

Besides Coulomb corrections E_0 is the direct band gap without field, $\epsilon_1(n)$ is related to the reduced effective mass, and $\epsilon_2(n)$ contains information about the band nonparabolicity. In [4.1] analytical expressions were derived for these coefficients based on a (8×8)-$\boldsymbol{k}\cdot\boldsymbol{p}$-model for electrons and light-mass valence Landau levels. This can be used for a direct and simple comparison of $\epsilon_1(n)$ and $\epsilon_2(n)$, determined from the fits, with material parameters. The fit of (5.1) to the data in Fig. 5.5 was performed by a least-squares method considering only the range between 1.53 and 1.57 eV in order to eliminate magneto-polaron effects which occur at higher energies. For $B \to 0$ the fitted fan lines converge at 1.5185 eV ($n \geq 4$) and are thus in very good agreement with the literature value of the E_0 band gap in GaAs [2.153].

The linear coefficient $\epsilon_1(n)$ describes the slope of the fan lines for $B \to 0$ and is therefore related to the effective mass of electrons and light holes at $\boldsymbol{k} = 0$. The electron part to the slope is given by $\epsilon_1^{\rm c}(n) = m_{\rm e}^{-1}(n + 1/2)$ (in units of $e\hbar/m_0$), while the light hole contribution can be calculated from the

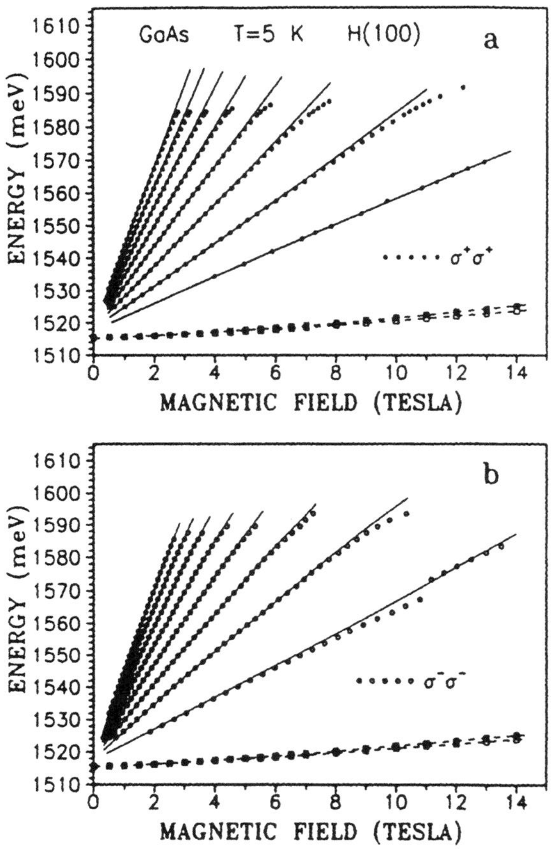

Fig. 5.5. Fan plots of magneto-optical transitions in GaAs observed by resonant magneto-luminescence. *Filled* and *open circles* are measured data for the $\bar{z}(\sigma^+, \sigma^+)z$ (electron Landau levels $1 \leq n \leq 8$) **(a)** and $\bar{z}(\sigma^-, \sigma^-)z$ configurations (electron Landau levels $2 \leq n \leq 9$) **(b)**. The *solid lines* are fits as discussed in the text. The lowest transitions (*open* and *filled squares, dashed lines*) indicate the change of the exciton ground state with B, determined from nonresonant luminescence measurements

(2×2)-matrices given by Luttinger (see (4.7a) and (4.7b)) which are quite accurate for the case of large n and small B [1.43, 3.74, 4.1, 4.9]. A detailed description of the transitions assigned to the fan lines in Fig. 5.5 is given in [1.43]. Values of $\epsilon_1(n)$ for transitions in σ^+ and σ^- polarization are shown in Fig. 5.6 vs. the Landau level index [1.43]. For the identification of the relevant hole components of the mixed wave functions which participate in the observed transitions we used the ($J_z = \pm 3/2$) admixtures of the light mass valence levels which, according to (4.3a), (4.3b), and Table 4.1, dominate for large n. The data points in Fig. 5.6 depend almost linearly on the Landau index. For $n \geq 5$ they are best described by

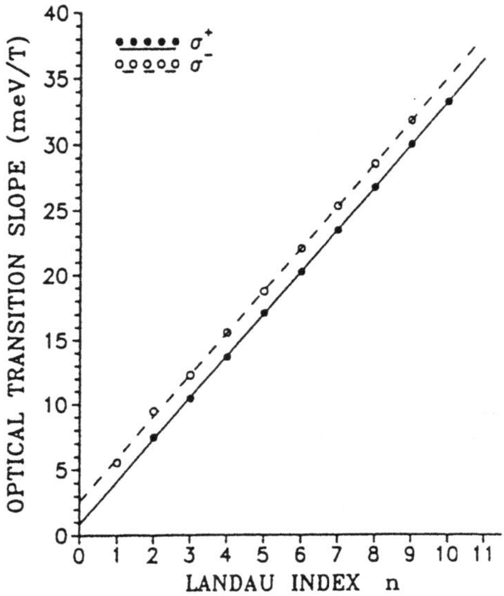

Fig. 5.6. Slopes $\epsilon_1(n)$ of the fan lines fitted to the magneto-optical interband transitions measured by resonant hot magneto-luminescence in Fig. 5.5 (*open (filled) circles:* σ^+ (σ^-) polarization)

$$\epsilon_1(n) = (3.242\,n + 0.78)\,\mathrm{meV/T} \quad (\sigma^+) \tag{5.2a}$$

$$\epsilon_1(n) = (3.246\,n + 2.54)\,\mathrm{meV/T} \quad (\sigma^-)\,. \tag{5.2b}$$

For large n one obtains the reduced effective mass μ from

$$\epsilon_1(n) - \epsilon_1(n-1) = \frac{1}{\mu} = \frac{1}{m_\mathrm{e}} + \frac{1}{m_\mathrm{lh}} \tag{5.3}$$

and finds a value of $\mu^{-1} = 28.02\,m_0^{-1}$. To determine μ the values of $\epsilon_1(n)$ have to be divided by $e\hbar/m_0 = 0.1158\,\mathrm{meV/T}$. Using an electron mass between $m_\mathrm{e} = 0.0653\,m_0$ and $0.0660\,m_0$, which was found in cyclotron resonance measurements taking the conduction band nonparabolicity into account [5.17–5.19], one obtains a light-hole effective mass for a magnetic field along the [001] direction of $m_\mathrm{lh} = (0.078 \pm 0.002)\,m_0$. This value is somewhat smaller than earlier results between $m_\mathrm{lh} = 0.082\,m_0$ and $0.087\,m_0$ from magneto-optical interband experiments and cyclotron resonance for $\boldsymbol{B} \parallel [001]$ [2.153]. In a five-band $\boldsymbol{k}\cdot\boldsymbol{p}$-model the light-hole effective mass at $\boldsymbol{k} = 0$ was calculated to be $m_\mathrm{lh} = 0.08082\,m_0$ ([100] direction) and $m_\mathrm{lh} = 0.07054\,m_0$ ([110] direction) [5.20]. When taking into account that magneto-optical experiments with the field along [001] yield a mass averaged over the perpendicular [100] and [110] directions, these calculations are closer to the new experimental value for m_lh than to the previous results [1.43]. For large n the light-hole g factor can be determined from the difference of the lines in Fig. 5.6 [4.9, 5.21]. One finds $g_\mathrm{lh} = -30.4 - g_\mathrm{e} = -30.0$, where the electron g factor is given by $g_\mathrm{e} = -0.44$ [2.153], and the sign results from the selection rules. The Luttinger parameter κ which describes the isotropic

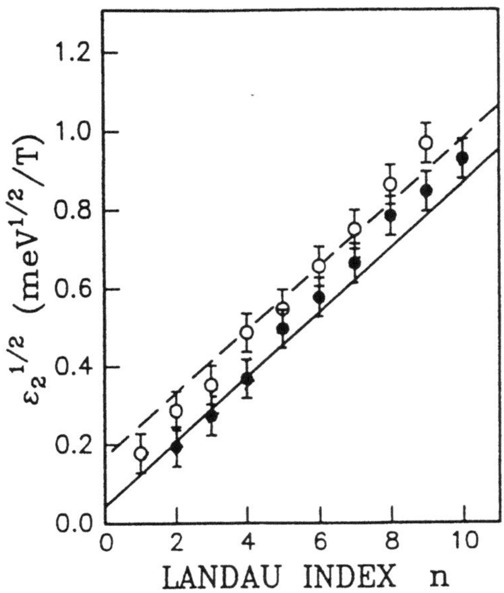

Fig. 5.7. Nonparabolicity coefficients $\sqrt{\epsilon_2(n)}$ of magneto-optical interband transitions with Landau index n in GaAs, determined from fits of (5.1) to the data in Fig. 5.5. *Filled (open) circles* are for measurements in $\bar{z}(\sigma^+,\sigma^+)z$ $(\bar{z}(\sigma^-,\sigma^-)z)$ polarizations. The *solid (dashed) lines* were calculated with (5.4) and parameters discussed in the text

part of the hole g factor is given by $\kappa = (|g_{\text{lh}}| - 2/m_{\text{lh}})/4$ [4.9, 5.21], and one obtains $\kappa = 1.09 \pm 0.2$, in good agreement with the literature value of $\kappa = (1.2 \pm 0.2)$ [2.153].

Near the polaron threshold around 1.590 eV the fitted fan lines in Fig. 5.5 devitate from the experiment. Due to the smaller Fröhlich coupling constant [4.18] in GaAs ($\alpha = 0.065$ [2.153]) which is only about half as large as for InP (see Sect. 4.2.2) resonant magneto-polarons are harder to observe. The resonant magneto-luminescence measurements in Fig. 5.5, however, yield data for the lower polaron branch even at high Landau indices. As for InP, the polaron splitting decreases with increasing n [4.33]. By comparison with calculations it was found that the influence of resonant magneto-polarons on the slopes of the fan lines for $B \to 0$ and thus on the effective mass can be neglected. However, they contribute about 20% to the nonparabolicity term $\epsilon_2(n)$ even below 1.570 eV. For a precise determination of this parameter they were therefore calculated according to [4.33] and, together with the exciton corrections mentioned above, considered in the fits [1.43].

The values of $\epsilon_2(n)$ found from fits to the corrected fan lines therefore only contain contributions due to the band nonparabolicity. Separating the effects of electrons and light holes $\epsilon_2(n)$ is a sum of two terms [1.43]

$$\epsilon_2(n) = \frac{\epsilon_1^{\text{c}}(n)^2}{\epsilon_{\text{c}}^*} + \frac{\epsilon_1^{\text{lh}}(n)^2}{\epsilon_{\text{lh}}^*} \tag{5.4}$$

where $\epsilon_1^{\text{c,lh}}(n)$ are the linear coefficients for these bands and $\epsilon_{\text{c,lh}}^*$ are effective parameters with the dimension of an energy. For $\epsilon_2(n)$ one expects a quadratic dependence on the Landau level index. Values of $\sqrt{\epsilon_2(n)}$ determined from the fits in Fig. 5.5 are shown in Fig. 5.7 [1.43, 5.22]. Note the almost linear increase with n which is in good agreement with calculations (solid and dashed lines) from (5.4) [1.43, 5.22]. The parameters $\epsilon_1^{\mathrm{c}}(n)$ and $\epsilon_1^{\mathrm{lh}}(n)$ in (5.4) were determined from the experimental results in (5.2a) and (5.2b) for $\epsilon_1(n) = \epsilon_1^{\mathrm{c}}(n) + \epsilon_1^{\mathrm{lh}}(n)$ using the coefficient $\epsilon_1^{\mathrm{c}}(n)$ calculated for an electron effective mass of $m_{\mathrm{e}} = 0.0660\, m_0$. One finds

$$\epsilon_1^{\mathrm{c}}(n) = 1.755 \left(n + \frac{1}{2}\right) \mathrm{meV/T} \tag{5.5a}$$

$$\epsilon_1^{\mathrm{lh}}(n) = 1.485 \left(n + \frac{1}{2}\right) \mathrm{meV/T}\,. \tag{5.5b}$$

The effective energies ϵ_{c}^* und ϵ_{lh}^* are obtained from expressions given by $\boldsymbol{k} \cdot \boldsymbol{p}$-theory [1.35, 1.43]. In a two-band model the conduction band dispersion without magnetic field can be expressed by

$$\epsilon(k) = \epsilon_1^{\mathrm{c}} \left(1 - \frac{\epsilon_1^{\mathrm{c}}}{\epsilon_{\mathrm{c}}^*}\right) \tag{5.6}$$

with $\epsilon_1^{\mathrm{c}} = \hbar^2 k^2/(2m_{\mathrm{e}})$ and the effective energy ϵ_{c}^*. Considering the band nonparabolicity via (3.17) with $\beta_0 = 0$ [1.35, 3.71] one obtains

$$\epsilon_{\mathrm{c}}^* = -\left(\frac{\hbar^2}{2m_{\mathrm{e}}}\right)^2 \frac{1}{\alpha_0}\,. \tag{5.7}$$

The nonparabolicity coefficient α_0 was determined experimentally for GaAs from the magneto-Raman measurements as $\alpha_0 = -2370\,\mathrm{eV\AA^4}$ (see Sect. 4.2.1) [1.35]. This yields an effective energy of $\epsilon_{\mathrm{c}}^* = 1.41\,\mathrm{eV}$. For light holes one obtains

$$\epsilon_{\mathrm{lh}}^* = \left(\frac{E_0 + 2\Delta_0}{2E_0\Delta_0}\right)^{-1} \tag{5.8}$$

from an expansion to different powers in the magnetic field [4.1] and finds $\epsilon_{\mathrm{lh}}^* = 0.47\,\mathrm{eV}$. The solid and dashed lines in Fig. 5.7 were calculated with these parameters from (5.4). They are in good agreement with the experiment. The stronger nonparabolicity of the transitions in the $\bar{z}(\sigma^-, \sigma^-)z$ configuration arises from the larger energies of the hole Landau levels with ($J_z = +3/2$) components involved in the transitions as compared to the $\bar{z}(\sigma^+, \sigma^+)z$ geometry where the ($J_z = -3/2$) admixtures contribute. This is a direct consequence of the mixed valence band wave functions (see (4.3a), (4.3b), and Table 4.1) where the Landau indices of these admixtures for the nth state always differ by two units. This effect was taken into account in the calculation of the hole contributions to $\epsilon_2(n)$. A simple expression for α_0 has also been derived in [1.35]. With

$$\alpha_0 = -\frac{P^4}{E_0^{*3}} - \frac{P^2}{E_0^{*2}} \left(\frac{Q^2}{E_0'^*} - \frac{P'^2}{E_0'^* - E_0^*}\right) + \frac{P'^2 Q^2}{E_0^*(E_0'^* - E_0^*)^2} \tag{5.9}$$

and matrix elements from [3.80] ($P = 9.957\,\mathrm{eV\AA}$, $P' = 4.763\,\mathrm{eV\AA}$, $Q = 6.763\,\mathrm{eV\AA}$) one finds $\alpha_0 = -2279\,\mathrm{eV\AA^4}$ and $\epsilon_{\mathrm{c}}^* = 1.46\,\mathrm{eV}$, in good agreement with the experimental value from [1.35] (see Sect. 4.2.1). The average band gaps used in (5.9) are $E_0^* = E_0 + \Delta_0/3 = 1.631\,\text{eV}$ and $E_0'^* = E_0' + 2\Delta_0'/3 = 4.602\,\text{eV}$ [3.80].

5.2 Inter-Landau-Level Scattering in Quantum Wells

In Sect. 5.1 we discussed the recombination of hot photoexcited carriers between electron and hole Landau levels of bulk GaAs. The observed fan lines were interpreted in terms of valence band mixing effects, effective masses and band nonparabolicities. These effects can also be studied in QWs and applied to determine their electronic structure. In order to identify the mechanisms which lead to the appearance of hot luminescence one has to understand the relaxation of carriers between the QW Landau levels. In this context acoustic phonons play an important role. From investigations in QWs, where the disorder-induced continuous emission discussed in Chap. 3 provides special access to these modes, one thus expects to gain important insight into these phenomena. The results obtained in QWs should also be relevant for the understanding of hot luminescence in the bulk where the acoustic modes cannot be observed as easily. In the following we discuss resonant hot magneto-luminescence experiments performed on a GaAs/$Al_{0.36}Ga_{0.64}As$ MQW [1.44]. We present possible scattering mechanisms involving acoustic phonons and discuss the role of crystal-momentum conservation. When excited in the vicinity of interband magneto-optical transitions the spectra also show lines due to electronic Raman scattering between conduction band Landau levels, populated by photoexcitation [1.15, 1.44, 5.8]. The resonance behavior of these peaks leads to further conclusions about the hot luminescence and the role played by crystal-momentum conservation. Hot magneto-luminescence between heavy- and light-hole excitons as well as from the lowest Landau level transition in QWs has also been reported in other investigations [5.23, 5.24].

Resonant Hot Magneto-Luminescence. Figure 5.8 shows a luminescence spectrum of the (100/103) Å GaAs/$Al_{0.36}Ga_{0.64}As$ MQW sample which has already been discussed in Sects. 3.1 and 3.7 [1.44]. The excitation density of the spectra discussed in this section is less than $5\,\text{W/cm}^2$ in order to avoid the build-up of a large photoexcited population. Different regimes can be distinguished in the spectrum: Superimposed on the incoherent, exponentially decaying luminescence background from the lowest exciton state around 1.555 eV one observes the continuous emission (indicated by dashed lines) within about 5 meV around the laser (ω_l). Emission maxima occur at lower-energy transitions between Landau levels (LL1, LL2) where the joint density of states has singularities. Finally, there are peaks with a Raman shift corresponding to the electron cyclotron energy (SRS and ARS). The features due to hot luminescence and electronic Raman scattering between Landau levels which are of interest here are shown by the shaded regions.

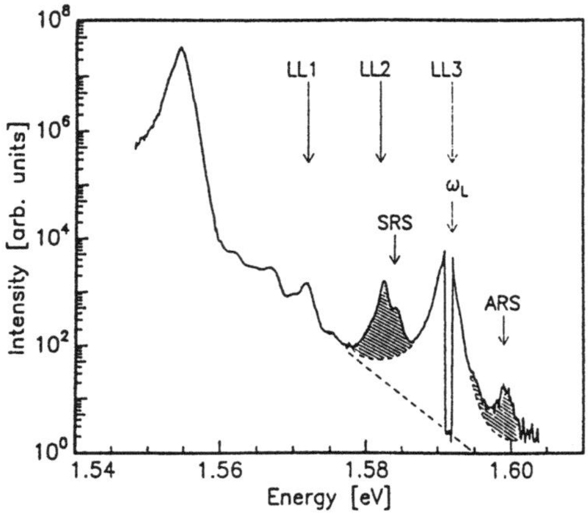

Fig. 5.8. Luminescence spectrum of a (100/103) Å GaAs/$Al_{0.36}Ga_{0.64}As$ MQW measured at 4.7 T in resonance with the (n = 3)-transition between electron and heavy-mass Landau levels (LL3) in $\bar{z}(\sigma^-, \sigma^-)z$ Faraday backscattering configuration (T = 10 K). See text for details

The maximum marked by LL2 in Fig. 5.8 corresponds to a transition between electron and heavy-mass Landau levels with index $n = 2$. It differs from those at lower energies in several aspects: Its intensity is much larger than expected by comparison with the exponentially decreasing luminescence tail from the lowest-energy exciton. The intensity of this line changes linearly with the excitation power while the lower ones have an approximately quadratic dependence. Furthermore, LL2 has a characteristic polarization behavior. As shown in Fig. 5.8 for $\bar{z}(\sigma^-, \sigma^-)z$ scattering, a strong signal is also found in the $\bar{z}(\sigma^+, \sigma^+)z$ geometry. For the $\bar{z}(\sigma^\pm, \sigma^\mp)z$ configurations, however, the LL2 intensity only reflects the Maxwell–Boltzmann distribution of thermalized carriers in the luminescence tail.

The polarization and power dependence of LL2 therefore corresponds to that of the continuous emission. This indicates that both effects might have the same origin, i. e., that acoustic phonons contribute to the LL2 signal. The polarization behavior of LL2 also corresponds to that of the resonant hot magneto-luminescence excited close to interband transitions in bulk GaAs (see Sect. 5.1.2). On the other hand, the hot-luminescence fine structure (see Sect. 5.1.1), excited about one LO phonon away from resonance, appears in all scattering geometries. It depends only on the polarization of the analyzer. This indicates that the hot luminescence at LL2 and in Sect. 5.1.2 has a more coherent character than the transitions discussed in Sect. 5.1.1. The reason for this difference may be found in the smaller number of intermediate steps participating in the energy relaxation, due to the smaller energy which has to be overcome before recombination occurs. It should thus be possible, within certain approximations, to treat these phenomena in a coherent picture, i. e., as Raman scattering.

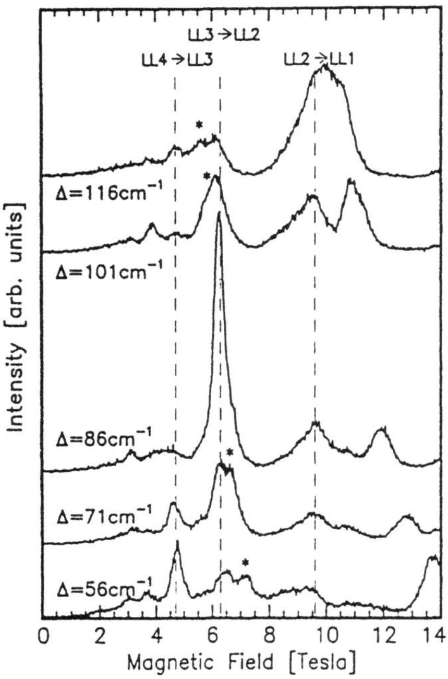

Fig. 5.9. Magneto-oscillations of the hot luminescence intensity in a (100/103) Å GaAs/$Al_{0.36}Ga_{0.64}$As MQW for different shifts of the detection Δ with respect to the laser ($\hbar\omega_l = 1.5999\,\mathrm{eV}$). Intensity oscillations with incoming *(dashed vertical lines)* and outgoing character *(asterisks)* can be identified. For $\Delta = 86\,\mathrm{cm}^{-1}$ in- and outgoing resonances at about 6 T coincide and a "double resonance" occurs

The interpretation of LL2 as a Raman process is also suggested by its resonance behavior. Figure 5.9 shows intensity profiles of magneto-oscillations measured for a fixed excitation energy $\hbar\omega_l = 1.5999\,\mathrm{eV}$ and different frequency shifts $\Delta = \hbar\omega_l - \hbar\omega_s$ with a double monochromator as spectral band pass of about $0.3\,\mathrm{cm}^{-1}$ width at $\hbar\omega_s$ [1.44]. The laser energy for which the intensity profiles were measured is shown by the horizontal dashed line in the $\bar{z}(\sigma^-, \sigma^-)z$ transition fan plot from [1.26] (see Fig. 3.33) given in Fig. 5.10 [1.44]. The resonances in Fig. 5.9 are attributed to magneto-optical interband transitions between heavy-mass valence and electron Landau levels LLn. Around 6 T the spectra show a constant incoming resonance with LL3 (dashed vertical line) and an outgoing resonance with LL2 (asterisks) whose position varies with Δ. The increase of the signal for the incoming LL3 resonance which occurs over a wide range of Δ is caused by enhanced LL2 and continuous-emission signals as well as by the luminescence background. In outgoing resonance one records the LL2 maximum. For $\Delta = 86\,\mathrm{cm}^{-1}$ the two resonances coincide and the intensity is strongly enhanced. This phenomenon can be interpreted as a "double resonance" where the excitation occurs into LL3 and emission is recorded from LL2. In Fig. 5.9 further "double resonances" occur for LL4 → LL3 and LL2 → LL1 at $\Delta = 56\,\mathrm{cm}^{-1}$ and

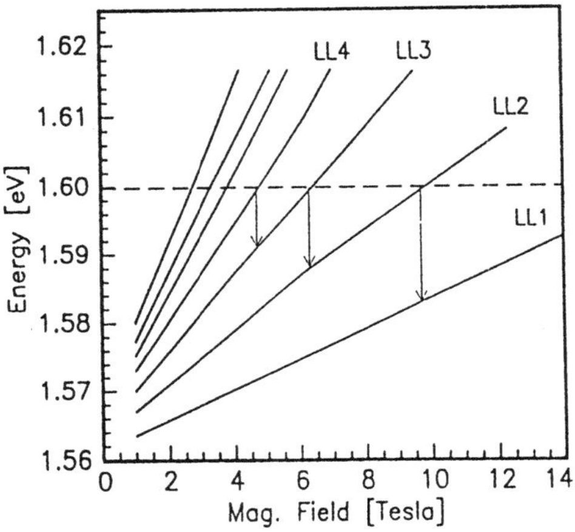

Fig. 5.10. Fan plot of the transitions between heavy-mass valence and electron Landau levels LLn considered in Fig. 5.9. The *horizontal dashed line* corresponds to the excitation energy. The *vertical arrows* denote possible "double resonances"

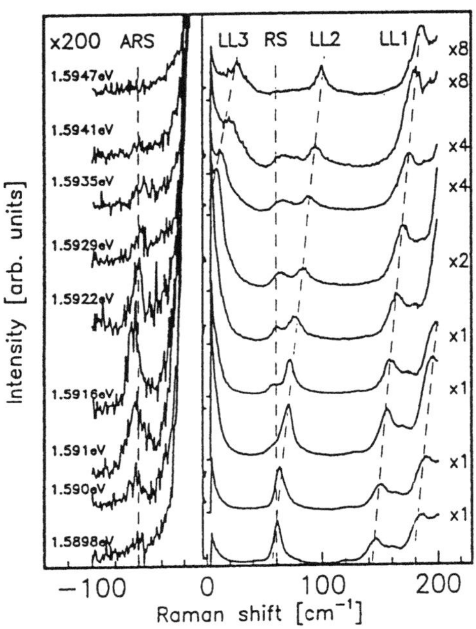

Fig. 5.11. Raman spectra of a (100/103) Å GaAs/$Al_{0.36}Ga_{0.64}As$ MQW at 4.7 T ($\bar{z}(\sigma^-, \sigma^-)z$ geometry) for different excitation energies, given next to each curve, around LL3 (1.5916 eV). Peaks due to LLn transitions shift with the laser energy while "true" Raman lines (SRS, ARS) do not depend on it. The spectra were scaled with the factors indicated

116 cm^{-1}, respectively. Note that these "double resonances" differ from those discussed in Sects. 4.1.3 and 4.2.1 since in a LLn → LL(n-1) transition the Landau oscillator indices of both, electron and hole levels, change.

The resonance behavior of the hot luminescence can also be investigated in series of Raman spectra measured for different excitation energies at a constant magnetic field. Figure 5.11 shows such spectra at 4.7 T, recorded around the incoming LL3 resonance ($\hbar\omega_l = 1.5916$ eV) [1.44]. For excitation close to the LL3 resonance, the LL2 intensity is larger than that of LL1 due to the double resonance. However, when the excitation moves away from resonance this situation is reversed, and the relative line intensities approach values expected from a Maxwell–Boltzmann distribution. Contrary to a "true" Raman line, LL2 in Fig. 5.11 depends on the excitation energy and shifts accordingly. As mentioned above and substantiated theoretically below it is the polarization and resonance behavior of this line as well as its dependence on the excitation power which justifies its description by a Raman-type process.

Photoexcited Raman Scattering. Other structures which are labeled in Fig. 5.8 and 5.11 by SRS (Stokes Raman scattering) and ARS (Anti-Stokes Raman scattering) behave like "true" Raman lines [1.15]. They have a fixed frequency shift which does not vary with the excitation energy. As can be seen in Fig. 5.11 they are strongly resonant at the LLn transitions. In fact, the ARS line is only observed in the immediate vicinity of the LL3 resonance. An important feature of these peaks is their magnetic-field dependence. Figure 5.12 shows the Raman shifts of the SRS and several LLn resonances vs. field for a fixed excitation energy $\hbar\omega_l = 1.5999\,\mathrm{eV}$ which corresponds to the horizontal dashed line in Fig. 5.10 [1.44]. The Raman shift of the LLn lines decreases with increasing field since the magneto-optical transitions approach the laser energy. However, the Raman shift of the SRS line increases linearly with field. By comparison with the calculations in [1.26] (see Sect. 3.7) one finds that SRS corresponds to the electron cyclotron energy. It is therefore due to electronic excitations between conduction band Landau levels. The difference between the separation of subsequent LLn lines and the energy of SRS in Fig. 5.12 thus corresponds to the hole contribution to the slope of the fan lines. The SRS and ARS peaks also obey selection rules which are, however, not as stringent as for the LLn transitions. In the $\bar{z}(\sigma^{\pm}, \sigma^{\pm})z$ configurations the signals are about three times stronger than in $\bar{z}(\sigma^{\pm}, \sigma^{\mp})z$.

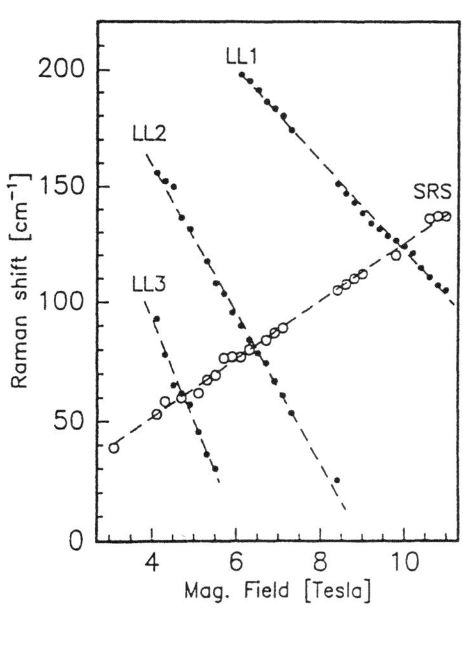

Fig. 5.12. Magnetic-field dependence of the Raman shifts for the SRS and LLn lines in $\bar{z}(\sigma^-, \sigma^-)z$ configuration ($\hbar\omega_l = 1.5999\,\mathrm{eV}$). The *dashed lines* are fits to the data

Mechanisms. In the following we discuss mechanisms which are responsible for the resonant electronic Raman (SRS, ARS) and hot magneto-luminescence (LLn) signals. Transitions between neighboring electron Landau levels by electronic Raman scattering, shown schematically in Fig. 5.13, can occur with the help of magneto-optical interband transitions [1.44]. In the ideal case (Fig. 5.13 (a)) an electron in the valence state m is promoted to the electron Landau level n by photon absorption. This creates a hole which subsequently recombines with an electron from Landau level p at a smaller (Stokes process, SRS) or larger energy than n (anti-Stokes process, ARS). Similar processes are also possible with additional steps mediated by interface roughness scattering, as shown in Fig. 5.13 (b). For the description of experimental spectra in nominally undoped QWs it has to be assumed that laser excitation creates a quasi-stationary electron population of the conduction band Landau levels. A problem for the calculation of Raman intensities from the processes of Fig. 5.13 (a) arises from the requirement to conserve the Landau quantum number in magneto-optical interband transitions. This

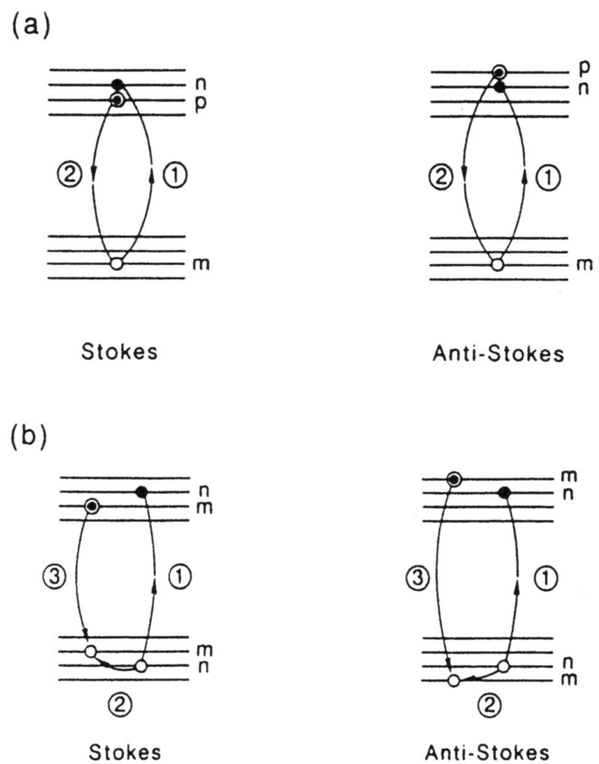

Fig. 5.13. Schematic Stokes and anti-Stokes processes for electronic Raman scattering of single-particle excitations between electron Landau levels due to direct interband transitions **(a)** and an additional step mediated by interface roughness scattering **(b)**

restriction can be circumvented, however, when additional steps mediated by interface roughness scattering, as shown in Fig. 5.13 (b), are taken into account. The change in the Landau index of the hole level due to these additional steps leads to the conservation of the Landau quantum number for each interband magneto-optical transition.

Another important point which was discovered in recent studies of electronic Raman scattering is the influence of the final-state linewidth on the spectrum [5.25]. If this broadening is neglected, one obtains from the scattering processes of Fig. 5.13 the Stokes and anti-Stokes lines at $\omega_s = \omega_l \pm (n-p)\omega_e$, where ω_e is the electron cyclotron frequency. The scattering intensity depends on a resonance denominator and the population factors of the Landau levels. The final-state broadening, i. e., the lifetime of the electron in level n and the hole in p, causes further interesting effects which are obtained from an extension of the theory for single-particle intersubband excitations in QWs [5.25]. Qualitatively speaking, due to their finite linewidth, states which do not exactly coincide with an interband transition can also contribute to resonances. Otherwise they would only be involved in virtual processes giving small signals. It could be shown that this effect causes further resonance denominators in the expression for the scattering cross section [1.44]. In addition to the signal enhancement in the incoming channel, excitation at LLn causes a resonance at $\omega_s = \omega_l - \omega_e$ which can be regarded as a double resonance for scattering with the cyclotron frequency. If $\hbar\omega_l$ does not coincide exactly with the energy of the LLn transition, E_{LLn}, a line appears in the Raman spectrum which corresponds to $E_{\mathrm{LLn}} - \omega_e$ on an absolute energy scale. This is an outgoing resonance which does not exist for electronic Raman scattering without final-state broadening. The frequency of this line is independent of the excitation energy. It therefore has the character of a luminescence, despite of its origin in a Raman-type process.

To illustrate these concepts Fig. 5.14 shows spectra calculated for the processes in Fig. 5.13 (b) with the inclusion of interface roughness scattering and parameters adapted to Fig. 5.11 [1.44]. The electronic structure (fan lines LLn) was calculated at $B = 4.7\,\mathrm{T}$ with respect to the energy of the lowest QW Landau level transition at 1.5575 eV using effective masses of $m_e = 0.076\,m_0$ and $m_h = 0.22\,m_0$. The Stokes/anti-Stokes ratio of the SRS and ARS lines corresponds to a carrier temperature of $T_e \sim 60\,\mathrm{K}$, obtained from the exponential decay of the luminescence in Fig. 5.8. The population n_e of the electron Landau levels n is therefore proportional to $\exp(-n(\omega_e/\omega_{\mathrm{LO}})/0.15)$. For the broadening of the transitions we assumed an empirical dependence on the Landau index given by $\Gamma_n = 0.05(n+1)^2\,\mathrm{meV}$.

The calculated cyclotron-scattering lines SRS and ARS are resonantly enhanced for excitation at LL3, in good agreement with the experiment. They are thus due to transitions between photoexcited electron Landau levels and can be observed in the Faraday backscattering geometry due to interface roughness scattering. These excitations were not observed in earlier inves-

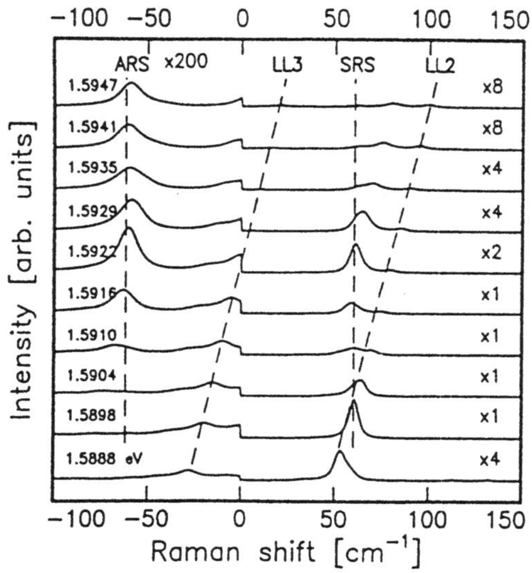

Fig. 5.14. Theoretical spectra of electronic Raman scattering for the experiments of Fig. 5.11. For excitation around LL3, resonances of cyclotron scattering occur (SRS, ARS). Recombination at LL2 is also observed. The energies and scaling factors are as in Fig. 5.11

tigations of electronic Raman scattering in undoped QWs [5.26]. Only for nonzero angles of the magnetic field with respect to the QW growth direction they appear as coupled cyclotron-intersubband excitations [1.15, 5.26]. This is another indication for the participation of steps mediated by interface roughness in the scattering processes. The QWs used in [5.26] are more than three times wider than those investigated here. Therefore the change in the QW energy corresponding to a monolayer fluctuation is about one order of magnitude weaker. Scattering by interface roughness is insignificant and a nonzero angle between growth direction and magnetic field is necessary in order to observe cyclotron resonance scattering. For resonant excitation of spectra as in Fig. 5.11 at LLn transitions with larger Landau indices and the same magnetic field one finds a strong decrease of the ARS signal while the SRS line is still observed. This is attributed to the lower photoexcited electron density in higher Landau states and their shorter lifetime. The spin polarization of the photoexcited carriers which is reflected in the above-mentioned selection rules (strong lines in the $\bar{z}(\sigma^{\pm}, \sigma^{\pm})z$ geometries) is consistent with other investigations. It was noticed in [5.12, 5.27] that the complete quantization of the two-dimensional QW electronic structure in a magnetic field along the growth direction favors spin-conserving relaxation processes. This leads to both faster carrier thermalization and recombination for these channels than for those with spin changes. In addition to the lines from cyclotron scattering the theoretical spectra of Fig. 5.14 also show structures at $E_{\mathrm{LL3}} - \omega_{\mathrm{e}}$, E_{LL3} and E_{LL2} which are due to final-state broadening effects. However, compared to the measured spectra in Fig. 5.11 these lines are weaker than observed. This indicates that further mechanisms contribute to the emission at LL2. Due to the small energy shift between the excitation and the interband transition acoustic phonons may play a role. For a discussion of several possible processes see [1.44].

Effective Masses. Finally, we would like to mention an aspect of resonant inter-Landau level scattering which extends the possibilities for the determination of electronic structure parameters compared to measurements of intensity oscillations of the continuous emission or the hot luminescence in undoped QWs. Note that the energy of the SRS and ARS lines directly reflects the effective mass of the electron Landau levels which in interband experiments is usually determined from fits to fan lines with the help of additional information, e. g., from calculations or other measurements [1.26, 1.39, 3.70]. Therefore electronic Raman scattering provides direct access to this parameter which can otherwise only be obtained from cyclotron resonance or, in the visible, optically-detected cyclotron resonance [5.28]. In contrast to cyclotron resonance, where transitions are always observed from the Fermi level to the next higher Landau states, the photoexcited carrier distribution in these experiments allows us to measure the separation between arbitrary Landau levels, within a certain range, which can be selected by choosing appropriate resonance conditions. The effective mass in the limit of small fields is then given by the separation between adjacent Landau levels: $m_\mathrm{e} = \hbar eB/\Delta E(B)$. In Sect. 3.7 it was shown that this value corresponds to the effective mass of the free dispersion perpendicular to the growth direction which is enhanced compared to bulk GaAs by nonparabolicity and anisotropy effects [3.72]. The magnetic-field dependence of the SRS line in Fig. 5.12, determined from measurements in resonance with several LLn transitions, is linear over the whole range. The slope of the dashed line fitted to the data gives $m_\mathrm{e} = (0.076 \pm 0.001)\, m_0$. For $B \to 0$ it extrapolates to a nonzero Raman shift of $(2.5 \pm 1.0)\,\mathrm{cm}^{-1}$. This phenomenon was also observed in cyclotron resonance measurements and explained by the localization effects of the electrons due to interface roughness [5.28]. A comparison of this effective mass with the calculation ($m_\mathrm{e}^{\|} = 0.074\, m_0$, from (3.18c)), the results of continuous emission ($m_\mathrm{e}^{\|} = (0.073 \pm 0.001)\, m_0$, same sample, see Sect. 3.7) [1.26], and optic-phonon magneto-Raman scattering ($m_\mathrm{e}^{\|} = 0.076\, m_0$, AlAs barriers, see Sect. 4.4) [1.39] allows an estimate of the remaining uncertainties in the determination of the electronic structure parameters by these methods.

Recent experiments on n-type modulation-doped QWs have shown that multiple electron-cyclotron resonances can also be observed by inelastic light scattering [5.29, 5.30]. Cascades with up to fifteen cyclotron lines which appear in all scattering geometries were attributed to the resonant relaxation of holes between heavily mixed valence Landau levels [5.30]. This opens new possibilities for investigations of the QW electronic structure and its renormalization due to electron–electron interactions.

6. Applications and Trends

Parallel to significant progress in the understanding of the basic effects, phonon Raman scattering in semiconductors, quantum wells and superlattices has also found a wide range of practical applications. Some of them have been presented in the previous chapters. In the following we give further examples to highlight the possibilities which Raman spectroscopy offers for modern materials research. We also discuss new developments which are presently gaining importance and show how they might affect future directions of Raman spectroscopy as well as the topics investigated by the method.

6.1 Raman Scattering in Applied Semiconductor Research

Progress in semiconductor epitaxy has made it possible to fabricate structures whose interfaces are defined with an accuracy of single atomic layers [1.8]. Consequently, quantum effects, such as the size-quantization of electronic or vibrational states, are now being exploited in a wide range of devices which cannot be realized with bulk semiconductors alone. These advances also pose a significant challenge for the analytical methods employed to determine the interface or device properties during and after production, as well as for new developments and applied research. The presence of many different compounds on very small length scales requires procedures which are material-specific, surface-sensitive and applicable with high lateral resolution. In order to be used on a routine basis, such materials characterization methods should also be simple and inexpensive. For these reasons optical techniques, most notably ellipsometry, luminescence, and Raman scattering have found widespread applications. While linear optical techniques are nowadays well established also in an industrial context, Raman scattering is only beginning to be used as an analytical method for manufacturing purposes. Its capabilities have been demonstrated in various examples related to materials research, and the number of commercial applications is increasing.

In the following we present some highlights of Raman scattering being used for materials characterization. Its application to monitor the properties of semiconductor structures in-situ, i. e., during their epitaxial growth, is also discussed. The next section is dedicated to different strategies for improvements of the method which are presently being actively pursued. The much higher spatial resolution in the near-field optical regime and simple spectrometers with larger throughput, based on new optical devices, are expected to have a profound impact on the applications of Raman scattering, also in semiconductor research and microstructure fabrication.

6.1.1 Materials Characterization

Many applications of Raman spectroscopy for the characterization of semiconductor interfaces, heterostructures and multilayer systems have been discussed in earlier reviews [1.7, 1.13]. The technological advances, which made possible the routine measurement of weak signals from very thin films, down to one atomic monolayer, such as the use of CCD and other detectors for the multichannel recording of spectra, have also been outlined [1.6, 1.7]. In this section we therefore concentrate on recent examples and emphasize new concepts and materials. The different topics were chosen in order to illustrate the broad range of applications which Raman scattering has found in the meantime. Issues close to basic research are covered as well as its use as an analytical tool for quality control in an industrial context.

Interface Quality: Optic and Acoustic Phonons. In the investigations of phonons in semiconductor QWs and SLs it was quickly realized that the interfaces play an important role for mode frequencies and scattering intensities. In fact, a mapping of the confined optic modes in GaAs/AlAs SLs onto the bulk dispersions can only be achieved when the finite penetration of the phonon displacement patterns into the adjacent material is taken into account (see Chap. 2 and (2.8)) [1.13, 2.10, 2.12, 2.13]. While this holds also for ideal interfaces, the strong dependence of the optic-mode frequencies on the QW width makes them a very sensitive probe of layer thickness fluctuations and interface roughness. Since these properties are influenced by the growth conditions and may even vary across individual wafers, Raman scattering by optic phonons can be applied to the determination of interface quality and sample homogeneity.

In high-quality samples using MBE with growth interruption or migration-enhanced epitaxy to obtain smooth interfaces, it has been found that the high-order confined phonons are more sensitive to roughness than the low-order ones, and even discrete confined modes from those parts of a QW which differ in thickness by one monolayer have been observed [2.36]. Together with photoluminescence data, Raman scattering by optic phonons has been applied to investigate both the large- and small-scale interface structure in GaAs/AlAs QWs grown with a considerable thickness gradient across a wafer [6.1].

The powerful ab-initio techniques for lattice-dynamical calculations developed in recent years allow one to perform detailed simulations of interface roughness and interdiffusion effects on the phonon frequencies [2.38–2.40, 6.2,

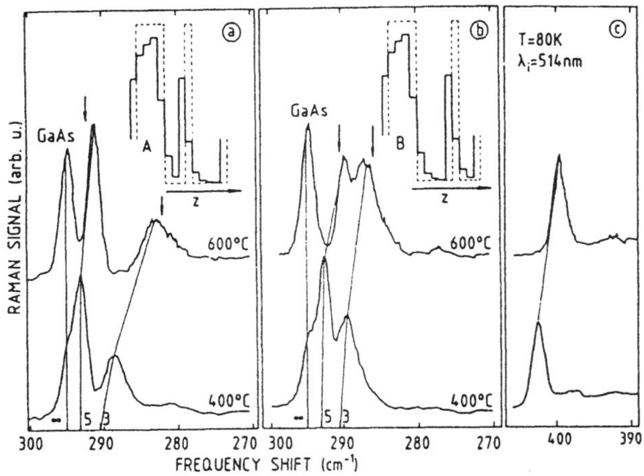

Fig. 6.1. Raman spectra of $(GaAs)_3/AlAs/(GaAs)_5/(AlAs)_4$ **(a)** and $(GaAs)_5/AlAs/(GaAs)_3/(AlAs)_4$ **(b)** SLs (GaAs energy range) grown at different temperatures as indicated. The insets show the nominal composition profiles (*dashed lines*) and, for the 600°C-grown samples, the ones deduced from a segregation model (solid lines). **(c)** Spectra in the AlAs energy range which are independent of the layer sequence. See text for details (from [2.44])

6.3] which can be compared to Raman measurements on samples made under different growth conditions and layer sequences. Figure 6.1 shows Raman spectra of GaAs/AlAs SLs with layer sequences $(GaAs)_3/AlAs/(GaAs)_5/(AlAs)_4$ ((a), series A) and $(GaAs)_5/AlAs/(GaAs)_3/(AlAs)_4$ ((b), series B) [2.44]. If these samples had ideal interfaces, their spectra would be identical since they only differ in the sequence with which the $(GaAs)_3$ and $(GaAs)_5$ QWs are grown with respect to the AlAs barriers. This case is realized to a good approximation for the two samples grown at a temperature of 400°C (Fig. 6.1 (a,b)) which show three peaks due to the bulk LO phonon from the GaAs substrate and the LO_1 modes in the $(GaAs)_3$ and $(GaAs)_5$ layers, respectively. A small deviation from the expected ideal behavior is found for the $(GaAs)_3$ layer. This can be seen from the thin lines connecting peaks of the same origin and the calculated frequencies, assuming perfect interfaces, which are indicated at the bottom of each plot.

For samples grown at the higher temperature of 600°C atomic intermixing at the interfaces becomes more important, and the Raman spectra of series-A and -B specimens show characteristic differences. As compared to the 400°C data the frequencies of both $(GaAs)_3$ and $(GaAs)_5$ shift towards lower values. This indicates a reduction of the GaAs layer thickness which leads to larger effective wavevectors and reduced frequencies, given by the bulk GaAs dispersion. This effect is strongest for the layer next to the $(AlAs)_4$ barrier, i.e., $(GaAs)_3$ in Fig. 6.1 (a) and $(GaAs)_5$ in Fig. 6.1 (b). The greater reduction of the effective layer thickness for the GaAs film next to $(AlAs)_4$ can also be seen in the interface profiles shown in the insets of Fig. 6.1, deduced from a segregation model (solid lines), as compared to the nominal layer sequence (dashed lines) [2.44]. The LO_1 frequencies of the $(GaAs)_3$ and $(GaAs)_5$ layers calculated with this profile are given by the vertical arrows in Fig. 6.1 (a,b). They are in good agreement with the experiment.

Further information about the intermixing at the interfaces can be obtained from Raman spectra in the AlAs region which do not depend on the layer sequence (Fig. 6.1 (c)). Due to the rather flat AlAs LO-phonon dispersion, the shift between the two peaks in Fig. 6.1 (c) is only due to different Ga contents in the $(AlAs)_4$ layers. By comparison with calculations it is found that the AlAs frequency in the sample grown at 400°C corresponds to an Al concentration of 92 % while that of the 600°C specimen reflects 83 % Al. These numbers are in good agreement with those obtained from the composition profiles determined independently [2.44]. This example highlights the sensitivity of optic-phonon Raman spectra in GaAs/AlAs QWs to interface properties. In combination with theoretical models detailed information about atomic intermixing at the interfaces and its dependence on the exact growth conditions can be obtained.

While confined optic phonons are sensitive to interface properties of individual QWs, the folded acoustic modes of a SL allow one to investigate the periodicity of the whole system as well as changes due to interdiffusion. In stacked SLs with different periods and length scales [6.4] as well as in mirror-plane SLs [3.18] with different numbers of building units, sometimes separated by bulk-like cavities, interference effects of the contributions from different QWs to the Raman intensity lead to a fine structure of the folded-phonon doublets. These effects have been demonstrated very recently and are now also being applied to determine the SL periodicity and length scales of interface roughness.

As has been shown in Sect. 2.4 [1.13,2.6] the intensity of the folded-phonon doublets in a SL depends on the modulation of the elasto-optic constants. The observation of higher-order doublets requires large Fourier components of their spatial profile which, in the ideal case, should therefore have a rectangular shape. In nonideal systems which are, for example, modified by interdiffusion or growth imperfections, one expects to observe folded-phonon doublets as long as the SL periodicity is not destroyed. However, as the modulation of the elasto-optic constants gets shallower, their intensities decrease with the higher-order compounds vanishing faster. This tendency is shown in Fig. 6.2 for a Sn_1Ge_{21} SL (α-Sn) which was annealed under different conditions [6.5]. The as-grown sample (bottom spectrum) shows in the low-frequency regime two folded acoustic-phonon doublets (FLA_1, FLA_2). With increasing annealing, performed incrementally on the same sample when the same temperature is given, the FLA_2 intensity vanishes rapidly since the interfaces are smeared out due to interdiffusion. The FLA_1 signal also decreases but vanishes only after longer annealing (top spectrum). At this point the SL has been de-

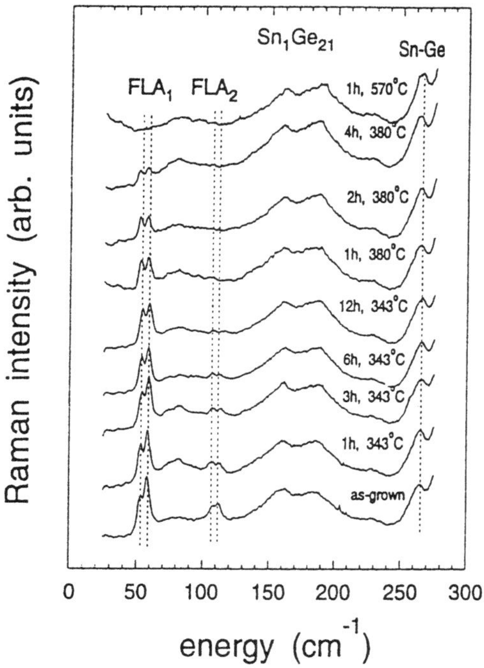

Fig. 6.2. Raman spectra of a Sn_1Ge_{21} SL measured after different annealing treatments. With increasing interdiffusion the intensities of the folded acoustic-phonon doublets (FLA_1, FLA_2) decrease and vanish. The spectra are normalized with respect to the Ge LO_1 phonon (not shown) (from [6.5])

stroyed and an α-$Sn_{0.05}Ge_{0.95}$ alloy is obtained. From such measurements the interdiffusion constant can be determined [6.5].

Interface Sensitivity: Resonance Effects. The strong dependence of sample and device properties on the quality of heterostructure interfaces makes it necessary to develop special techniques which allow one to access this important region. Even though Raman scattering provides high chemical selectivity to compositional changes at interfaces which are, for example, caused by diffusion or growth imperfections, the signal intensities are sometimes rather small due to the vanishing sampling volume. Solutions to this problem, based on more sensitive detection schemes and instrumental improvements, have been summarized in [1.7] with the observation of spectra from single Ge films with a thickness down to two monolayers embedded in Si/Ge/Si QWs being a special highlight. However, in recent years another approach has become more and more important. It is based on the resonant enhancement of the Raman intensity at interband critical points of the constituent materials [3.1,6.6].

As an example, Fig. 6.3 shows Raman spectra of a single InAs QW with a thickness of 15 nm embedded between AlSb layers [3.1]. Since these two semiconductors have no common anions or cations, the heterointerfaces can be either InSb- or AlAs-like, depending on the conditions and shutter sequences chosen during MBE growth. Peaks in the Raman spectrum, which can be attributed to either of these materials, should thus allow one to determine the character of the interface. In addition to the strong lines at 235 and 350 cm^{-1},

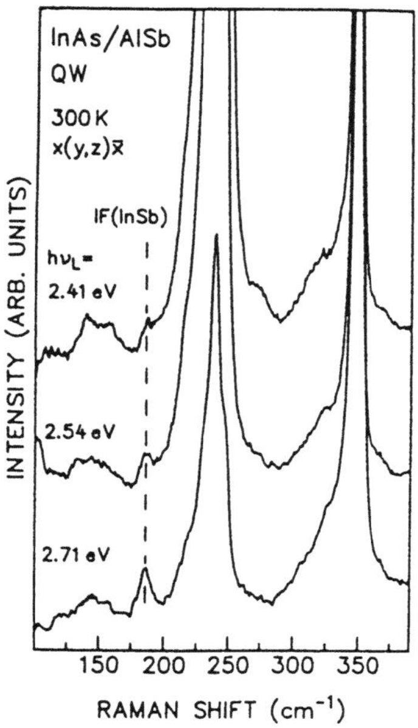

Fig. 6.3. Raman spectra of a single InAs/AlSb QW measured for different laser energies $h\nu_L$. The longitudinal InSb interface mode which is strongly resonant at 2.71 eV is denoted by IF(InSb) (from [3.1])

which are due to the LO phonons of the InAs and AlSb layers, respectively, the spectra in Fig. 6.3 have an additional peak at 190 cm^{-1} which can be assigned to a longitudinal InSb interface mode. The intensity of this phonon increases for larger excitation energies. It is strongest at $h\nu_L = 2.71$ eV, a photon energy which matches the InAs $E_1/E_1+\Delta_1$ resonance. At $h\nu_L = 2.41$ eV, the $E_1 + \Delta_1$ gap of InSb, however, the line is much weaker. This appears unusual at first sight since one would expect an InSb interface mode to resonate at the gap of this material and not at that of the InAs well. It was shown in the SiGe system, however, that a minimum of three to four monolayers is necessary in order to observe such a resonance behavior [3.1,6.6]. For thinner layers the interface phonon intensity follows the resonance profile of the well material. In addition to the chemical sensitivity to single interfaces which has been achieved using resonant excitation, this phenomenon therefore also yields an upper limit for the extent of the interface region. In the resonance measurements in [3.1] it was further possible to distinguish between the LO phonon of the InAs QW and that from a 10 nm protective GaSb cap layer on top of the InAs/AlSb structure whose frequency is quite close.

An extensive study of the two different interfaces in InAs/AlSb SLs, and their influence on optical and structural properties, has been reported in [6.7, 6.8]. Differences in the Raman spectra and shifts of interband critical points, measured by ellipsometry and photoluminescence, could be related to structural information obtained from X-ray diffraction and transmission electron microscopy. It was found that the type of interface strongly affects the strain state of the samples and the critical layer thickness which is much smaller for AlAs-like interfaces than for those with InSb [6.7, 6.8]. This leads to different defect densities and strain relaxation, a fact which influences the optical and electrical properties. In samples with AlAs-like interfaces there is also considerable diffusion of As into the AlSb layers which could be quantified by Raman spectroscopy [6.7, 6.8].

Interfaces of Isotopic Superlattices: Self-Diffusion. One of the most fundamental processes of matter transport in semiconductors, the self-diffusion of atomic species, has been investigated for a long time almost exclusively by radioactive tracer methods [2.86, 6.9]. However, the results are very sensitive to surface effects such as contamination and oxidation. This makes the development of other procedures highly desirable.

New approaches to study self-diffusion have emerged with the advent of isotopic heterostructures and SLs. Secondary-ion mass spectroscopy (SIMS) measurements of carefully annealed thick heterostructures with different isotopes have been performed recently and the parameters for self-diffusion in Ge as well as those of Ga in GaAs and GaP have been determined [2.108, 2.109, 6.10]. An example is given in Fig. 6.4 which shows the atomic fractions vs. layer depth for ^{69}Ga and ^{71}Ga in as-grown and annealed isotopic GaP heterostructures [2.109]. The samples were grown by MBE with a thickness of 2000 Å for each layer using undoped GaP substrates.

As has been demonstrated in [2.86] such investigations can also be performed on isotopic SLs using Raman scattering which has the advantage of being a nondestructive probe. Figure 6.5 compares the Raman spectra of a $^{70}Ge_{16}/^{74}Ge_{16}$ SL after different annealing treatments (left panel) with the results of a calculation (right panel) [2.86]. In the modeling of the spectra, one first derives depth profiles for the Ge isotope distribution by solving Fick's second diffusion equation for a one-dimensional system using an expression for the self-diffusion coefficient which corresponds to the thermally activated behavior associated with the so-called vacancy mechanism [2.86]. From these

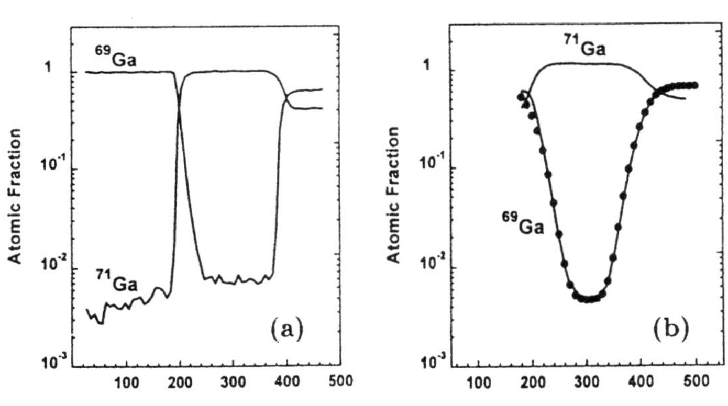

Fig. 6.4. Self-diffusion in GaP studied by SIMS measurements of the Ga isotope fractions for an **(a)** as-grown and **(b)** annealed (231 min at 1111°C) ^{69}GaP/^{71}GaP heterostructure. The *filled circles* in (b) represent a calculated ^{69}Ga profile

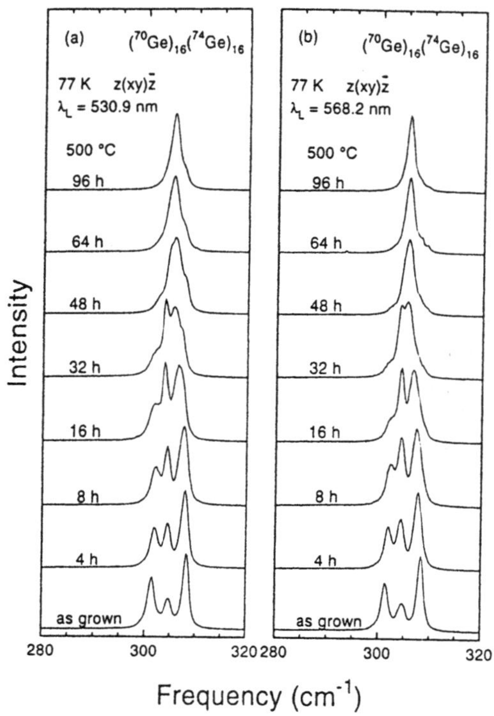

Fig. 6.5. Raman spectra of a $^{70}Ge_{16}/^{74}Ge_{16}$ SL for different annealing times at 500°C. **(a)** Experimental data, **(b)** theory (from [2.86])

depth profiles one determines average masses which are then used together with the planar bond-charge model (see Sect. 2.3.1) to obtain the phonon frequencies and, by means of bond polarizabilities, the Raman spectra. Adjusting the self-diffusion parameters, good agreement between theory and experiment is obtained. Apart from being nondestructive, the Raman method is sensitive to changes at the interfaces on an atomic scale. Unlike SIMS, whose resolution is limited to about 40 Å (160 Å) per decade of the measured atomic fraction at the leading (falling) edge of a layer [6.10], Raman spectroscopy can therefore be applied to investigate self-diffusion at much lower temperatures. The Raman spectra also consist of several modes, all of which can be monitored. Their relative intensities provide an even more critical test to determine self-diffusion parameters.

Mechanical Stress in Microelectronics Devices. The stress-dependence of phonon frequencies has been investigated intensively for many semiconductors, mainly by Raman scattering [6.11, 6.12], but also by infrared reflectivity [6.13], in order to determine fundamental material properties such as phonon deformation potentials or mode Grüneisen parameters which allow one to perform critical tests of models for the underlying electronic and lattice-dynamical structure [6.14,6.15]. As an example, Fig. 6.6 shows the behavior of the optical phonon in silicon for compressive uniaxial stress along the [111] and [001] directions [6.12]. The threefold degenerate zone-center mode splits into a singlet (frequency shift $\Delta\omega_s$) with atomic displacements

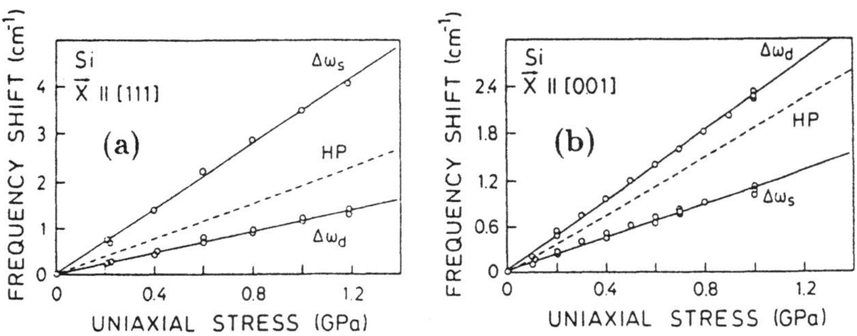

Fig. 6.6. Frequency shifts of the singlet and doublet components of the optic phonon in Si for uniaxial stress along the [111] **(a)** and [001] directions **(b)**. The *solid lines* are linear least-squares fits to the experimental data (*open circles*) obtained by Raman scattering at 110 K using the 1.064 μm line of a Nd-YAG laser (from [6.12])

along the stress direction and a doublet ($\Delta\omega_d$) where the eigenvectors are perpendicular to the stress. The dashed lines (HP) are calculated shifts due to the hydrostatic component of the applied stress. They are the same for both directions. However, the uniaxial stress component leads to different shifts for the singlet and doublet phonons which are also different for the two stress directions shown. These shifts result from a secular equation containing the phonon deformation potentials as parameters and the respective nonvanishing strain tensor components which depend on the stress direction [6.11,6.14]. While the phonon frequencies increase for compressive stress, tensile stress causes negative shifts.

In connection with the increasing importance of semiconductor microstructures in basic research and applications, these results, obtained for bulk materials, have become very useful since they also allow one to determine the strain state of thin films, as was demonstrated early by Raman measurements of Si grown on sapphire [6.16]. Since then Raman scattering has found a variety of applications for the analysis of devices and even integrated circuits, mostly in Si technology, which have been reviewed in [6.14, 6.15].

Figure 6.7 shows a frequency-shift profile of the Si optic phonon measured by micro-Raman scattering across a 10 μm wide line of 240 nm Si_3N_4 on 50 nm polycrystalline Si and a 10 nm SiO_2 pad oxide covering the whole Si substrate [6.14,6.17]. Such layer sequences are used for insulation purposes in the "local oxidation of silicon" (LOCOS) technique. The Si_3N_4 acts as a mask and protects the areas where active devices will be formed during a thermal oxidation process [6.14]. On approaching the Si_3N_4 line in Fig. 6.7, the phonon frequency shifts to lower values, indicating the presence of tensile stress just outside the mask. Underneath, however, the shifts are positive, and the stress is compressive. Such stress inhomogeneities arise, for example, from different lattice constants or thermal expansion coefficients of the materials involved and generate dislocations which affect the device performance [6.14]. Frequency-shift profiles obtained by Raman spectroscopy can be used as an analytical tool for the minimization of such effects. This procedure requires

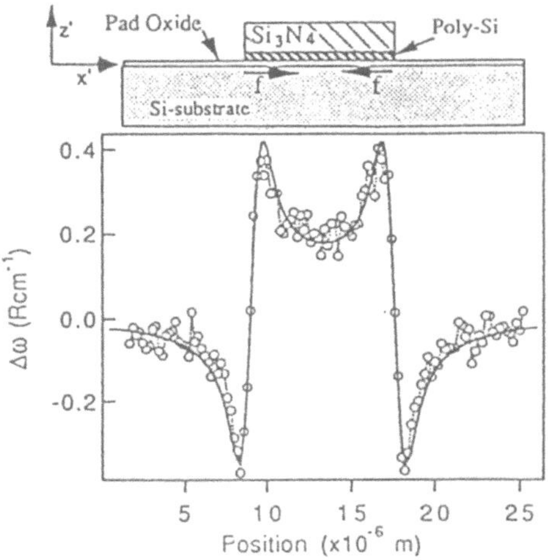

Fig. 6.7. Raman frequency shift of the Si optic phonon across a line of Si_3N_4 on polycrystalline Si. The device structure is given above the profile. The shifts $\Delta\omega$ refer to the stress-free value. The *open circles* are experimental data; the *solid line* is a fit using a specific model for the force distribution causing the strain (from [6.14, 6.17])

finding the appropriate layer dimensions and processing parameters. The precise determination of stress values from the observed phonon-frequency shifts, however, requires theoretical models which are based on assumptions about the origin of the stress for the specific geometry investigated [6.17].

Epitaxy of New Optoelectronic Materials. The advantages of Raman spectroscopy for the investigation of thin films are exploited extensively for the characterization of new materials, such as the wide-gap II-VI or III-V semiconductors whose growth by various epitaxial methods is presently a very active field of research.

In some cases, layer growth is difficult because of the large lattice mismatch between the crystal of interest and the substrate. This causes strain, and the resulting shifts in the electronic structure as well as defects or dislocations at the interface affect the device properties. Such effects have been investigated recently with spin-flip Raman scattering (SFRS) from nitrogen-doped p-type ZnSe grown on a GaAs substrate [6.18, 6.19]. Being an electronic Raman scattering process [1.14, 1.15, 6.20], SFRS is usually performed in resonance with defect-bound exciton states. It yields precise information about g factors whose magnitude and angular variation in a magnetic field depend strongly on the strain and the symmetry of the electronic states involved. Figure 6.8 shows the angular dependence of the hole spin-flip Raman line for acceptor-bound excitons in nitrogen-doped p-type ZnSe epilayers measured at 1.5 K and 6 T [6.18]. In one case (curve "A") the SFRS line has a nonzero value for a magnetic field parallel to the surface normal (Faraday geometry). It increases when the field is tilted in the sample plane (Voigt geometry). This is typical for spin-flip processes between hole states with $|3/2, \pm 1/2\rangle$ quasi-angular-momentum components measured with resonant excitation at the light-hole acceptor-bound exciton. This state contributes to the lowest-energy interband transition, observed also in photoluminescence,

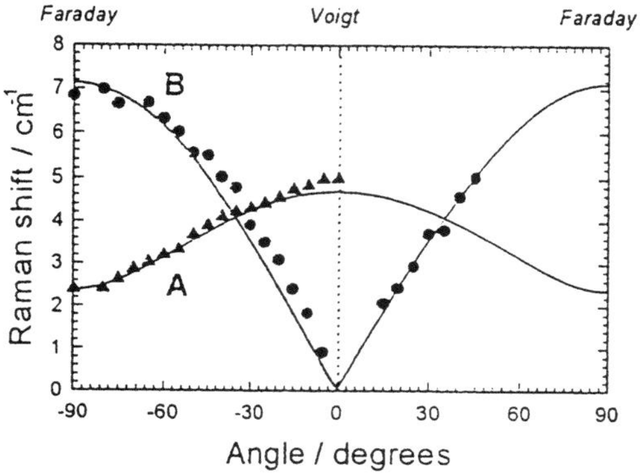

Fig. 6.8. Angle-dependent spin-flip Raman shifts of acceptor-bound excitons in nitrogen-doped p-type ZnSe under biaxial tensile (A) and compressive stress (B). The *solid lines* are fits to the data (*filled triangles and circles*). See text for details

for ZnSe layers experiencing a biaxial tensile strain due to their different thermal expansion coefficient compared to the GaAs substrate. For a second measurement (curve "B") a ZnSe film was glued onto a glass slide, and the substrate was removed by polishing and etching. In this case the different thermal contraction upon cooling to 1.5 K causes a compressive strain, and the $|3/2, \pm 3/2\rangle$ state contributes to the lowest-energy heavy-hole acceptor-bound exciton. The corresponding SFRS lines were observed for resonant excitation at this transition which also dominates the luminescence spectrum. A typical signature of these states is a cosine-like angular dependence of the SFRS shift. It is nonvanishing in Faraday geometry but zero in Voigt configuration [6.21]. The solid lines in Fig. 6.8 are fits to the data with a spin Hamiltonian for the magnetic-field dependence of the acceptor states. This allows one to determine the sample strain and valence band parameters [6.18, 6.19].

Another problem in the growth of, for example, GaN, is the existence of two different phases with zincblende and wurtzite structure which differ in their physical properties. For materials development, analytical tools are required to identify the presence of these modifications even as a minority phase, since they could affect the optical properties in an undesirable way. The polarization selection rules of Raman scattering, different in these two cases, have recently been used to exclude the presence of wurtzite GaN in an epilayer with zincblende structure down to the 1% level [6.22]. This approach is therefore rather suitable for the routine characterization of samples [6.22].

The doping of wide-gap semiconductors also presents a technological challenge [6.23]. In the II-VI materials self-compensation effects are very strong and often permit easy doping only for one type of carrier. New approaches have to be found in order to incorporate the other. In ZnSe, which can easily be made n-type by doping with chlorine, holes have to be introduced in a more complicated way which, for MBE growth, involves plasma-activated N_2. The properties of holes and compensating donors, created during the doping process, have been investigated intensively by spin-flip Raman scattering [6.24, 6.25]. For GaN it has been found that the presence of carriers

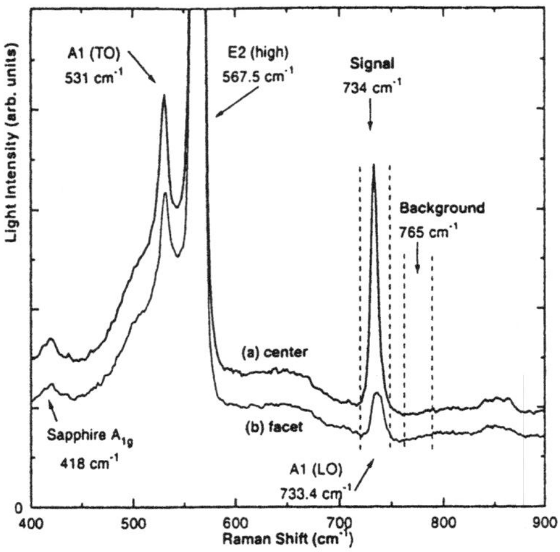

Fig. 6.9. Raman spectra at different positions of a hexagonal GaN crystallite. The intensity of the A_1(LO) peak is inversely correlated to the presence of few **(a)** or many **(b)** free carriers introduced by Si doping. The dashed regions denote the spectral ranges used to determine the A_1(LO) signal and the background in an imaging experiment. See text for details (from [6.27])

due to doping strongly affects the A_1(LO) phonon via plasmon–phonon coupling [6.20,6.26,6.27]. Lineshape fits to the renormalized mode yield the same electron concentration as Hall effect measurements, and Raman scattering is therefore used as a simple and contact-free method for the routine characterization of doped samples [6.26]. The dependence of the A_1(LO) phonon intensity on doping has been exploited in a Raman imaging experiment (see Sect. 6.2.2), to determine the free-carrier concentration across a 20 μm diameter area of a Si-doped GaN crystal ($n \sim 2 \times 10^{17}\,\mathrm{cm}^{-3}$ at 300 K) with a resolution of about 0.5 μm [6.27]. As can be seen from the Raman spectra in Fig. 6.9 the A_1(LO) phonon signal is strongly reduced on the triangular side faces (b) of a hexagonal dome-shaped crystallite compared to the center (a) where the different facets touch each other [6.27]. This difference is attributed to the presence of free carriers on the facets, which suppress the signal due to plasmon–phonon coupling, and their absence at the center, indicating some mechanism of preferential donor incorporation during growth.

Chemical vapor deposition of diamond often yields granular films with a crystallite size around 100 μm. Raman mapping has been performed on such grains in order to investigate their fine structure. Figure 6.10 shows plots of the diamond phonon intensity (A), its width (B), and position (C) over a $40 \times 40\,\mu\mathrm{m}^2$ area rastered in steps of $1\,\mu\mathrm{m}^2$ across a grain on the back surface of a free standing diamond film (thickness 225 μm) grown by chemical vapor deposition on a Si substrate [6.28]. These results show that the diamond properties vary considerably even on length scales smaller than the grain size. The intensity (A) and linewidth plots (B) are almost complementary to each other, indicating strong scattering and narrow linewidths from "high-quality domains". In the area with enhanced linewidth (B) one also observes

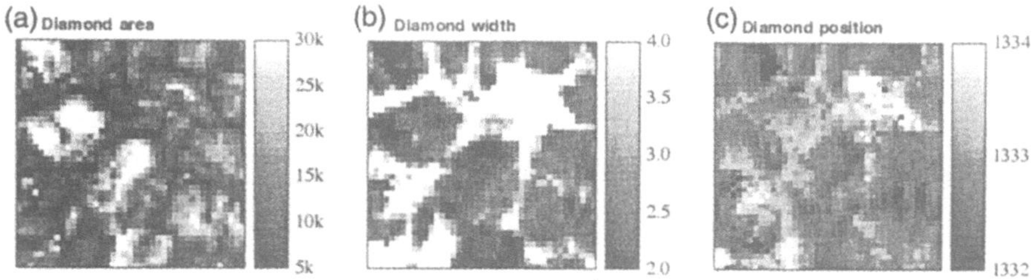

Fig. 6.10. Raman maps of the integrated peak area (arb. units) **(a)**, the HWHM (in cm^{-1}) **(b)**, and the frequency (in cm^{-1}) **(c)** of the diamond phonon measured on a grain of a thin diamond film. See text for details (from [6.28])

significant frequency shifts (C) indicating stress or other inhomogeneities [6.28].

Local-Mode Spectroscopy of Defects in GaAs. Another application of Raman scattering to determine doping effects in semiconductors uses the local vibrational modes of substitutional impurity atoms at host lattice sites. Figure 6.11 shows Raman spectra of Be-doped MBE-grown GaAs with different dopant concentrations as determined from SIMS and infrared absorption measurements [6.29]. With its much smaller atomic number Be (100% ^{9}Be), substituting for Ga and thus acting as an acceptor, forms a light mass defect in GaAs. The frequency of the associated local vibrational mode is 482 cm^{-1}, much higher than that of the intrinsic LO phonon (294 cm^{-1}) [2.153]. The spectra in Fig. 6.11 show this peak ($^{9}Be_{Ga}$) superimposed on the intrinsic second-order Raman spectrum of GaAs. Its intensity increases monotonically with the Be concentration. To facilitate the observation of the usually weak local modes a laser energy of 3.00 eV was chosen, thereby exploiting their resonant enhancement at the GaAs E_1 gap. Before calibrating the dependence of the local-mode intensity on the Be concentration the measured signal was normalized to the second-order GaAs feature at 540 cm^{-1}. This avoids the difficulties associated with the determination of absolute scattering efficiencies. Figure 6.12 shows the dependence of the normalized local-mode intensity on the Be concentration [6.29]. The detection limit of this method has been estimated to be $\approx 3 \times 10^{18}\,cm^{-3}$ for layer thicknesses exceeding 100 Å [6.29]. Similar studies have been performed for Si-doped GaAs where, due to its amphoteric character, local modes of Si on either Ga or As sites occur along with Si_{Ga}–Si_{As} pairs and other defects [6.30, 6.31]. In combination with infrared absorption, Raman scattering yields a detailed insight on doping effects in semiconductors as well as practical information which can be used in materials science and applications.

Quality Control of Diamond-Like Carbon Coatings. Thin carbon layers are used widely as protective coatings, for example, for hard disks which can be severely damaged when the read/write head, floating less than a micrometer above the surface, crashes into the magnetic recording film [6.32].

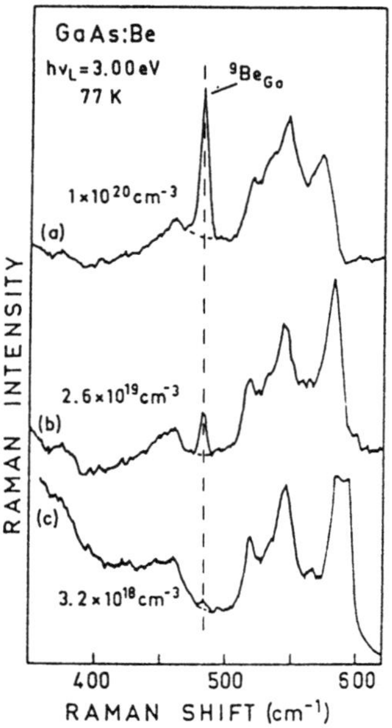

Fig. 6.11. Raman spectra of GaAs with different amounts of Be doping. Superimposed on the second-order GaAs spectrum a local mode ($^9Be_{Ga}$) occurs whose intensity is proportional to the hole concentration (from [6.29])

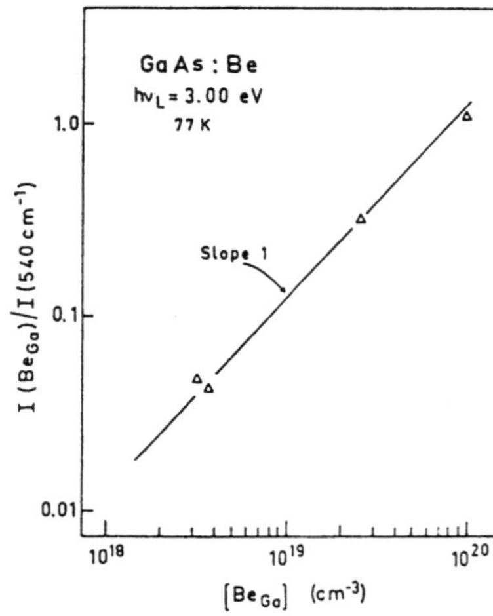

Fig. 6.12. Normalized local-mode intensity from the Raman spectra in Fig. 6.11 compared to the Be concentration. The *solid line* indicates the direct proportionality between these two quantities (from [6.29])

While pure diamond would provide optimum hardness and wear resistance, diamond-like (amorphous) carbon thin films are being used in practice since they are much easier to manufacture by sputtering or other deposition techniques. Despite its close relationship to graphite, which as a layered crystal is very soft, diamond-like carbon (DLC), consisting of randomly oriented graphitic microcrystallites, has a much greater hardness and mechanical strength which can be comparable to diamond. In recent years Raman scattering has emerged as the technique of choice to control the quality of such layers, and it is now even used in an industrial context.

The physical origin for this can be seen in Figs. 6.13 and 6.14 which show Raman spectra of polycrystalline graphite for different crystallite sizes as well as those of ion-beam sputtered amorphous carbon (DLC) films annealed at increasing temperatures [6.32–6.34]. In the high-frequency region the Raman spectrum of large graphite single crystals exhibits one sharp line at $1575\,\mathrm{cm}^{-1}$ due to an E_{2g} phonon for which scattering is allowed [6.35]. This so-called G (graphite) line dominates the spectra in Fig. 6.13. In samples with small crystallites, however, an additional peak (D (disorder) line) appears around $1355\,\mathrm{cm}^{-1}$. It can be attributed to disordered graphitic clusters consisting

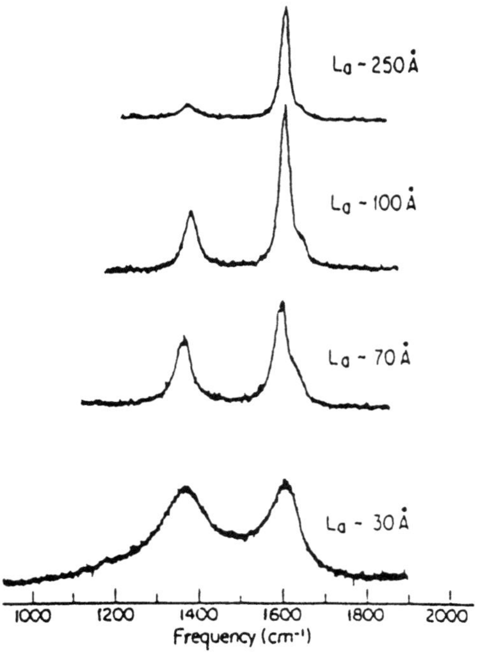

Fig. 6.13. Raman spectra of polycrystalline graphite with different crystallite sizes L_a in the basal plane (from [6.33])

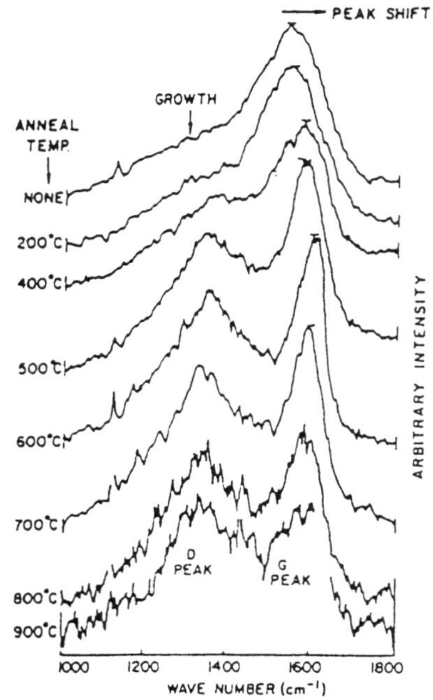

Fig. 6.14. Raman spectra of amorphous carbon (DLC) films annealed at different temperatures (from [6.34])

only of a few hexagonal building blocks. Raman scattering from A_{1g}-type modes of these entities becomes possible due to their finite-size which causes a relaxation of the crystal-momentum conservation law [6.35]. The spectrum of the as-grown sample in Fig. 6.14 only shows a broad G peak at $1536\,\mathrm{cm}^{-1}$. Its overall shape reflects the graphite phonon density of states, a feature typical of many amorphous semiconductors [6.36]. With increasing annealing temperature the G line sharpens and shifts to $1598\,\mathrm{cm}^{-1}$ [6.34]. At the same time the D line appears and its strength increases relative to the G line. This behavior is caused by the recrystallization of amorphous carbon into randomly oriented graphitic microcrystallites which, for the intermediate annealing temperatures shown, have a lateral extension of about 20 Å, consistent with the spectra of Fig. 6.13 [6.34]. The D line signal increases as long as the number of these islands is growing. However, a decrease of the intensity ratio $I(D)/I(G)$ has been reported when crystal-momentum conservation is restored in larger domains formed after annealing at higher temperatures [6.34]. X-ray diffraction measurements in this regime show that $I(D)/I(G)$ is inversely proportional to the crystallite size [6.35]. As can be seen from Fig. 6.13, the position of the G line for larger L_a shifts back towards the lower value of single-crystal graphite.

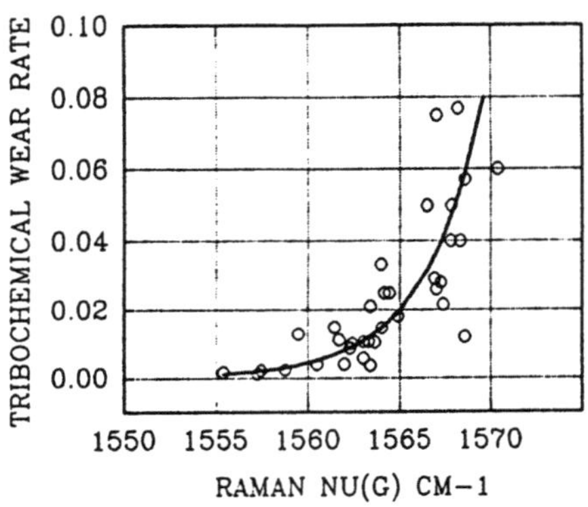

Fig. 6.15. Tribochemical wear rate of DLC coatings vs. the *G* line position (from [6.37])

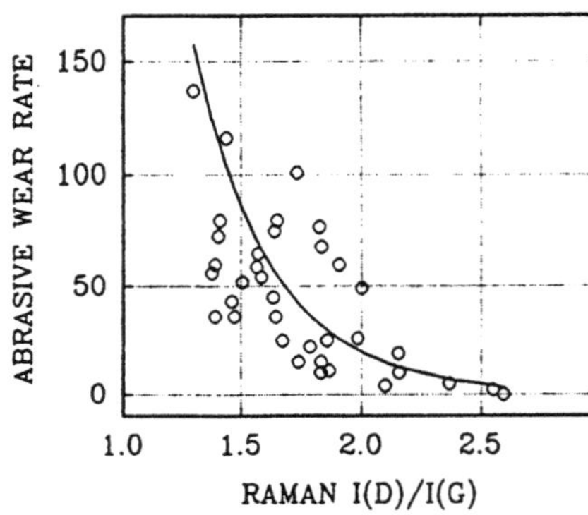

Fig. 6.16. Abrasive wear rate of DLC coatings vs. the intensity ratio I(*D*)/I(*G*) (from [6.37])

In order to optimize the protective properties of DLC coatings and to apply Raman spectroscopy for the purpose of quality control, the variations in the spectral features just discussed have to be correlated with tribological quantities. Among the relevant wear mechanisms for the head/disk interface of thin film media which involve physical properties of the coating are adhesive, impact, and abrasive wear [6.37]. Another mechanism is tribochemical wear in the oxygen-containing air which leads to smoother surfaces and higher friction forces [6.37]. Figures 6.15 and 6.16 show the measured tribochemical and abrasive wear rates of different magnetron-sputtered DLC films in ambient air which depend on the *G* line position and the ratio I(*D*)/I(*G*), obtained from Raman spectra, respectively [6.37]. The tribochemical wear rate was estimated in a drag test from the increase of the friction coefficient per cycle when a rotating film was exposed to a slider with a certain weight load. The abrasive wear rate was estimated from the number of scratches on the thin film surface generated by a SiC abrasive tape. Based on these results Raman scattering is nowadays applied routinely to determine tribological quantities and to perform quality control during hard disk manufacturing [6.38, 6.39].

The data in Figs. 6.15 and 6.16 can be understood with a structural model of amorphous carbon which assumes small graphitic microcrystallites randomly embedded in a three-dimensional network of sp^3- and sp^2-bonded carbon [6.37]. Theoretically, a decrease of the *G*-line frequency has been predicted when the sp^3/sp^2 ratio increases [6.37]. A larger fraction of tetrahedrally-bonded carbon atoms strengthens the connections between different crystallites. This enhances the durability of the coating and results in a lower tribochemical wear rate. The correlation of the sp^3 content (*G* line frequency) with tribochemical wear reflects the different affinities of graphite and diamond for oxygen chemisorption. While graphite reacts readily with molecular oxygen and forms surface oxides, diamond is inert. The increased presence of sp^3 bonds in DLC films has analogous consequences. As discussed above, $I(D)/I(G)$ depends on the number of small graphitic microcrystals. It has been suggested that the reduced abrasive wear of films with a large $I(D)/I(G)$ ratio is due to the denser packing of the microcrystals when their number increases [6.37].

Damage Mechanisms in QW Lasers. One of the major limiting factors for the lifetime of GaAs/$Al_xGa_{1-x}As$ QW laser structures is the so-called catastrophic optical damage, which manifests itself by a sudden degradation of the output power during operation [6.40]. A thorough understanding and control of the mechanisms responsible for this effect is essential in order to design reliable devices for optoelectronics and communications as well as for high-power applications such as materials processing. The laser resonator facets, usually cleaved surfaces terminating a structure, play a crucial role in the occurrence of catastrophic damage processes [6.40]. Using the ability to measure temperatures from Stokes/anti-Stokes intensity ratios, which can also be done with high spatial resolution, Raman scattering has provided valuable information about changes in these critical areas during laser operation and breakdown. It has thus helped to identify different damage mechanisms [6.41, 6.42].

Figure 6.17 shows the facet-temperature and output-power variation with time for different operating currents of a GRINSCH (graded-index separate confinement heterostructure) laser diode with a 70 Å $Al_{0.04}Ga_{0.96}As$ single QW embedded between 2000 Å thick regions of parabolic Al composition ($0.22 < x < 0.36$) [6.41]. At a current of $I = 115\,\mathrm{mA}$ (Fig. 6.17 (a)), corresponding to a slowly decreasing initial output power of 46 mW, it takes about 10 min until the laser output rapidly drops by a factor of three. The facet temperature, as determined by Raman scattering, is seen to increase slowly before the breakdown and to rise suddenly to 800 K above room temperature. For lower currents, this catastrophic breakdown occurs after much longer operation times. It takes about 1 h for $I = 106\,\mathrm{mA}$ (b), while for $I = 95\,\mathrm{mA}$ (c) no effect is observed even after 3 h of operation. However, a sudden drop of the output power always occurs after the facet has reached a critical temperature of 120–140°C, independent of the operating current [6.41].

A suggested explanation of the phenomenon comes from the initial slow rise of the facet temperature whose rate rapidly decreases with the operating current (10°/min (a), 0.7 °/min (b), 0.15 °/min (c)) [6.41]. After breakdown, the temperature reached at the laser surface also depends on the operating current. Measurements of the laser emission wavelength along the resonator, however, yield a homogeneous temperature distribution which is only slightly higher than room temperature [6.42]. These results indicate that catastrophic optical damage occurs due to processes near the facets. It could be demonstrated that external influences such as oxidation, erosion or contamination of the facets contribute to the initial slow power decrease and rise in temperature [6.41]. On a microscopic level this creates defects and enhances the nonradiative recombination at the facets which generates heat. Consequently,

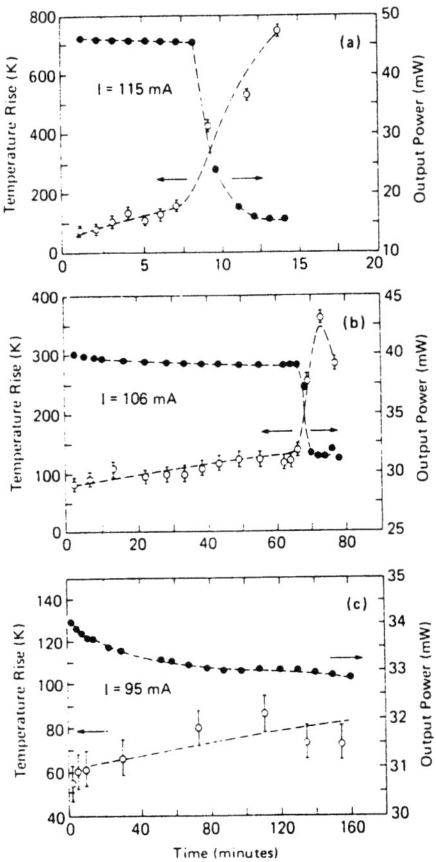

Fig. 6.17. Facet temperature rise with respect to room temperature (*open circles*) and output power (*filled circles*) of a QW laser in air vs. time for different constant operating currents as indicated. The temperatures were determined by Raman scattering from the Stokes/anti-Stokes intensity ratio of the GaAs-like TO phonon (from [6.41])

the absorption coefficient in the facet region increases, and further heating occurs from nonradiative recombination of absorbed laser photons. At some critical temperature these processes feed back positively and thermal runaway occurs [6.40, 6.41]. The surface region can be heated up to the melting point, a fact which severely degrades the laser output power. It has been noted, however, that thermal runaway can occur even at defect-free cleaved surfaces where local heating is favored by heat flow considerations [6.40].

Raman thermometry was also applied to optimize the QW laser design [6.43]. It was found that the active region can be kept away from the laser facets by using segmented electrodes which provide separate contacts to the main lasing area and the device edges. For appropriate voltages processes leading to catastrophic breakdown are suppressed. The facet temperatures are significantly reduced and the device performance is improved [6.43].

Applications in Other Areas. Progress in Raman instrumentation (see Sect. 6.2.2) has already led to a broad range of applications in very different fields of science and engineering. A good overview can be found in the proceedings of the International Conference on Raman Spectroscopy (ICORS) which is held biannually [6.44–6.46]. Without aiming at completeness, a few of the new approaches introduced there over the past 5 years include such diverse topics as the Raman-profiling of stress distributions and dopant concentrations in optical fibers, the use of Raman scattering in chemical agent identification sets for toxic substances, and on-line monitoring of aromatics separation and other processes in the chemical industry. Remote Raman-sensing with optical fibers for applications in difficult environments as well as multi-sensor concepts for the simultaneous control of many process parameters have recently attracted a lot of attention [6.44–6.46]. Raman scattering in the near-infrared spectral range using diode lasers allows measurements from a wide range of organic materials such as polymers or rubber which cannot be investigated with excitation in the visible due to their strong luminescence. Biological applications include the imaging of cells and even subcellular components such as chromosomes, and Raman studies of the carotenoid distribution in different lymphozytes are relevant for cancer research [6.47]. Raman scattering is intensively applied to a wide range of inorganic compounds, for example, for the compositional analysis of rocks in geology and prospecting, the determination of color pigments in paintings and manuscripts, and the analysis of ceramic materials [6.44–6.46].

6.1.2 In-Situ Monitoring of Epitaxial Growth

Exploiting the short penetration depths of photons and the strong signal enhancement obtained for resonant excitation, Raman scattering has recently joined the group of surface- and interface-sensitive optical techniques, such as reflectance-difference spectroscopy, laser light scattering and surface photoabsorption, which are being applied for the on-line analysis of the epitaxial growth of semiconductors [6.48–6.50]. In addition to its capability to detect phonons already from less than one atomic monolayer of material deposited on a semiconductor substrate, Raman scattering has a very high chemical sensitivity which allows one to study interdiffusion phenomena and the formation of new compounds or contacts at surfaces and interfaces [6.48]. As with the other optical techniques, its application is not restricted to growth processes under the ultra-high vacuum (UHV) conditions used in MBE, but it can also be applied to study semiconductor formation from the vapor phase [6.48].

The potential of Raman spectroscopy for the investigation of semiconductor epitaxy can be seen in Figs. 6.18 and 6.19 which show Raman spectra from a (110)-oriented InP surface for different coverages with Sb [6.51]. The InP surface was prepared by cleaving in UHV, and the Sb was evaporated with the substrate at room temperature. Raman spectra were recorded in-situ after each deposition step using the 5145 Å line of an Ar-ion laser. The clean surface shows only two peaks at $305\,\mathrm{cm}^{-1}$ and $345\,\mathrm{cm}^{-1}$ which are due to the TO and LO phonons of bulk InP, respectively. Only the TO mode is expected according to the selection rules for a (110) surface [1.11]. The much weaker LO line might appear due to slight deviations from the exact

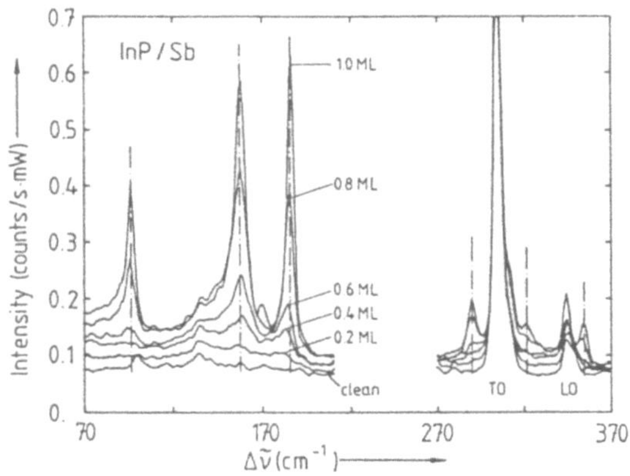

Fig. 6.18. Raman spectra of an InP (110) surface for different coverages with Sb up to 1 ml. The *dashed-dotted vertical lines* indicate "InSb" interface phonons (from [6.51])

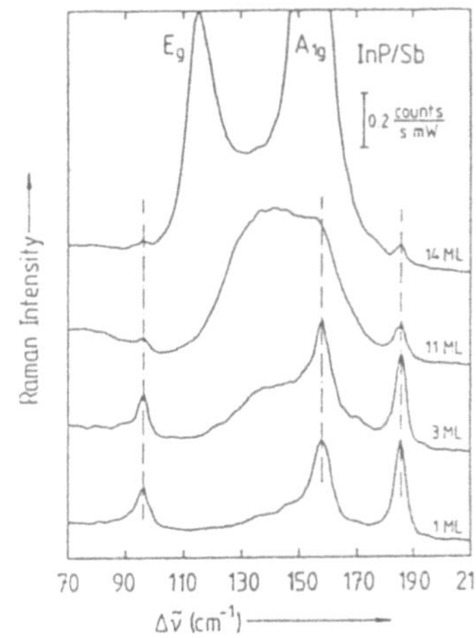

Fig. 6.19. Raman spectra of InP (110) with Sb coverages between 1 and 14 ml. See text for details (from [6.51])

backscattering geometry. When more than about 0.5 ml Sb is deposited on the surface six new peaks appear in the spectra whose intensity increases with the coverage. At 1 ml their scattering strength is comparable to that of the TO line. They are due to interface phonons of the substrate–Sb-monolayer system and reflect an ordered Sb structure grown epitaxially on the InP surface [6.51]. A confirmation of this picture comes from resonance Raman measurements. Signal enhancements for these modes with changing excitation energy occur neither at critical points of the Sb nor of the bulk InP band structure but at surface resonances of the two-dimensional electronic structure of the Sb monolayer [6.52]. Ab-initio calculations predict the interface modes with good accuracy [6.53]. As can be seen in Fig. 6.19, a broad peak around 140 cm^{-1}, characteristic of amorphous Sb, develops for coverages between 3 and 11 ml, and the interface-phonon intensity decreases [6.51]. Above 13 ml the deposited film crystallizes abruptly, and the E_g and A_{1g} phonons of crystalline Sb appear at 113 cm^{-1} and 152 cm^{-1}, respectively.

Figure 6.20 shows series of Raman spectra for the room temperature deposition of different group-III elements on Sb substrates [6.54, 6.55]. Annealing spectra for the same systems but with initial low-temperature deposition are given in Fig. 6.21 [6.54, 6.56]. These results give an impression about the kind of information which is obtained when deposition processes and epitaxial growth are monitored by Raman spectroscopy [6.54–6.56]. In principle, the reaction of Sb substrates with the group-III metals In, Ga, or Al could lead to the formation of III-V semiconductors, i. e., InSb, GaSb, or AlSb. However, metallic layers or clusters might also appear at the surface. The spectra in Fig. 6.20 which monitor several hours of deposition time on (111)-oriented

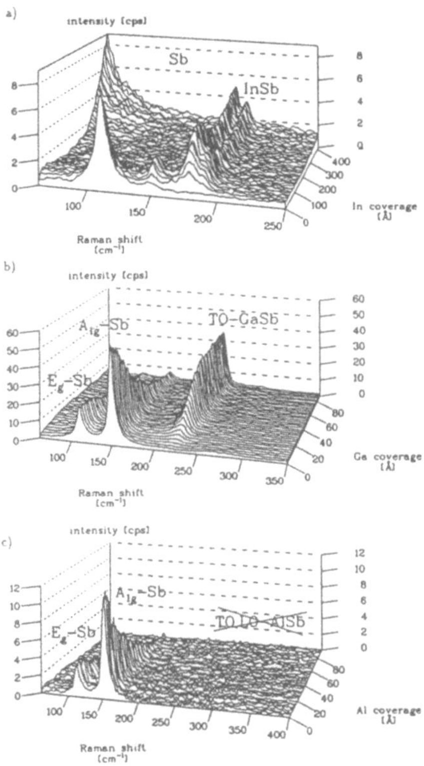

Fig. 6.20. In-situ Raman spectra taken during room temperature deposition of (**a**) In, (**b**) Ga, and (**c**) Al on Sb (111) substrates (from [6.54, 6.55])

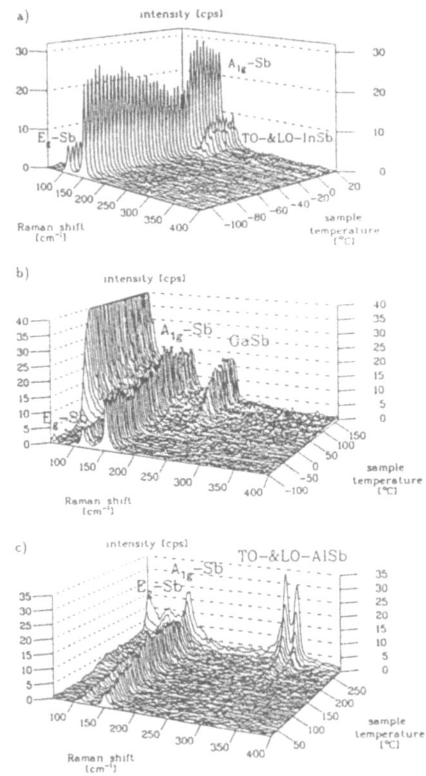

Fig. 6.21. In-situ Raman spectra taken while heating (**a**) In, (**b**) Ga, and (**c**) Al films on Sb substrates, initially deposited at lower temperatures (from [6.54, 6.56])

Sb substrates, prepared by cleaving of single crystals under UHV conditions, show that both cases are realized, depending on the element [6.54, 6.55]. To obtain the highest surface sensitivity, the Raman spectra were excited in resonance with the $E_1/E_1 + \Delta_1$ interband critical points of the semiconductor compounds expected to form, using laser wavelengths of 5145 Å for InSb, 4965 Å for GaSb, and 4579 Å for AlSb, respectively. The clean substrates only show the E_g and A_{1g} peaks of Sb. The A_{1g} intensity for Fig. 6.20 (a) is suppressed due to the different polarization geometry (crossed) used in this case as compared to Figs. 6.20 (b,c) (parallel). These peaks vanish with increasing coverage, indicating the formation of strongly absorbing overlayers. For In and Ga deposition the characteristic optic phonons of InSb and GaSb appear in the spectra. The intensity variation of the InSb lines can be described by a theoretical model assuming planar growth [6.54, 6.55]. As discussed below for ZnSe, the Raman signal for such homogeneous thin films is modulated by layer-thickness-dependent interference effects which lead to oscillations of the scattering intensity. For coverages exceeding 300 Å the InSb spectra develop a background at low frequencies. It is due to diffuse reflection from the surface and indicates a roughening of the layer. The intensity variations of the GaSb phonon cannot be explained by planar growth of a homogeneous film [6.54, 6.55]. The substrate signal is well observed even after the GaSb has appeared. This suggests the formation of GaSb islands. In the case of Al deposition no AlSb peaks appear in the spectra. An overlayer of elemental Al forms and suppresses the substrate signal [6.54, 6.55].

These results provide detailed insight about the relevant growth mechanisms. From thermodynamic arguments an increasing likelihood for the formation of III–V compounds is expected when going from In to Ga and Al deposition on Sb [6.54–6.56]. However, the Raman results of Fig. 6.20 indicate that the opposite is happening. While layer-by-layer growth at room temperature is observed for InSb, GaSb forms islands, and Al does not react at all. Another effect which influences the growth is the interdiffusion of the deposited and the substrate atoms across the III-V layer at the interface [6.54–6.56]. If this layer acts as a strong diffusion barrier at room temperature, semiconductor growth may be prohibited for kinetic reasons.

In order to determine the importance of these effects, Fig. 6.21 shows in-situ Raman spectra obtained when In, Ga or Al layers, deposited on Sb substrates at low temperatures to avoid any reaction, are annealed during a heating ramp [6.54, 6.56]. For this purpose the group-III elements were evaporated on UHV-cleaved (111)-oriented Sb substrates at $-110\,^\circ$C (In), $-100\,^\circ$C (Ga), and $+23\,^\circ$C (Al), respectively. These films had a thickness of about 40 Å (In, Al) or 70 Å (Ga). The ramp velocity was about 1°C/min, and the Raman spectra were measured at a rate of one per minute. With rising temperature the formation of III–V compounds is observed at threshold temperatures of -15°C (In), +40°C (Ga), and +165°C (Al), respectively. This reflects an increasing diffusion barrier height which dominates the growth of these materials. From the behavior of the substrate-mode intensities and the development of a low-frequency background for Ga and Al it was concluded that InSb grows as planar homogeneous layers while GaSb and AlSb tend to island formation [6.54, 6.56]. These results are consistent with the room-temperature observations of Fig. 6.20. Planar growth of InSb at 300 K (Fig. 6.20 (a)) was observed for overlayers as thick as 700 Å. This requires quite strong interdiffusion which has been attributed to the formation of cracks in the InSb due to the large lattice mismatch of 6% to the Sb substrate [6.54, 6.56].

In addition to a continuously rising intensity with increasing layer thickness during growth, the Raman signals of epitaxial films exhibit characteristic oscillations which are due to Fabry–Perot interferences of laser and scattered photons in the film–substrate system [6.57, 6.58]. An example for this interference-enhanced Raman scattering is shown in Fig. 6.22 for the growth

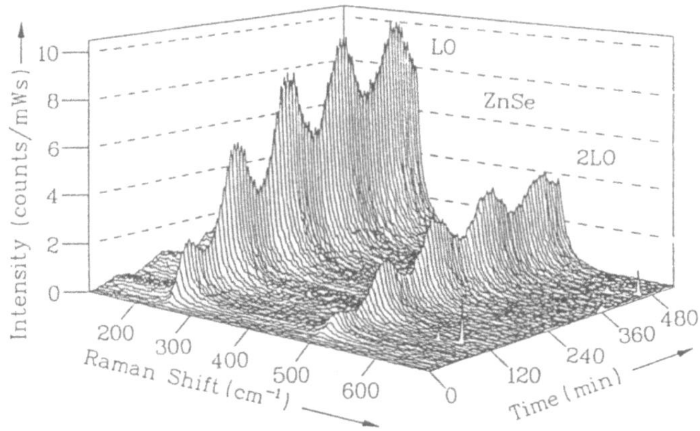

Fig. 6.22. In-situ Raman spectra monitoring the growth of ZnSe on GaAs. The intensity of one- and two-LO-phonon scattering oscillates due to Fabry–Perot interferences (from [6.58])

of ZnSe on GaAs (100) at a substrate temperature of 150°C [6.58]. The Raman spectra were recorded with parallel polarizations along the [001] direction of the substrate, and no GaAs phonons are observed for the clean surface. With ZnSe deposition the spectra show the one- and two-LO-phonon lines of that material, resonantly enhanced due to the laser energy of 2.60 eV (4765 Å), close to the E_0 gap of ZnSe at this temperature. The oscillations of the Raman signals with increasing layer thickness can be used to determine the layer thickness and the growth rate [6.58]. Information about surface roughness can be derived from the magnitude of the interferences [6.58]. The frequency of the Raman lines depends on the layer thickness and the substrate temperature. This yields information about strain [6.58]. For mixed crystals like ZnS_xSe_{1-x} the composition can be determined from the frequencies of the ZnSe- and ZnS-like phonons [6.59].

6.2 Recent and Future Developments

Optical spectroscopy has experienced significant progress in recent years due to the development of near-field optical techniques which allow one to overcome the spatial resolution limit given by diffraction [6.60–6.65]. This breakthrough has had a strong impact on many different fields of science such as physics, chemistry, and biology. An overview of the various technical developments and the broad range of topics investigated can be obtained from the proceedings of conferences dedicated to near-field optics [6.66–6.69]. In semiconductor physics the new possibility of obtaining subwavelength spatial resolution coincided with the trend to investigate dimensionality and confinement effects in nanoscale structures where major developments, such as the self-organized growth of quantum wires and dots [2.135–2.137,6.70,6.71] and nanocrystals prepared in a glass matrix [6.72, 6.73] or with the methods of colloidal chemistry [6.73, 6.74] have also occured in the past few years. The combination of the achievements in both, near-field spectroscopy and sample preparation, has made it possible in semiconductor physics to investigate the optical properties of individual nanoscopic objects [6.75–6.77]. This exciting perspective has also stimulated the development of other, sometimes easier, techniques which are used intensively for this purpose [3.33, 6.78–6.81].

The very high spatial resolution offered by near-field optics is also attractive for Raman scattering which, in addition to the linear optical properties or luminescence usually probed, has high chemical selectivity. It is hoped that many applications in basic research and analytics which are presently based on micro-Raman scattering and confocal microscopy can be extended to smaller length scales. However, the two factors of Raman scattering being a rather week effect (typically only one of about 10^7 photons is scattered inelastically) and a lot of light being lost when operating under near-field or similarly constrained conditions (usually only one of about 10^5 photons is transmitted through a 50 nm aperture) which have to be combined for this purpose make this a seemingly impossible undertaking. Nevertheless, several encouraging results have recently appeared [6.82–6.90].

In Section 6.2.1 we illustrate the basics of near-field optics and related techniques and discuss their present use and possible potential for Raman scattering. Section 6.2.2 is about progress in instrumentation which might help to further develop Raman spectroscopy in this and other directions.

6.2.1 Near-Field and Nano-Raman Scattering

Near-Field Optics (NSOM). Although near-field optics has become popular as an experimental method only within the last few years, the basic concept was proposed and discussed many decades ago [6.91, 6.92]. It was realized already in 1928 that subwavelength features can be imaged when a small aperture is scanned in the immediate vicinity of an object [6.91]. Unlike conventional optics, in near-field optics the resolution is determined by the diameter of the aperture and its distance to the sample rather than the wavelength of the light. Since, in principle, both parameters can be made arbitrarily small, resolution beyond the diffraction barrier is limited by the capability to make small apertures and to position them with high enough accuracy rather than by any fundamental physical laws. In other words, the key to very high spatial resolution in optics is to probe the evanescent non-radiating fields with high spatial frequencies which occur in the near-field regime of a sample exposed to electromagnetic radiation, while the well-known diffraction limit holds for the detection of radiating field components which propagate in the far-field [6.62–6.65]. This concept is quite general, and a resolution of $\lambda/60$ was demonstrated already 25 years ago using microwaves ($\lambda = 3$ cm) [6.93].

The advent of scanning tunneling microscopy also stimulated efforts to use optical probes [6.94]. A resolution exceeding $\lambda/20$ in the visible was soon reported for an "optical stethoscope" consisting of a metal-coated quartz tip with a small aperture, operated with 4880 Å radiation from an Ar-ion laser [6.60]. A strong impulse for near-field scanning optical microscopy (NSOM) came from the invention of tapered optical-fiber tips which can be produced reliably by conventional micropipette pulling techniques and Al-coated to achieve aperture sizes down to a few nanometers [6.62, 6.63]. Due to their conical shape at the end, light in these probes is transmitted with high efficiency. Nevertheless, since all waveguide modes are exponentially cut-off for such small diameters, the incident laser power is still attenuated by several orders of magnitude on the last few hundred nanometers approaching the aperture [6.95]. Therefore the resolution in NSOM is limited by the practical question of how much light is needed in an experiment and, ultimately, by the smallest possible aperture size which is about 10 nm due to the finite skin depth of the Al coating [6.62, 6.63]. Tapered optical-fiber tips are presently used almost exclusively in scanning near-field optics. Many adaptations of the basic NSOM principle have been developed for different experimental situations in recent years [6.64–6.69]. Some typical possibilities are given in Fig. 6.23 [6.96]. Further NSOM configurations can be found in [6.64].

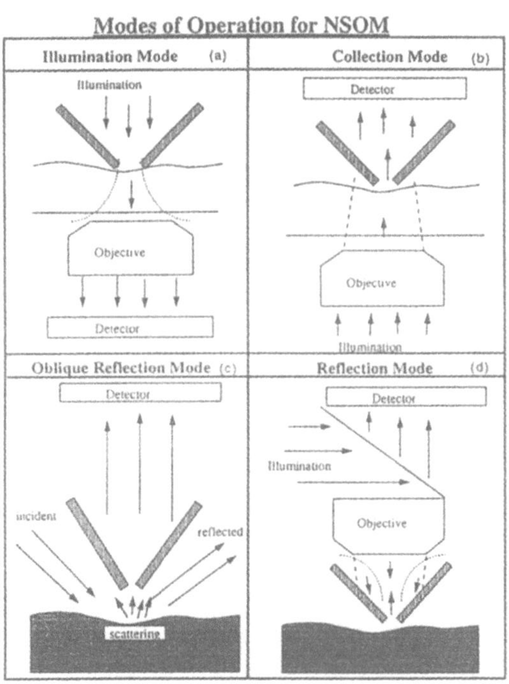

Fig. 6.23. Different operation modes for NSOM: Common to all techniques is the passage of photons through a small aperture, schematically indiated by the *dashed rectangles*, in the vicinity of the sample. The illumination (excitation) **(a)** and collection modes **(b)** are used with transparent samples. The reflection modes with external **(c)** (photons are collected through the aperture) and internal illumination **(d)** (exciting and detected photons go through the aperture) are suitable for absorbing specimen (from [6.96])

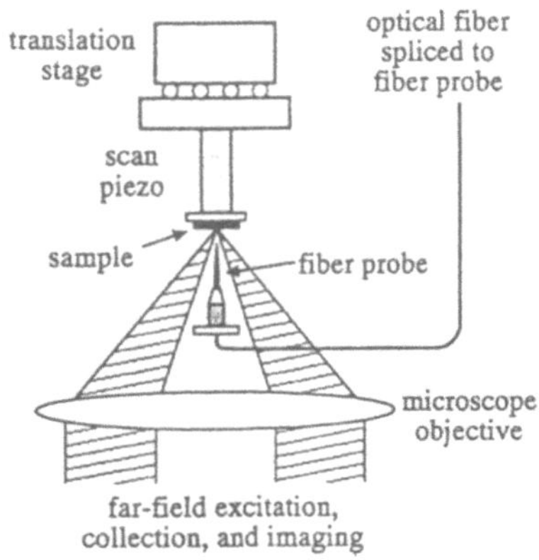

Fig. 6.24. Experimental setup for NSOM experiments which can be used also at low temperatures. The translation stage is for the coarse positioning of the sample mounted on the piezoelectric scanning tube by which it can be moved with three degrees of freedom. The fiber probe can also be moved vertically. The microscope objective provides far-field access to the sample. It is used for the excitation (collection) of light collected (excited) through the fiber tip in spectroscopic NSOM experiments as well as to obtain an overview of the sample during alignment (from [6.75])

The power of near-field optics and its suitability for the investigation of a wide range of phenomena derives from the fact that basically all methods known from conventional optics can be transferred, i. e., a plethora of contrast mechanisms such as absorption, reflectivity, refractive index, and polarization, as well as ways to obtain spectroscopic information are readily available [6.63]. To give one example we discuss the application of NSOM to investigate the optical properties of quantum wells and wires by low-temperature luminescence spectroscopy [6.75, 6.97]. Figure 6.24 shows a schematic sketch of the apparatus used. It allows one to perform NSOM measurements in both, a collection (excitation from the far-field, collection through the fiber probe, see Fig. 6.23 (c)) and an excitation mode (excitation through the fiber, collection in the far-field, inverted version of Fig. 6.23 (c)) [6.75]. Figure 6.25 gives a cross section of the sample as exposed to the fiber probe. A far-field spectrum and NSOM images of this specimen are shown in Fig. 6.26. The measurements were performed at 1.5 K in the collection mode. The sample was excited with photons from a Ti : sapphire laser (3 mW, $\lambda = 7500$ Å), focused to a spot with 30 μm diameter. The far-field luminescence was obtained with the fiber 1.0 μm above the sample. The intensity distribution of the three peaks observed in this case (Fig. 6.26 (a)) was then imaged in the near-field by scanning the 250 nm-diameter fiber probe 100 nm above the sample over a $2 \times 2\,\mu m^2$ area. The pixel size of the images is 0.1 μm. Even though the GaAs QWs in the MQW array and the overgrown SQW were designed to have the same thickness (70 Å, see Fig. 6.25), the different luminescence energies in Fig. 6.26 (a) show that they differ by 3 ml [6.75]. The energy of the quantum wire emission is smaller than both these values due to the more extended character of the wave function at the joints. The near-field intensity distributions prove the different origin of the three luminescence peaks observed in the far-field spectrum. The images in Fig. 6.26 (c,d) show that the QW luminescence decreases strongly about 0.6 μm away from the wire region. This has been attributed to diffusion of photoexcited carriers in the lower-energy wire states [6.75].

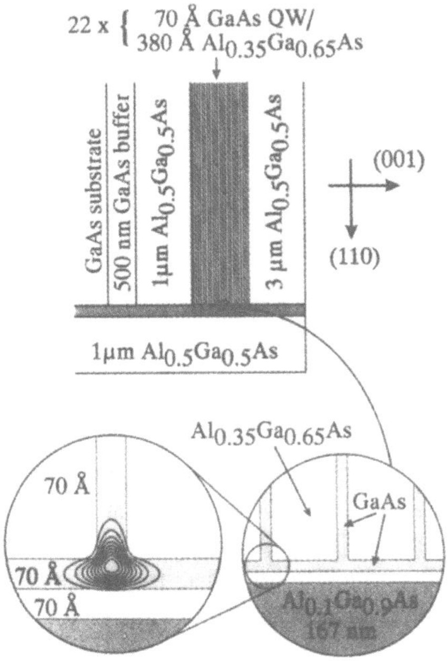

Fig. 6.25. Sketch of a quantum wire sample made by cleaved-edge overgrowth of a single 70 Å GaAs quantum well (SQW) on a GaAs/$Al_{0.35}Ga_{0.65}As$ MQW array. As shown in the magnified areas, the quantum wires are formed where the SQW and the MQWs touch each other. The contour plot indicates the constant density profiles of a quantum wire wave function (from [6.75])

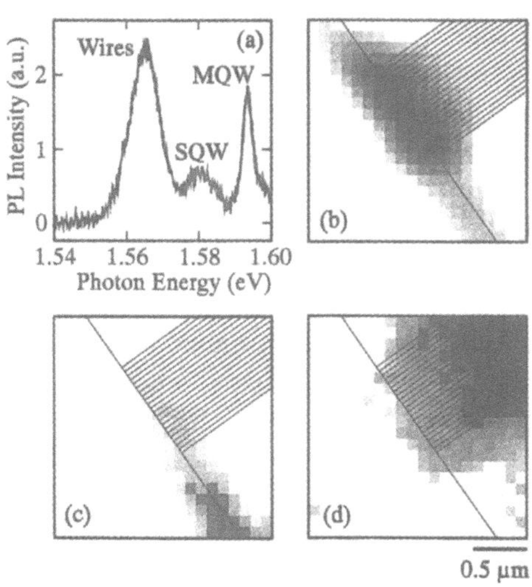

Fig. 6.26. **(a)** Far-field luminescence spectrum of the sample in Fig. 6.25. The three peaks at different energies correspond to recombination from the quantum wires, the SQW, and the MQW array. Greyscale images of the luminescence intensity distribution at these energies reveal the local origin of the emission from the quantum wires **(b)**, the SQW **(c)**, and the MQWs **(d)**. For comparison, the sample structure is overlaid on each image (from [6.75])

Despite the high spatial resolution of the experiments just discussed, the individual wires are not resolved due to their much smaller separation compared to the aperture size [6.75]. Emission from isolated confined structures has been observed, however, in near-field measurements on single GaAs/$Al_{0.32}Ga_{0.68}As$ QWs where excitons can also be localized laterally by interface roughness potentials [6.76]. In this case the inhomogeneously broadened far-field luminescence spectrum splits into a multitude of extremely sharp lines which arise from the emission of such zero-dimensional "quantum dot" excitons. Individual quantum dots with larger diameters (270 nm) in etched samples have also been imaged directly by NSOM [6.77]. Beyond these applications in semiconductors, near-field optics is used intensively for single-molecule detection and studies of their fluoresence dynamics [6.98–6.102].

Other High-Resolution Techniques. In addition to NSOM, several other methods have been developed in recent years which allow one to image individual luminescing entities and to investigate their properties spectroscopically. These new schemes have been pioneered mainly in the area of low-dimensional semiconductors where the inhomogeneous broadening of spectral features by interface roughness often makes detailed studies of intrinsic properties, such as carrier lifetimes and relaxation mechanisms, difficult (see Chap. 3).

In one of the first spectroscopic studies of single quantum dots, zero-dimensional systems were made by laser-induced local interdiffusion of a GaAs/$Al_{0.35}Ga_{0.65}As$ single QW structure [6.78]. The rapid thermal interdiffusion between the well and the barriers which takes place under the laser spot increases the band gap in the QW plane. This can be used for the lateral confinement of electrons and holes in untreated areas enclosed by an interdiffused region. Due to the easy controllability of the laser position this is an elegant way to fabricate dots with different submicron diameters. The luminescence of such structures shows several sharp lines which arise from the recombination between different quantum-dot states. The spectra are unaffected by inhomogeneous broadening and reveal the reduced electron relaxation times expected for zero-dimensional systems [6.78].

In another approach, sharp photoluminescence lines from individual QW excitons were obtained by systematically reducing the diameter of the laser spot exciting a sample [3.33]. While the spectrum of a high-quality 35 Å GaAs/$Al_{0.35}Ga_{0.65}As$ QW still shows an inhomogeneously broadened line with a FWHM of about 2 meV when excited with a spot diameter of 25 μm, a series of sharp lines (FWHM less than 0.1 meV) is observed when the laser is focused to 1.5 μm. Similar to near-field optics [6.76], the sharp lines arise from the recombination of single "quantum dot" excitons which are localized laterally by interface roughness [3.33]. For larger spot diameters, too many of these lines overlap and one obtains the usual inhomogeneously broadened spectrum. From intensity-distribution images at different recombination energies a map of the position and size of interface defects was obtained [3.33]. In coupled GaAs/AlAs QWs the electric-field tunable indirect X-conduction states in AlAs have been used to map the distribution of GaAs excitons localized by interface roughness [6.79]. For a small enough laser spot, sharp lines occur around the indirect recombination peak when carriers accumulated in the X-state are resonantly reinjected into GaAs "quantum dot" exciton states [6.79].

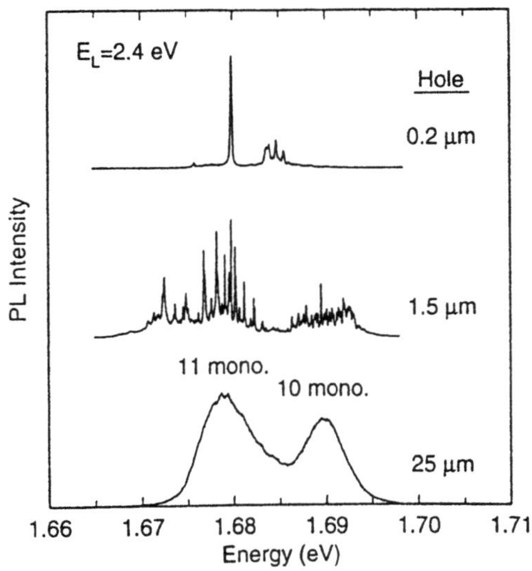

Fig. 6.27. Luminescence spectra of a 10 ml $GaAs/Al_{0.3}Ga_{0.7}As$ QW excited and detected through apertures with different diameters as indicated (from [6.103])

A third method to investigate individual nanoscopic objects uses small apertures in nontransparent Al films deposited directly on the sample surface [6.80,6.103,6.104]. Submicrometer pinholes can be produced by electron-beam lithography and subsequent metal liftoff. Figure 6.27 shows luminescence spectra of a single 28 Å (10 ml) $GaAs/Al_{0.3}Ga_{0.7}As$ QW measured with different apertures [6.103]. In addition to the inhomogeneously broadened peak at the nominal QW thickness of 10 ml, the 25 μm-spectrum in Fig. 6.27 shows another broad emission corresponding to a layer thickness of 11 ml. This indicates the existence of areas with that QW width where excitons can get trapped. The spectra taken with smaller apertures show a fine structure of many lines which are considerably sharper. They are due to QW excitons localized at different lateral potential fluctuations caused by interface roughness whose extension has been estimated to be around 100 nm [6.103]. These systems are therefore completely confined and represent individual "naturally occuring" quantum dots which are not affected by inhomogeneous broadening. Their zero-dimensional nature has been demonstrated in temperature-dependent measurements of the luminescence linewidth which increases significantly more slowly than for QWs or bulk GaAs [6.104]. Below 10 K homogeneous linewidths of $(23 \pm 10)\ \mu$eV (FWHM) have been obtained, corresponding to an exciton lifetime of 29 ps [6.104].

The spectra in Fig. 6.27 are typical also for those obtained by the other methods illustrated above. It should be noted that the aperture defined by the diameter of the laser spot or the hole in a metal film needs not to be located directly at the sample. There are reports where an external pinhole, positioned at an intermediate image, or even the spectrometer entrance slit were used to select a spatial region small enough for the observation of single quantum dot effects [6.81,6.105].

Raman Results from Solids and Liquids. Although the possibility to perform Raman spectroscopy with subwavelength resolution has been suggested early in the development of near-field optical techniques [6.61], it took several years before some of the formidable obstacles, arising mainly from the weak light levels in both, the Raman process and the detection through small apertures, could be partly overcome. The principle was demonstrated for diamond, where phonon spectra were obtained with a fiber probe of 100 nm diameter even though subwavelength resolution was not shown [6.82]. Using an NSOM setup designed to minimize the sample drift during long-time signal accumulations, it has recently been possible to record nano-Raman spectra and even mode-intensity images of phonons in potassium titanyl phosphate (KTP) under near-field conditions [6.83, 6.84, 6.86]. Figure 6.28 gives a comparison of Raman spectra in KTP taken under different conditions [6.84]. Most obvious is the large intensity difference between the micro-Raman scan of Fig. 6.28 (a) and the nano-Raman spectra, measured in excitation mode (excitation through a tapered fiber tip, collection via a microscope objective), shown in Fig. 6.28 (b,c). Since these spectra were recorded using single-channel detection (λ = 5145 Å, aperture diameter ≈ 250 nm, spectral resolution 4 cm^{-1}, room temperature), they require long accumulation times, a point which could be substantially improved (see Sect. 6.2.2). The reduction of the sampling volume between micro- and nano-Raman spectra is evident from the reduced Rayleigh-scattering background in the low-frequency region [6.84]. While the two far-field spectra in Fig. 6.28 are quite similar, several changes occur under near-field conditions. Most obvious is a splitting of the line at 700 cm^{-1} which belongs to a group of TiO_6 stretch-

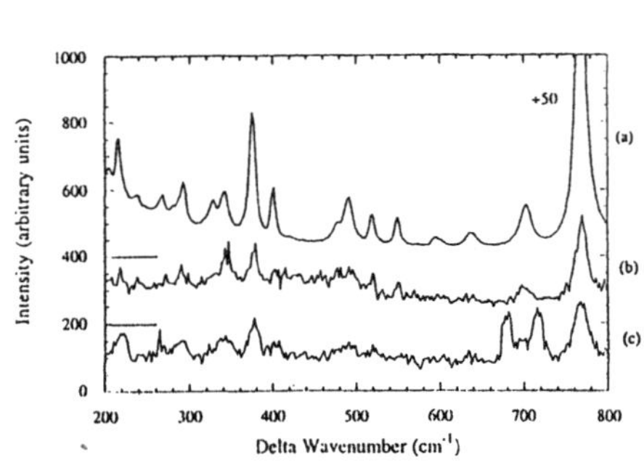

Fig. 6.28. Raman spectra of KTP: **(a)** conventional micro-Raman spectrum; **(b)** far-field spectrum using a tapered fiber tip; **(c)** near-field spectrum. Note the intensity scaling factor between micro- and nano-Raman spectra and the peak splitting of the 700 cm^{-1} line under near-field conditions (from [6.84])

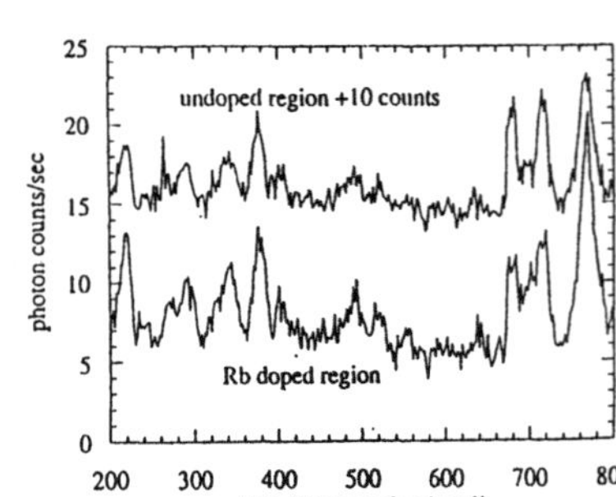

Fig. 6.29. Near-field Raman spectra of a KTP sample in areas with and without Rb doping. The most prominent difference is an almost twofold intensity change of the mode at 767 cm^{-1}. The spectral resolution is 2 cm^{-1} (from [6.84])

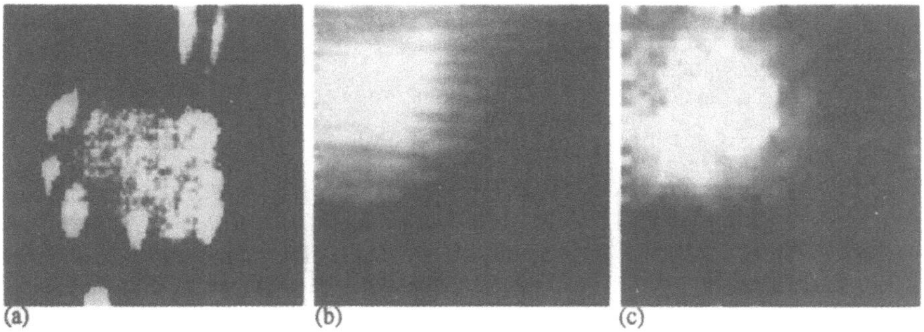

Fig. 6.30. Images of a KTP sample containing $5 \times 5\,\mu m^2$ Rb-doped areas: **(a)** topographic image ($10 \times 10\,\mu m^2$, gray scale: 17 nm); **(b)** transmission image ($4 \times 4\,\mu m^2$, gray scale: 460–550 nW); **(c)** Raman image ($4 \times 4\,\mu m^2$, gray scale: 18–22cts/s) (from [6.84])

ing modes of KTP found in this region [6.86]. It has been suggested that this effect arises from an enhanced surface sensitivity due to the near-field probe which therefore allows one to observe mode-splittings caused by surface stress [6.84]. However, these shifts are very large compared to those known from semiconductors (see, e. g., Fig. 6.6), a point which requires further investigation. Another explanation invokes components of the electromagnetic near-field normal to the surface, induced by the metal aperture, which could couple to phonons otherwise not Raman-active in the scattering geometry used [6.84,6.86]. Such polarization effects, known from NSOM single-molecule polarization experiments [6.98], might also explain some other differences between the near- and far-field spectra in the 200–400 cm^{-1} region which are most pronounced for the peak at 220 cm^{-1} [6.84,6.86]. A different polarization behavior than for the far-field was also reported recently in NSOM Raman spectra from liquid CCl_4 [6.90]. In these measurements the sampling volume of 70 nm diameter and 10 nm depth contained about 240 000 molecules.

The analytical capabilities of NSOM Raman spectroscopy were demonstrated by comparing pure and rubidium-doped KTP [6.84,6.86]. Figure 6.29 shows the respective near-field spectra [6.84]. The KTP sample used in these experiments contains an array of $5 \times 5\,\mu m^2$ Rb-doped areas where the intensity of the phonon at 767 cm^{-1} is found to be about twice as strong as in the undoped material. This intensity difference has been used as a contrast mechanism in an imaging experiment. Figure 6.30 shows the resulting nano-Raman image together with topographic and transmission information [6.84]. The Rb-doped regions have a slightly larger volume than the bulk and are therefore raised above the KTP surface. This leads to the topographic contrast in Fig. 6.30 (a), detected via the atomic shear-force mechanism used to control the tip position during the optical measurements [6.84]. The change in refractive index between the two materials ($\Delta n = 0.027$) leads to contrast in

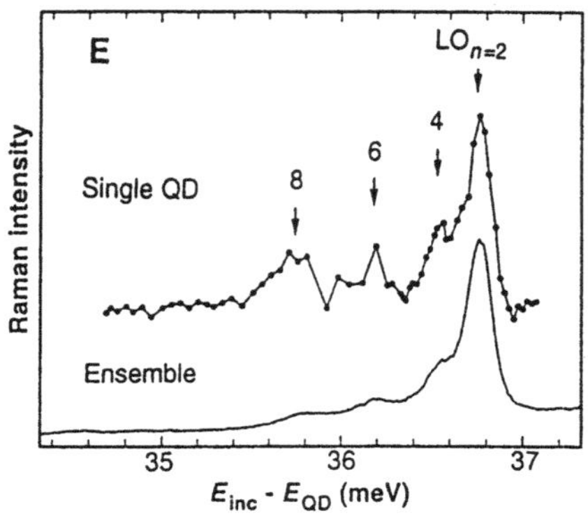

Fig. 6.31. Optic-phonon Raman spectra of a 19 ml GaAs/AlAs QW structure measured with aperture diameters of 25 μm (*Ensemble*) and 0.8 μm (*Single QD*) through an Al film deposited directly on the sample surface (from [6.87])

the transmission image of Fig. 6.30 (b) which shows an enlarged view around one of the corners of the Rb-doped square. The Raman image in Fig. 6.30 (c), accumulated during 10 h, reflects the intensity distribution of the 767 cm^{-1} phonon in the same area.

Using a fixed pinhole on a 19 ml GaAs/AlAs QW structure an optic-phonon nano-Raman spectrum has very recently been reconstructed from the intensity variations of a sharp single-quantum-dot localized-exciton emission line with the laser excitation energy [6.87]. This result, obtained with a narrow aperture, is compared to a micro-Raman spectrum, measured in the conventional way through a wide pinhole, in Fig. 6.31. In the GaAs region different "confined" phonons LO_n are observed in this strongly resonant luminescence-excitation experiment. Both spectra are quite similar which indicates the absence of lateral confinement for the phonons. From the spectral shape one can conclude that some roughness-induced coupling to mixed confined-optic and interface phonons is present (see Fig. 2.15 [2.31, 2.32]).

These results give rise to the expectation that Raman scattering can be established as one of the NSOM spectroscopies and that it is possible to exploit the very localized character of individual excitonic intermediate states for high resolution experiments. This should allow one to obtain detailed analytical information on the nature and distribution of defects as well as the local variation of parameters such as stress, alloy composition or sample inhomogeneity. In principle, nano-Raman scattering could be used in most of the examples in Sect. 6.1, and applications for the characterization of semiconductors and microelectronic materials and devices are being discussed in the literature [6.83, 6.85]. However, a large potential for improvements should be exploited in order to make the method more versatile. In experiments using optical fibers, care has to be taken to suppress the strong Raman signals from the fiber material itself [6.90], for example, by avoiding the reflection mode geometry of Fig. 6.23 (d). Advanced fiber design concepts could help to increase the power throughput and raise the signal levels such that much shorter accumulation times become feasible [6.106]. Similar improvements should be possible when higher-throughput or parallel detection schemes are employed for the imaging of Raman intensity distributions or the measurement of Raman spectra (see Sect. 6.2.2).

In addition to the resonant signal enhancement which is already used in some cases, intrinsically stronger Raman processes than the phonon scattering discussed so far could be considered. Spin-flip Raman scattering [6.20,6.21,6.107,6.108], an electronic process which is typically four to six orders of magnitude stronger than phonon Raman scattering is already used intensively for semiconductor research [6.109–6.112] as well as materials characterization (see Sect. 6.1.1 [6.18,6.19,6.24,6.25]). Despite the requirement of a magnetic field being present, this technique could be a very useful method to determine the origin of different impurities and their distribution, especially in new optoelectronic materials where doping is not yet fully understood. In fact, spin-flip Raman scattering relies heavily on bound excitons or states localized by interface roughness [6.109–6.112]. It thus occurs in an environment which is very similar to the "quantum dot" excitons discussed above. In angle-dependent resonant spin-flip Raman experiments using NSOM or other small-aperture techniques it might thus be possible to directly observe the different ground and excited state quantum-dot-exciton spin splittings reported in [6.105] and to separate the electron and hole contributions, respectively.

Surface-Enhanced Raman Scattering. Another technique where strong signals are obtained is surface-enhanced Raman scattering (SERS) [6.113–6.116]. In these processes, the Raman signal from molecules is enhanced by five to seven orders of magnitude when they are in the vicinity of metallic, mostly silver, nanoparticles or "rough" metal-coated surfaces. The presence of nanoscopic metal protrusions is essential for the effect. Although the origin of the signal enhancement is not yet understood in all detail, two main mechanisms are usually considered [6.113–6.116]. The so-called electromagnetic or classical effects arise from the local-field enhancement of the laser radiation near a small metallic object. The Raman intensity of a molecule in its vicinity increases by $10^4 - 10^5$, approximately proportional to the fourth power of the electric field enhancement ratio. So-called chemical enhancement ($\approx 10^2 - 10^3$) is attributed to resonant Raman processes involving a charge transfer between the metal and the molecule. The importance of the metal coating of tapered optical fibers for a modification of the electromagnetic field in the tip–surface region and the possibility of SERS has been pointed out early during the development of NSOM techniques [6.63]. In addition, we suggest that it should also be possible to obtain SERS signals from single molecules on nonmetallic or flat surfaces when they are approached by a metallic tip.

Encouraging results in this direction have been reported recently from SERS experiments on dilute colloidal Ag nanoparticles coated with dye molecules [6.88, 6.89]. SERS spectra of rhodamine-6G molecules revealed the existence of individual "hot particles" which yield especially strong Raman signals. The scattering from single rhodamine-6G molecules adsorbed at these particles is $10^{14} - 10^{15}$ times enhanced, indicating that in conventional SERS measurements, which perform an ensemble average, the vast majority of the molecules does not contribute much to the signal at all [6.88]. Figure 6.32 shows single-molecule SERS spectra for two Ag nanoparticles oriented differently with respect to the polarization of the incident laser light (5145 Å) [6.88]. Strong SERS signals occur when the exciting polarization is parallel to the elongated axis of the Ag particles. In this case the laser excites a strong electronic surface-plasmon resonance which contributes to the signal enhancement. The orientation of the rhodamine-6G molecules with respect to the Ag particles can be determined from an analysis of the scattered light polarization using unpolarized excitation [6.88]. Single-molecule conditions in these experiments are ensured by an overall occupation probability per particle of 10%.

In a colloidal Ag solution with an average of 0.6 crystal violet molecules in a sampling volume of 30 pl, containing about 100 Ag clusters, a series of SERS spectra shows a discrete intensity distribution for one of the dye molecule's Raman lines which corresponds to the simultaneous presence of zero, one, two or three molecules in the individual measurements [6.89]. The probability to find Raman lines with the different intensities follows a Poisson distribution which is evidence for single-molecule detection. The observed SERS signals are comparable to those from a reference sample with about 10^{14} (not SERS-active) methanol molecules in the sampling volume [6.89]. This provides additional evidence for the enormous signal enhancement of SERS from single molecules.

These results could help to establish Raman scattering as an additional method to detect and investigate single molecules. In contrast to the widely

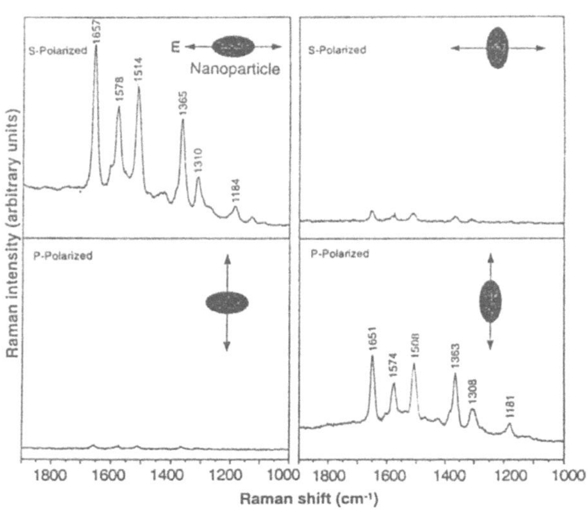

Fig. 6.32. Surface-enhanced Raman spectra of single rhodamine-6G molecules adsorbed on two different Ag nanoparticles. Their relative orientation with respect to the laser polarization is indicated schematically (from [6.88])

used laser-induced fluorescence, SERS spectra contain rich spectral structures which makes them very valuable for analytical purposes [6.88, 6.89]. Further advantages include enhanced count rates due to faster vibrational relaxation times which make photobleaching less likely and the less destructive character of the probe which is not in resonance with molecular transitions [6.88, 6.89]. In future applications such entities could be used as markers or in nanoscopic sensors, possibly even in connection with semiconductor devices.

6.2.2 Raman Instrumentation

Many applications of Raman scattering (see, for example, Sect. 6.1) would benefit from the availability of more efficient ways to measure the usually weak signals. Especially for the high-resolution methods described in Sect. 6.2.1, but also for many commercial purposes, progress is very closely related to advances in Raman instrumentation. Emerging new concepts in this area should help to make Raman spectroscopy an even more versatile tool for basic research and applications.

While resonances of the Raman effect can be used to obtain stronger signals in some cases, this approach is generally rather impractical since it often requires the use of special laser wavelengths which are not always at hand, especially in an industrial context. As far as the recording of scattered light is concerned, charge-coupled device (CCD) detectors have become very widespread during the past 10 years and are now being used in most multichannel Raman spectrometers. Among their favorable properties are a very high quantum efficiency over a wide spectral range, low dark counts when cooled, a long lifetime, and a good resistance to overexposure from high signal or stray-light levels [1.7]. Another area where significant improvements have been demonstrated concerns the optical components used for spectroscopy itself. New methods for the efficient suppression of stray light and the generation of spectra are presently becoming more and more popular. Their basic features and potential are now discussed.

High-Throughput Devices. The separation of Raman signals from the much stronger elastically scattered laser light is usually achieved by bandpass filter stages with two diffraction gratings arranged to obtain subtractive dispersion [1.6]. The steepness of the bandpass edges determines how close to the laser line spectra can be recorded. In holographic notch filters a different approach is used [6.117–6.119]. These structures suppress the incident light by more than six orders of magnitude over a narrow bandpass of about 10 nm while the rest of the spectrum is transmitted with a high efficiency of about 80%. With typical dimensions of 25 mm diameter and a thickness around 5 mm, holographic notch filters are readily available for most ion laser lines [6.119]. They consist of a thick ($> 50\,\mu$m) gelatin film with a small but very uniform sinusoidal modulation of the refractive index ($\Delta n \approx 0.02$) generated by holographic techniques [6.117–6.119]. These filters are a very attractive alternative for replacing the double-monochromator filter stage of conventional Raman triple-spectrometers with multichannel detection. Their spectral edgewidth of less than 4 nm allows one to measure signals at frequency shifts as low as 50 cm^{-1} [6.119]. Small Raman systems consisting of a single grating or prism spectrograph and a holographic notch filter for stray light rejection are presently gaining in popularity [6.38, 6.39, 6.119]. In addition to a five to ten times larger throughput, they are more compact, very robust, and less expensive than triple monochromators. This opens the way to many applications, for example, in the areas of semiconductors, superconductors, pharmaceuticals, chemicals and polymer science, forensic science, geology, materials and coatings, corrosion and electrochemistry for which Raman scattering was not widely used before [6.38, 6.39, 6.119]. In connection with fiber-optic sampling heads for remote sensing, such systems are even employed for real-time in-line process monitoring and control [6.38, 6.39, 6.119].

Progress in the manufacturing of volume holographic optical elements makes it nowadays possible to produce highly efficient holographic transmission gratings [6.119, 6.120]. Very compact Raman systems with apertures as low as f/1.8 and a focal length of 85 mm have been built using these devices in the spectrograph stage and holographic notch filters for stray light rejection [6.119, 6.120]. In addition to their significantly enhanced throughput, they cover a broad spectral range of about 4000 cm^{-1} with 5 cm^{-1} resolution. This is achieved with truly parallel signal detection by superimposing different index modulations in the same hologram such that the resulting grating diffracts two different spectral regions onto separate areas of a CCD detector [6.119, 6.120]. Such wide-range measurements are needed, for example, for the identification of chemical substances which can have a "fingerprint" region with many lines at frequencies up to about 2000 cm^{-1} and characteristic bands such as C–H or O–H stretching modes around 3000 cm^{-1} and beyond [6.120]. Applications include the monitoring of the bath composition used for cleaning silicon wafers in the microelectronics industry, the verification of the chemical composition of tablets in the pharmaceutical industry, and the determination of the saturation degree of fats in the food industry [6.120]. The large throughput of these instruments is not only advantageous for the detection of signals from intrinsically weak Raman scatterers such as some bulk materials or thin films. It also translates directly into shorter accumulation times (< 0.1 s) and reduced laser powers ($1 - 10$ mW) being required to obtain spectra in many other situations. This makes real-time process monitoring feasible and allows much smaller lasers to be used. Another advantage of high-throughput instruments is that they still work in the near infrared where CCD's are less efficient. In combination with semiconductor diode lasers ($\lambda \approx 800$ nm) they are an alternative to obtain Raman spectra from many chemical substances which are usually investigated by Fourier-transform Raman techniques and Nd:YAG lasers ($\lambda = 1.06\,\mu$m) in order to avoid a strong luminescence background in the visible [6.38, 6.39, 6.119].

Beyond Grating Monochromators. The wavelength selection by diffraction gratings in conventional spectrometers has the disadvantages that these instruments are quite large when high resolution is desired and rather inflexible in applications which require the measurement of many peaks over a wide spectral range. These problems can be circumvented with the use of an acousto-optic tunable filter (AOTF). Although known for many years, such devices have recently reached performance levels which make them suitable as narrow-band wavelength selectors in spectroscopy [6.121–6.125]. The basic principle of AOTF operation is the Bragg diffraction of light by the acousto-optic effect [6.126]. Traveling acoustic waves in suitable crystals like TeO_2 or quartz, generated by applying radio-frequency (rf) signals in the 100 MHz range via a piezoelectric transducer bonded to the crystal, periodically modulate the refractive index and create a moving phase grating [6.121–6.125]. For a given acoustic-wave frequency only light in a narrow spectral range fulfils a wavevector-conservation condition and is diffracted with an efficiency exceeding 80%. When the frequency of the rf signal is changed, phase-matching and diffraction is obtained for another optical wavelength. For practical reasons AOTFs are based on anisotropic Bragg diffraction in birefringent crystals, where light from a wide acceptance angle is diffracted into the same direction, independent of its frequency [6.121–6.125]. This process involves a change of the light polarization by 90°. Depending on whether a collinear or noncollinear configuration is chosen to obtain phase-matching, the diffracted light can be separated by polarizers or spatial filtering, respectively. One of the advantages of AOTFs is that their transmission wavelength can be tuned easily by varying the frequency of the rf signal, for example, in TeO_2 a range between 400 and 1900 nm can be covered with acoustic frequencies from 40 to 170 MHz using rf powers of 1–2 W, and operation up to 5500 nm has been achieved [6.125]. More importantly, AOTF wavelengths can be switched randomly in about 10 μs, limited by the propagation time of the acoustic waves through the device [6.121–6.125]. In contrast to grating monochromators, AOTFs are compact solid-state devices with no moving parts. The resolution of an AOTF depends on the length of the interaction region. While filters with a moderate bandpass of about 50 cm^{-1} at 630 nm are commonly used, the technology limit was estimated to be below 2 cm^{-1} [6.122]. AOTF filters with a resolution of 8 cm^{-1} have become available recently [6.127].

Due to these special properties, AOTFs have been introduced in many different applications in recent years. To mention only a few, they are employed as fast tunable wavelength selectors and modulators in combination with incoherent or laser white-light sources [6.123], as well as in emission, fluorescence and luminescence spectroscopy over a very wide spectral range from the visible to the near infrared [6.128, 6.129]. Reports on AOTFs used for remote sensing [6.124], in remote multisensor systems [6.130], in plasma diagnostics and environmental analysis [6.127], in waste recycling, or in spectroscopic ellipsometry [6.131] can be found in the literature. In combination

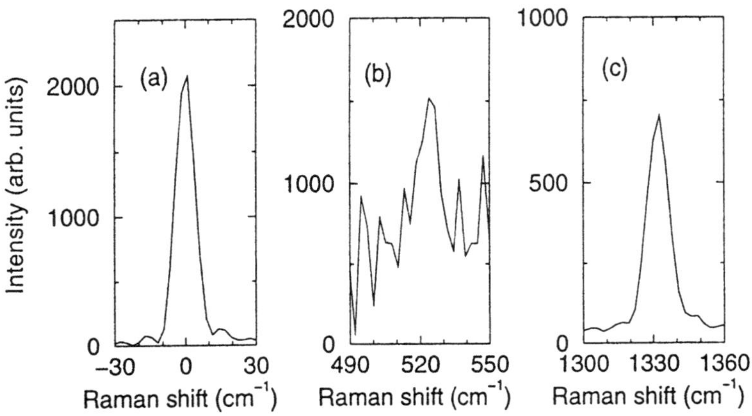

Fig. 6.33. Laser line at 5145 Å **(a)** and Raman spectra of Si **(b)** and diamond **(c)** measured with a spectrometer consisting of an AOTF and a holographic notch filter. See text for details

with holographic notch filters, AOTFs have been implemented in miniaturized high-throughput Raman spectrometers which are used for a wide variety of analytical applications in chemistry and biochemistry as well as in chemical-process monitoring and control [6.132–6.135]. The potential of such instruments for the spectroscopy of semiconductors and other solids has been demonstrated [6.135, 6.136]. Figure 6.33 shows the Raman spectra of silicon (b) and diamond (c) measured with a resolution between 8 and 10 cm^{-1} using 5145 Å laser excitation (a) and an improved narrow-band AOTF operated in a quasi-collinear configuration [6.136]. In all spectra we used a 50× microscope objective to collect the signal and a holographic notch filter [6.119] to suppress the laser light (laser power: ≈ 20 mW, accumulation time per point: 2 s). The spectrum of Si was magnified 50 times before plotting. Note the side-lobes of the lines in Fig. 6.33 (a,c). They arise from the phase-matching condition for anisotropic Bragg diffraction and give an impression about the quality of the AOTF bandpass. From these results we expect that AOTF-based spectrometers will soon find widespread applications in semiconductor characterization and research which exploit the vibrational, luminescent, and linear optical properties of these materials.

Raman Imaging. An important topic where Raman and luminescence spectroscopy are used intensively is the imaging of physical and chemical properties over a sample region. While point-to-point mapping is applied in many cases, true imaging systems are presently gaining in importance [6.38, 6.39, 6.119]. In most Raman maps (see, for example, Fig. 6.10) ordinary wide-range spectra with the laser focused to about 1 μm are first recorded for a narrow raster of points using multichannel and micropositioning techniques [6.28]. The desired spectral information is then retrieved from fits to the lines of interest. This procedure can yield very precise information. However, it is rather slow and involves a large computational effort.

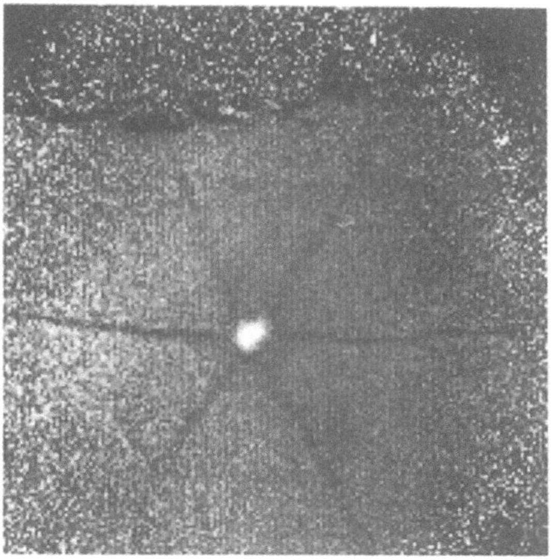

Fig. 6.34. Raman image of the A_1(LO) phonon intensity in wurtzite GaN over a $20 \times 20\,\mu m^2$ region. Dark (bright) areas indicate high (low) carrier concentration (from [6.27])

In Raman imaging the laser illuminates the whole sample area of interest, typically up to $100\,\mu m$ in diameter. The scattered light is sent through a passband filter before being projected onto a two-dimensional CCD detector. The resulting images directly show the desired signal intensity distribution. This concept is very fast and effective. It has significantly benefitted from the recent progress in filter technology, and very narrow dielectric, holographic, and acousto-optic filters are now used in Raman imaging instruments [6.38,6.39,6.119]. Various compact designs with holographic notch filters and AOTFs, often based on a microscope or fiber optic platform, have been described [6.125,6.128,6.132,6.134,6.135,6.137].

In order to illustrate the possibilities for Raman imaging in semiconductors, Fig. 6.34 shows the intensity distribution of the GaN A_1(LO) phonon over a hexagonal facet on the surface of a Si-doped crystal (see Sect. 6.1.1, Fig. 6.9) [6.27]. The intensity of this mode is strongly influenced by plasmon–phonon coupling [6.20,6.26,6.27]. In the bright area at the center the Raman signal is strong and the carrier concentration is low. Darker areas on the triangular side faces of the crystal indicate the presence of donor electrons, and the vanishing signal at the facet edges (dark lines, highest carrier concentration) reveals a preferential incorporation of the Si impurities at these locations during growth [6.27]. Raman imaging thus yields detailed information about parameters which are important for the growth of optoelectronic materials.

References

Chapter 1

1.1 A. Jayaraman and A. K. Ramdas, Physics Today **41**, 57 (1988).
1.2 G. Venkataraman, *Journey into Light*, (Indian Academy of Sciences, Bangalore, 1988).
1.3 A. Jayaraman, *C. V. Raman*, (Affiliated East-West Press, Madras, 1989)
1.4 *The Scientific Papers of Sir C. V. Raman*, Vol. 1 *Scattering of Light*, edited by the Indian Academy of Sciences, (Indian Academy of Sciences, Bangalore, 1978).
1.5 *The Scientific Papers of Sir C. V. Raman*, Vol. 1 *Scattering of Light*, Vol. 2 *Acoustics*, Vol. 3 *Optics*, Vol. 4 *Miscellaneous Papers, Colour, Optics of Minerals, Diamond*, Vol. 5 *Crystal Physics*, Vol. 6 *Colour and its Perception*, edited by the Indian Academy of Sciences, (Indian Academy of Sciences, Bangalore, 1988).
1.6 R. K. Chang and M. B. Long in *Light Scattering in Solids II*, edited by M. Cardona and G. Güntherodt, Vol. 50 of *Topics in Applied Physics*, (Springer-Verlag, Berlin, 1982), Chap. 3, p. 179.
1.7 J. C. Tsang in *Light Scattering in Solids V*, edited by M. Cardona and G. Güntherodt, Vol. 66 of *Topics in Applied Physics*, (Springer-Verlag, Berlin, 1989), Chap. 6, p. 233.
1.8 K. Ploog and G. Döhler, Adv. in Physics **32**, 285 (1983).
1.9 J. Faist, F. Capasso, D. L. Sivco, C. Sirtori, A. L. Hutchinson, and A. Y. Cho, Science **264**, 553 (1994).
1.10 A. Pinczuk and E. Burstein in *Light Scattering in Solids I*, edited by M. Cardona, Vol. 8 of *Topics in Applied Physics*, (Springer-Verlag, Berlin, 1982), Chap. 2, p. 23.
1.11 M. Cardona in *Light Scattering in Solids II*, edited by M. Cardona and G. Güntherodt, Vol. 50 of *Topics in Applied Physics*, (Springer-Verlag, Berlin, 1982), Chap. 2, p. 19.
1.12 M. Cardona, Surf. Sci. **37**, 100 (1973).
1.13 B. Jusserand and M. Cardona in *Light Scattering in Solids V*, edited by M. Cardona and G. Güntherodt, Vol. 66 of *Topics in Applied Physics*, (Springer-Verlag, Berlin, 1989), Chap. 3, p. 49.
1.14 G. Abstreiter, M. Cardona and A. Pinczuk in *Light Scattering in Solids IV*, edited by M. Cardona and G. Güntherodt, Vol. 54 of *Topics in Applied Physics*, (Springer-Verlag, Berlin, 1984), Chap. 2, p. 5.
1.15 A. Pinczuk and G. Abstreiter in *Light Scattering in Solids V*, edited by M. Cardona and G. Güntherodt, Vol. 66 of *Topics in Applied Physics*, (Springer-Verlag, Berlin, 1989), Chap. 4, p. 153.
1.16 J. Spitzer, I. Gregora, T. Ruf, M. Cardona, K. Ploog, F. Briones, and M. I. Alonso, Solid State Commun. **84**, 275 (1992).

1.17 J. Spitzer, T. Ruf, M. Cardona, W. Dondl, R. Schorer, G. Abstreiter, and E. E. Haller, Phys. Rev. Lett. **72**, 1565 (1994).
1.18 Z. Popović, J. Spitzer, T. Ruf, M. Cardona, R. Nötzel, and K. Ploog, Phys. Rev. B **48**, 1659 (1993).
1.19 V. F. Sapega, V. I. Belitsky, T. Ruf, H. D. Fuchs, M. Cardona, and K. Ploog, Phys. Rev. B **46**, 16005 (1992).
1.20 V. F. Sapega, V. I. Belitsky, A. J. Shields, T. Ruf, M. Cardona, and K. Ploog, Solid State Commun. **84**, 1039 (1992).
1.21 T. Ruf, V. I. Belitsky, J. Spitzer, V. F. Sapega, M. Cardona, and K. Ploog, Phys. Rev. Lett. **71**, 3035 (1993).
1.22 V. I. Belitsky, T. Ruf, J. Spitzer, and M. Cardona, Phys. Rev. B **49**, 8263 (1994).
1.23 T. Ruf, J. Spitzer, V. F. Sapega, V. I. Belitsky, M. Cardona, and K. Ploog, Phys. Rev. B **50**, 1792 (1994).
1.24 T. Ruf, J. Spitzer, V. F. Sapega, V. I. Belitsky, M. Cardona, and K. Ploog in *Festkörperprobleme/Advances in Solid State Physics*, Vol. 34, edited by R. Helbig, (Vieweg, Braunschweig/Wiesbaden, 1994), p. 237.
1.25 T. Ruf, J. Spitzer, M. Cardona, W. Braun, and K. Ploog, in *Proceedings of the 22nd International Conference on the Physics of Semiconductors*, edited by D. J. Lockwood, (World Scientific, Singapur, 1995), p. 923.
1.26 G. Goldoni, T. Ruf, V. F. Sapega, A. Fainstein, and M. Cardona, Phys. Rev. B **51**, 14542 (1995).
1.27 V. F. Sapega, T. Ruf, M. Cardona, H. T. Grahn, and K. Ploog, Phys. Rev. B **56**, 1041 (1997).
1.28 G. Ambrazevičius, M. Cardona, and R. Merlin, Phys. Rev. Lett. **59**, 700 (1987).
1.29 T. Ruf, A. Cantarero, M. Cardona, and U. Rössler, in *Proceedings of the 19th International Conference on the Physics of Semiconductors*, edited by W. Zawadzki, (Polish Academy of Sciences, Warsaw, 1988), p. 1473.
1.30 T. Ruf, R. T. Phillips, A. Cantarero, G. Ambrazevičius, M. Cardona, J. Schmitz, and U. Rössler, Phys. Rev. B **39**, 13378 (1989).
1.31 T. Ruf, C. Trallero-Giner, R. T. Phillips, and M. Cardona, Solid State Commun. **72**, 67 (1989).
1.32 C. Trallero-Giner, T. Ruf, and M. Cardona, Phys. Rev. B **41**, 3028 (1990).
1.33 T. Ruf, R. T. Phillips, C. Trallero-Giner, and M. Cardona, Phys. Rev. B **41**, 3039 (1990).
1.34 T. Ruf and M. Cardona, Phys. Rev. Lett. **63**, 2288 (1989).
1.35 T. Ruf and M. Cardona, Phys. Rev. B **41**, 10747 (1990).
1.36 F. Iikawa, T. Ruf, and M. Cardona, in *High Magnetic Fields in the Physics of Semiconductors*, edited by D. Heimann, (World Scientific, Singapore, 1995), p. 350.
1.37 V. López, F. Comas, C. Trallero-Giner, T. Ruf, and M. Cardona, Phys. Rev. B **54**, 10502 (1996).
1.38 S. I. Gubarev, T. Ruf, and M. Cardona, Phys. Rev. B **43**, 1551 (1991).
1.39 A. Cros, T. Ruf, J. Spitzer, M. Cardona, and A. Cantarero, Phys. Rev. B **50**, 2325 (1994).
1.40 F. Iikawa, T. Ruf, and M. Cardona, in *Proceedings of the 20th International Conference on the Physics of Semiconductors*, edited by E. M. Anastassakis and J. D. Joannopoulos, (World Scientific, Singapur, 1990), p. 2013.
1.41 F. Iikawa, T. Ruf, and M. Cardona, Phys. Rev. B **43**, 4849 (1991).
1.42 S. I. Gubarev, T. Ruf, M. Cardona, and K. Ploog, Solid State Commun. **85**, 853 (1993).

1.43 S. I. Gubarev, T. Ruf, M. Cardona, and K. Ploog, Phys. Rev. B **48**, 1647 (1993).
1.44 A. Fainstein, T. Ruf, M. Cardona, V. I. Belitsky, and A. Cantarero, Phys. Rev. B **51**, 7064 (1995).

Chapter 2

2.1 G. Bastard, *Wave mechanics applied to semiconductor heterostructures*, (Les éditions de physique, Les Ulis, 1988).
2.2 C. Weisbuch and B. Vinter, *Quantum semiconductor structures*, (Academic Press, Boston, 1991).
2.3 G. Bastard and J. Brum, IEEE J. of Quantum Electronics **22**, 1625 (1986).
2.4 C. Kittel, *Introduction to Solid State Physics*, (Wiley, New York, 1986).
2.5 T. Suemoto, G. Fasol, and K. Ploog, Phys. Rev. B **37**, 6397 (1988).
2.6 C. Colvard, T. A. Gant, M. V. Klein, R. Merlin, R. Fischer, H. Morkoç, and A. C. Gossard, Phys. Rev. B **31**, 2080 (1985).
2.7 M. V. Klein, IEEE J. of Quantum Electronics **22**, 1760 (1986).
2.8 M. Cardona, Superlattices and Microstructures **5**, 27 (1989).
2.9 M. Cardona, in *Surface Science*, edited by F. A. Ponce and M. Cardona, Vol. 62 of *Springer Proceedings in Physics*, (Springer-Verlag, Berlin, 1992), p. 319.
2.10 A. K. Sood, J. Menéndez, M. Cardona, and K. Ploog, Phys. Rev. Lett. **54**, 2111 (1985).
2.11 H. Rücker, E. Molinari, and P. Lugli, Phys. Rev. B **45**, 6747 (1992).
2.12 A. Fasolino, E. Molinari, and K. Kunc, Phys. Rev. Lett. **56**, 1751 (1986).
2.13 B. Jusserand and D. Paquet, Phys. Rev. Lett. **56**, 1752 (1986).
2.14 D. J. Mowbray, M. Cardona, and K. Ploog, Phys. Rev. B **43**, 1598 (1991).
2.15 M. Cardona, in *Optics of Excitons in Confined Systems*, edited by A. D. D'Andrea, R. Del Sole, R. Girlanda, and A. Quattropani, Vol. 123 of *Institute of Physics Conference Series*, (Institute of Physics, Bristol, 1992), p. 219.
2.16 S. M. Rytov, Akust. Zh. **2**, 71 (1956); [Sov. Phys. – Acous. **2**, 68 (1956)].
2.17 H. Brugger, H. Reiner, G. Abstreiter, H. Jorke, H. J. Herzog, and E. Kasper, Superlattices and Microstructures **2**, 451 (1986).
2.18 A. K. Sood, J. Menéndez, M. Cardona, and K. Ploog, Phys. Rev. Lett. **54**, 2115 (1985).
2.19 R. Fuchs and K. L. Kliewer, Phys. Rev. **140**, A2076 (1965).
2.20 J. J. Licari and R. Evrard, Phys. Rev. B **15**, 2254 (1977).
2.21 R. Lassnig, Phys. Rev. B **30**, 7132 (1984).
2.22 R. E. Camley and D. L. Mills, Phys. Rev. B **29**, 1695 (1984).
2.23 B. K. Ridley, Phys. Rev. B **39**, 5282 (1989).
2.24 C. Trallero-Giner, F. García-Moliner, V. R. Velasco, and M. Cardona, Phys. Rev. B **45**, 11944 (1992).
2.25 M. P. Chamberlain, M. Cardona, and B. K. Ridley, Phys. Rev. B **48**, 14356 (1993).
2.26 K. Huang and B. Zhu, Phys. Rev. B **38**, 13377 (1988).
2.27 H. Rücker, E. Molinari, and P. Lugli, Phys. Rev. B **44**, 3463 (1991).
2.28 Z. V. Popović, M. Cardona, E. Richter, D. Strauch, L. Tapfer, and K. Ploog, Phys. Rev. B **41**, 5904 (1990).
2.29 A. Huber, T. Egeler, W. Ettmüller, H. Rothfritz, G. Tränkle, and G. Abstreiter, Superlattices and Microstructures **9**, 309 (1991).

2.30 M. Zunke, R. Schorer, G. Abstreiter, W. Klein, G. Weimann, and M. P. Chamberlain, Solid State Commun. **93**, 847 (1995).
2.31 A. J. Shields, M. Cardona, and K. Eberl, Phys. Rev. Lett. **72**, 412 (1994).
2.32 A. J. Shields, M. P. Chamberlain, M. Cardona, and K. Eberl, Phys. Rev. B **51**, 17728 (1995).
2.33 M. Nakayama, K. Kubota, H. Kato, S. Chika, and N. Sano, Solid State Commun. **53**, 493 (1985).
2.34 M. Cardona, T. Suemoto, N. E. Christensen, T. Isu, and K. Ploog, Phys. Rev. B **36**, 5906 (1987).
2.35 Z. P. Wang, D. S. Jiang, and K. Ploog, Solid State Commun. **65**, 661 (1988).
2.36 G. Fasol, M. Tanaka, H. Sakaki, and Y. Horikoshi, Phys. Rev. B **38**, 6056 (1988).
2.37 S. Baroni, P. Giannozzi, and E. Molinari, Phys. Rev. B **41**, 3870 (1990).
2.38 E. Molinari, S. Baroni, P. Giannozzi, and S. de Gironcoli, Phys. Rev. B **45**, 4280 (1992).
2.39 P. Giannozzi, S. de Gironcoli, P. Pavone, and S. Baroni, Phys. Rev. B **43**, 7231 (1991).
2.40 S. Baroni, P. Pavone, P. Giannozzi, S. de Gironcoli, and E. Molinari, in *Semiconductor Superlattices and Interfaces*, edited by A. Stella and L. Miglio, Vol. 117 of *Proceedings of the International School of Physics "Enrico Fermi"*, (North-Holland, Amsterdam, 1993), p. 243.
2.41 S. Baroni, P. Giannozzi, and A. Testa, Phys. Rev. Lett. **58**, 1861 (1987).
2.42 A. Ishibashi, M. Itabashi, Y. Mori, K. Kawado, and N. Watanabe, Phys. Rev. B **33**, 2887 (1986).
2.43 T. Toriyama, N. Kobayashi, and Y. Horikoshi, Jpn. J. Appl. Phys. **25**, 1895 (1986).
2.44 B. Jusserand, F. Mollot, R. Planel, E. Molinari, and S. Baroni, Surf. Science **267**, 171 (1992).
2.45 F. Briones and A. Ruiz, J. Crystal Growth **111**, 194 (1991).
2.46 R. Cingolani, K. Ploog, L. Baldassarre, M. Ferrara, M. Lugará, and C. Moro, Appl. Phys. A **50**, 189 (1990).
2.47 D. Strauch and B. Dorner, J. Phys.: Condens. Mat. **2**, 1457 (1990).
2.48 M. Cardona, P. Etchegoin, H. D. Fuchs, and P. Molinàs-Mata, J. Phys.: Condens. Matter **5**, A61 (1993).
2.49 M. Cardona, in *Festkörperprobleme/Advances in Solid State Physics*, Vol. 34, edited by R. Helbig, (Vieweg, Braunschweig/Wiesbaden, 1994), p. 35.
2.50 H. D. Fuchs, C. H. Grein, C. Thomsen, M. Cardona, W. L. Hansen, E. E. Haller, and K. Itoh, Phys. Rev. B **43**, 4835 (1991).
2.51 H. D. Fuchs, C. H. Grein, R. I. Devlen, J. Kuhl, and M. Cardona, Phys. Rev. B **44**, 8633 (1991).
2.52 H. D. Fuchs, C. H. Grein, M. Bauer, and M. Cardona, Phys. Rev. B **45**, 4065 (1992).
2.53 H. D. Fuchs, P. Etchegoin, M. Cardona, K. Itoh, and E. E. Haller, Phys. Rev. Lett. **70**, 1715 (1993).
2.54 H. D. Fuchs, P. Molinàs-Mata, and M. Cardona, Superlattices and Microstructures **13**, 447 (1993).
2.55 J. Spitzer, P. Etchegoin, M. Cardona, T. R. Anthony, and W. F. Banholzer, Solid State Commun. **88**, 509 (1993).
2.56 T. Strach, T. Ruf, E. Schönherr, and M. Cardona, Phys. Rev. B **51**, 16460 (1995).
2.57 A. Göbel, T. Ruf, M. Cardona, K. Tötemeyer, and K. Eberl, unpublished (1995).

2.58 A. Göbel, T. Ruf, M. Cardona, C. Lin, K. Matsumoto, R. Lauck, and E. Schönherr, unpublished (1995).
2.59 A. K. Ramdas, Solid State Commun. **96**, 111 (1995).
2.60 E. E. Haller, J. Appl. Phys. **77**, 2857 (1995).
2.61 T. Ruf, H. D. Fuchs, and M. Cardona, Phys. Bl. **52**, 1115 (1996).
2.62 J. Emsley, *The Elements*, (Clarendon Press, Oxford, 1991); see also: *CRC Handbook of Chemistry and Physics*, edited by R. C. Weast, 70th Edition, (CRC Press, Boca Raton, 1989).
2.63 P. Etchegoin, J. Weber, M. Cardona, W. L. Hansen, K. Itoh, and E. E. Haller, Solid State Commun. **83**, 843 (1992).
2.64 A. T. Collins, S. C. Lawson, G. Davies, and H. Kanda, Phys. Rev. Lett. **65**, 891 (1990).
2.65 H. V. Klapdor-Kleingrothaus, Nucl. Phys. B **28A**, 207 (1992).
2.66 R. Moreh, O. Beck, I. Bauske, W. Geiger, R. D. Heil, U. Kneissl, J. Margraf, H. Maser, and H. H. Pitz, Phys. Rev. C **48**, 2625 (1993).
2.67 O. Beck, U. Kneissl, T. Ruf, and M. Cardona, to be published (1997).
2.68 Yu. V. Shvyd'ko, private communication (1994); A. I. Chumakov, G. V. Smirnov, A. Q. R. Baron, J. Arthur, D. E. Brown, S. L. Ruby, G. S. Brown, and N. N. Salashchenko, Phys. Rev. Lett. **71**, 2489 (1993).
2.69 J. M. Rowe, R. M. Nicklow, D. L. Price, and K. Zanio, Phys. Rev. B **10**, 671 (1974).
2.70 A. Debernardi, N. M. Pyka, A. Göbel, T. Ruf, R. Lauck, S. Kramp, and M. Cardona, Solid State Commun. **103**, 297 (1997).
2.71 I. Broser and K.-H. Franke, J. Phys. Chem. Sol. **26**, 1013 (1965).
2.72 R. Magerle, A. Burchard, M. Deicher, T. Kerle, W. Pfeiffer, and E. Recknagel, Phys. Rev. Lett. **75**, 1594 (1995).
2.73 G. Davies, Physica Scripta **T66**, 113 (1996).
2.74 H. D. Fuchs, *Spektroskopische Untersuchungen von ungeordneten und niederdimensionalen Halbleitern*, Thesis, Universität Stuttgart, (1993).
2.75 A. Ishibashi, in *Spectroscopy of Semiconductor Microstructures*, edited by G. Fasol, A. Fasolino, and P. Lugli, Vol. 206 of *NATO Advanced Study Institute, Series B: Physics*, (Plenum, New York, 1989), p. 21.
2.76 E. E. Haller, Semicond. Sci. and Technol. **5**, 319 (1990).
2.77 J. Spitzer, *Untersuchung der Grenzflächeneigenschaften von Halbleiter-Übergittern mit Ramanstreuung*, Thesis, Universität Stuttgart, (1994).
2.78 P. Molinàs-Mata, and M. Cardona, Phys. Rev. B **43**, 9799 (1991).
2.79 P. Molinàs-Mata, A. J. Shields, and M. Cardona, Phys. Rev. B **47**, 1866 (1993).
2.80 W. Weber, Phys. Rev. Lett. **33**, 371 (1974).
2.81 W. Weber, Phys. Rev. B **15**, 4789 (1977).
2.82 P. Etchegoin, H. D. Fuchs, J. Weber, M. Cardona, L. Pintschovius, N. Pyka, K. Itoh, and E. E. Haller, Phys. Rev. B **48**, 12661 (1993).
2.83 S. Go, H. Bilz, and M. Cardona, Phys. Rev. Lett. **34**, 580 (1975).
2.84 B. Jusserand, D. Paquet, and A. Regrency, Superlattices and Microstructures **1**, 61 (1985).
2.85 R. Schorer, *Ramanstreuung an α-Sn-, Si- und Ge-Übergittern*, Thesis, Technische Universität München, (1994).
2.86 E. Silveria, W. Dondl, G. Abstreiter, and E. E. Haller, Phys. Rev. B **56**, 2062 (1997).
2.87 A. Magerl, private communication (1996).

2.88 J. Spitzer, T. Ruf, M. Cardona, C. Grein, W. Dondl, R. Schorer, G. Abstreiter, and E. E. Haller, in *Proceedings of the 22nd International Conference on the Physics of Semiconductors*, edited by D. J. Lockwood, (World Scientific, Singapore, 1995), p. 971.
2.89 M. I. Bansal, A. K. Sood, and M. Cardona, Solid State Commun. **78**, 579 (1991).
2.90 D. J. Mowbray, H. Fuchs, D. W. Niles, M. Cardona, C. Thomsen, and B. Friedl, in *Proceedings of the 20th International Conference on the Physics of Semiconductors*, edited by J. D. Joannopoulos and E. M. Anastassakis, (World Scientific, Singapore, 1990), p. 2017.
2.91 K. C. Hass, M. A. Tamor, T. R. Anthony, and W. F. Banholzer, Phys. Rev. B **44**, 12046 (1991).
2.92 K. C. Hass, M. A. Tamor, T. R. Anthony, and W. F. Banholzer, Phys. Rev. B **45**, 7171 (1992).
2.93 M. P. D'Evelyn, C. J. Chu, R. H. Hange, and J. L. Margrave, J. Appl. Phys. **71**, 1528 (1992).
2.94 A. K. Ramdas, S. Rodriguez, M. Grimsditch, T. R. Anthony, and W. F. Banholzer, Phys. Rev. Lett. **71**, 189 (1993).
2.95 R. Vogelgesang, A. K. Ramdas, S. Rodriguez, M. Grimsditch, and T. R. Anthony, Phys. Rev. B **54**, 3989 (1996).
2.96 T. Y. Tan, H. M. You, S. Yu, U. M. Goesele, W. Jäger, D. W. Boeringer, F. Zypman, R. Tsu, and S.-T. Lee, J. Appl. Phys. **72**, 5206 (1992).
2.97 A. Göbel, T. Ruf, M. Cardona, and C. Lin, in *Proceedings of the 4th International Conference on Phonon Physics and the 8th International Conference on Phonon Scattering in Condensed Matter*, Physica B **219+220**, 511 (1996).
2.98 C. T. Lin, E. Schönherr, A. Schmeding, T. Ruf, A. Göbel, and M. Cardona J. Crystal Growth **167**, 612 (1996).
2.99 A. Göbel, T. Ruf, M. Cardona, C. T. Lin, and J. C. Merle, Phys. Rev. Lett. **77**, 2591 (1996).
2.100 A. Göbel, T. Ruf, C. T. Lin, M. Cardona, J. C. Merle, and M. Joucla, Phys. Rev. B **56**, 210 (1997).
2.101 N. Garro, A. Cantarero, M. Cardona, T. Ruf, A. Göbel, C. Lin, K. Reimann, S. Rübenacke, and M. Steube, Solid State Commun. **98**, 27 (1996).
2.102 T. Ruf, A. Göbel, M. Cardona, N. Garro, A. Cantarero, J. Wrzesinski, K. Reimann, S. Rübenacke, M. Steube, C. Lin, K. Eberl, and J. C. Merle, *15th General Conference of the Condensed Matter Division of the European Physical Society*, Baveno-Stresa, unpublished (1996).
2.103 T. Ruf, A. Göbel, M. Cardona, C. Lin, J. Wrzesinski, M. Steube, K. Reimann, N. Garro, A. Cantarero, and J. C. Merle, in *Proceedings of the 23rd International Conference on the Physics of Semiconductors*, edited by M. Scheffler and R. Zimmermann, (World Scientific, Singapore, 1996), p. 185.
2.104 J. M. Zhang, T. Ruf, A. Göbel, R. Lauck, and M. Cardona, in *Proceedings of the 23rd International Conference on the Physics of Semiconductors*, edited by M. Scheffler and R. Zimmermann, (World Scientific, Singapore, 1996), p. 201.
2.105 N. Garro, A. Cantarero, M. Cardona, A. Göbel, T. Ruf, and K. Eberl, in *Highlights of Light Spectroscopy on Semiconductors*, edited by A. D'Andrea, L. G. Quagliano, and S. Selci, (World Scientific, Singapore, 1996), p. 21.
2.106 N. Garro, A. Cantarero, M. Cardona, A. Göbel, T. Ruf, and K. Eberl, Phys. Rev. B **54**, 4732 (1996).
2.107 A. Debernardi and M. Cardona, Phys. Rev. B **54**, 11305 (1996).
2.108 L. Wang, L. Hsu, E. E. Haller, J. W. Erickson, A. Fischer, K. Eberl, and M. Cardona, Phys. Rev. Lett. **76**, 2342 (1996).

2.109 L. Wang, J. A. Wolk, L. Hsu, E. E. Haller, J. W. Erickson, M. Cardona, T. Ruf, J. P. Silveira, and F. Briones, Appl. Phys. Lett. **70**, 1831 (1997).
2.110 C. Thomsen, in *Light Scattering in Solids VI*, Vol. 68 of *Topics in Applied Physics*, edited by M. Cardona and G. Güntherodt, (Springer-Verlag, Berlin, 1991), p. 285.
2.111 E. Liarokapis, N. Poulakis, D. Palles, K. Conder, E. Kaldis, and K. A. Müller, in *Proceedings of the 4th International Conference on Phonon Physics and the 8th International Conference on Phonon Scattering in Condensed Matter*, Physica B **219+220**, 139 (1996).
2.112 V. G. Ivanov, M. N. Iliev, and C. Thomsen, Phys. Rev. B **52**, 13652 (1995).
2.113 E. T. Heyen, R. Wegerer, and M. Cardona, Phys. Rev. Lett. **44**, 10195 (1991).
2.114 T. Ruf, in *Proceedings of the 4th International Conference on Phonon Physics and the 8th International Conference on Phonon Scattering in Condensed Matter*, Physica B **219+220**, 132 (1996).
2.115 R. Henn, T. Strach, E. Schönherr, and M. Cardona, Phys. Rev. B **55**, 3285 (1996).
2.116 T. Strach, T. Ruf, M. Cardona, C. T. Lin, S. Jandl, V. Nekvasil, D. I. Zhigunov, S. N. Barilo, and S. V. Shiryaev, Phys. Rev. B **54**, 4276 (1996).
2.117 S. Jandl, M. Iliev, C. Thomsen, T. Ruf, M. Cardona, B. M. Wanklyn, and C. Chen, Solid State Commun. **87**, 609 (1993).
2.118 F. I. Kreĭngol'd, K. F. Lider, and L. E. Solov'ev, Pis'ma Zh. Eksp. Teor. Fiz. **23**, 679 (1976); [JETP Lett. **22**, 624 (1976)].
2.119 F. I. Kreĭngol'd, K. F. Lider, and V. F. Sapega, Fiz. Tverd. Tela **19**, 3158 (1977); [Sov. Phys. Solid State **19**, 1849 (1977)].
2.120 A. I. Bobrysheva, I. I. Jeru, and S. A. Moskalenko, phys. stat. sol. **113**, 439 (1982).
2.121 F. I. Kreĭngol'd, Fiz. Tverd. Tela **27**, 2839 (1985); [Sov. Phys. Solid State **27**, 1712 (1985)].
2.122 J. Menéndez, J. B. Page, and S. Guha, Philosophical Magazine B **70**, 651 (1994).
2.123 B. Eckert, R. Bini, H. J. Jodl, and S. Califano, J. Chem. Phys. **100**, 912 (1994).
2.124 J. Wagner, J. Schmitz, R. C. Newman, and C. Roberts, J. Raman Spectrosc. **27**, 231 (1996).
2.125 M. D. Sciacca, A. J. Mayur, N. Shin, I. Miotkowski, A. K. Ramdas, and S. Rodriguez, Phys. Rev. B **51**, 6971 (1995).
2.126 S. H. Kwok, R. Merlin, W. Q. Li, and P. K. Bhattacharya, J. Appl. Phys. **72**, 285 (1992).
2.127 L. T. P. Allen, E. R. Weber, J. Washburn, and Y. C. Pao, Appl. Phys. Lett. **51**, 670 (1987).
2.128 D. Kirillov and Y. C. Pao, in *Epitaxy of Semiconductor Layer Structures*, Mat. Res. Soc. Symp. Proc. No. 102, edited by R. T. Tung, L. R. Dawson, and R. L. Gunshor, (MRS, Pittsburgh, 1988), p. 169.
2.129 Z. V. Popović, M. Cardona, E. Richter, D. Strauch, L. Tapfer, and K. Ploog, Phys. Rev. B **40**, 1207 (1989).
2.130 Z. V. Popović, M. Cardona, E. Richter, D. Strauch, L. Tapfer, and K. Ploog, Phys. Rev. B **40**, 3040 (1989).
2.131 T. Hayakawa, K. Takahashi, M. Kondo, T. Suyama, S. Yamamoto, and T. Hijikata, Phys. Rev. Lett. **60**, 349 (1988).
2.132 L. W. Molenkamp, G. E. W. Bauer, R. Eppenga, and C. T. Foxon, Phys. Rev. B **38**, 6147 (1988).
2.133 T. Fukunaga, T. Takamori, and H. Nakashima, J. Crystal Growth **81**, 85 (1987).

2.134 S. Subbana, H. Kroemer, and J. L. Merz, J. Appl. Phys. **59**, 488 (1986).
2.135 R. Nötzel, N. N. Ledentsov, L. Däweritz, M. Hohenstein, and K. Ploog, Phys. Rev. Lett. **67**, 3812 (1991).
2.136 R. Nötzel, N. N. Ledentsov, L. Däweritz, K. Ploog, and M. Hohenstein, Phys. Rev. B **45**, 3507 (1992).
2.137 R. Nötzel, L. Däweritz, and K. Ploog, Phys. Rev. B **46**, 4736 (1992).
2.138 Z. V. Popović, M. Cardona, L. Tapfer, K. Ploog, E. Richter, and D. Strauch, Appl. Phys. Lett. **54**, 846 (1989).
2.139 Z. V. Popović, H. J. Trodahl, M. Cardona, E. Richter, D. Strauch, and K. Ploog, Phys. Rev. B **40**, 1202 (1989).
2.140 Z. V. Popović, M. Cardona, E. Richter, D. Strauch, L. Tapfer, and K. Ploog, Phys. Rev. B **43**, 4925 (1991).
2.141 F. Calle, D. J. Mowbray, D. W. Niles, M. Cardona, J. M. Calleja, and K. Ploog, Phys. Rev. B **43**, 9152 (1991).
2.142 Z. V. Popović, E. Richter, J. Spitzer, M. Cardona, A. J. Shields, R. Nötzel, and K. Ploog, Phys. Rev. B **49**, 7577 (1994).
2.143 A. J. Shields, Z. V. Popović, M. Cardona, J. Spitzer, R. Nötzel, and K. Ploog, Phys. Rev. B **49**, 7584 (1994).
2.144 J. Spitzer, Z. V. Popović, T. Ruf, M. Cardona, R. Nötzel, and K. Ploog, Solid-State Electronics **37**, 753 (1994).
2.145 A. B. Talochkin, V. A. Markov, I. G. Neizvestnyĭ, O. P. Pchelyakov, M. P. Sinyukov, and S. I. Stenin, Pis'ma Zh. Eksp. Teor. Fiz. **50**, 21 (1989), [JETP Lett. **50**, 24 (1989)].
2.146 A. B. Talochkin, V. A. Markov, Yu. A. Pusep, O. P. Pchelyakov, and M. P. Sinyukov, Superlattices and Microstructures **10**, 179 (1991).
2.147 B. A. Auld, *Acoustic Fields and Waves in Solids*, (Wiley, New York, 1973).
2.148 Z. V. Popović, Univ. Beograd, Publ. Elektrotechn. Fak. Ser. Tek. Fiz. **1**, 5 (1992).
2.149 S. Tamura and J. P. Wolfe, Phys. Rev. B **35**, 2528 (1987).
2.150 S. Tamura, D. C. Hurley, and J. P. Wolfe, Phys. Rev. B **38**, 1427 (1988).
2.151 P. V. Santos, L. Ley, J. Mebert, and O. Koblinger, Phys. Rev. B **36**, 4858 (1987).
2.152 F. Calle, M. Cardona, E. Richter, and D. Strauch, Solid State Commun. **72**, 1153 (1989).
2.153 *Physics of Group IV Elements and III-V Compounds*, edited by O. Madelung, M. Schulz, and H. Weiss, Landolt-Börnstein, New Series, Group III, Vol. 17a (Springer-Verlag, Berlin, 1982); *Intrinsic Porperties of Group IV Elements and III-V, II-VI, and I-VII Compounds*, edited by O. Madelung and M. Schulz, Landolt-Börnstein, New Series, Group III, Vol. 22a (Springer-Verlag, Berlin, 1987).
2.154 B. Jusserand, D. Paquet, F. Mollot, F. Alexandre, and G. Le Roux, Phys. Rev. B **35**, 2808 (1987).
2.155 J. Sapriel, J. He, B. Djafari-Rouhani, R. Azoulay, and F. Mollot, Phys. Rev. B **37**, 4099 (1988).
2.156 J. He, B. Djafari-Rouhani, and J. Sapriel, Phys. Rev. B **37**, 4086 (1988).
2.157 J. He, J. Sapriel, and H. Brugger, Phys. Rev. B **39**, 5919 (1989).
2.158 P. Etchegoin, J. Kircher, M. Cardona, C. Grein, and E. Bustarret, Phys. Rev. B **46**, 15139 (1992).
2.159 S.-Y. Ren and W. A. Harrison, Phys. Rev. B **23**, 762 (1981).
2.160 Note that the values of p_{44} used here are smaller than those in [1.18, 2.144] by a factor of two. It arises from different definitions for the relation between the change in the dielectric function and the piezo-optic tensor (see (2.27))

used in the stress-dependent measurements of [2.158] (Eq. (6)) and the photoelastic mechanism of Brillouin scattering. The π_{44} data in Fig. 3 (c) of [2.158] thus have to be divided by two when used to derive the elasto-optic constants p_{44}. The values of p_{11} and p_{12} used here are from Fig. 4 in [2.158] (see *Note added in proof* of P. Etchegoin et al., Phys. Rev. B **47**, 10292 (1993)). In [2.158] Fig. 4 (a) is the calculated difference of the spectra in Figs. 3 (a,b), while Fig. 4 (b) is from a measurement with stress along [111] (P. Etchegoin, private commun., 1997).
2.161 F. Canal, M. Grimsditch, and M. Cardona, Solid State Commun. **29**, 523 (1979).
2.162 K. Strössner, S. Ves, and M. Cardona, Phys. Rev. B **32**, 6614 (1985).
2.163 The improvement in the agreement between the calculation and the experiment which is obtained in [1.18] and [2.144] when the elasto-optic constants of GaP are used instead of the unknown ones for AlAs is due to a computational error. The Brillouin tensor of the QT [$\bar{2}$33]-mode in Eq. (6) of [1.18] has to be replaced by that given in Table 2.3.

Chapter 3

3.1 J. Wagner, J. Schmitz, M. Maier, J. D. Ralston, and P. Koidl, Solid-State Electronics **37**, 1037 (1994).
3.2 R. Schorer, G. Abstreiter, H. Kibbel, H. Presting, C. Tserbak, and G. Theodorou, Solid State Commun. **93**, 1025 (1995).
3.3 F. Agulló-Rueda, E. E. Mendez, and J. M. Hong, Phys. Rev. B **38**, 12720 (1988).
3.4 H. Schneider, J. Wagner, K. Fujiwara, and K. Ploog, Phys. Rev. B **42**, 11430 (1990).
3.5 A. J. Shields, C. Trallero-Giner, M. Cardona, H. T. Grahn, K. Ploog, V. A. Haisler, D. A. Tenne, N. T. Moshegov, and A. I. Toropov, Phys. Rev. B **46**, 6990 (1992).
3.6 F. Cerdeira, E. Anastassakis, W. Kauschke, and M. Cardona, Phys. Rev. Lett. **57**, 3209 (1986).
3.7 A. Alexandrou, and M. Cardona, Solid State Commun. **64**, 1029 (1987).
3.8 A. Alexandrou, Y. Pusep, and M. Cardona, Phys. Rev. B **39**, 8308 (1989).
3.9 R. C. Miller, D. A. Kleinman, and A. C. Gossard, Solid State Commun. **60**, 213 (1986).
3.10 R. C. Miller, D. A. Kleinman, C. W. Tu, and S. K. Sputz, Phys. Rev. B **34**, 7444 (1986).
3.11 D. A. Kleinman, R. C. Miller, and A. C. Gossard, Phys. Rev. B **35**, 664 (1987).
3.12 A. Alexandrou, M. Cardona, and K. Ploog, Phys. Rev. B **38**, 2196 (1988).
3.13 D. J. Mowbray, unpublished (1990); see also M. Cardona, in *Proceedings of the 22nd International Conference on the Physics of Semiconductors*, edited by D. J. Lockwood, (World Scientific, Singapore, 1995), p. 3.
3.14 D. N. Mirlin, I. A. Merkulov, V. I. Perel', I. I. Reshina, A. A. Sirenko, and R. Planel, Solid State Commun. **82**, 305 (1992).
3.15 T. Ruf, V. I. Belitsky, J. Spitzer, V. F. Sapega, M. Cardona, and K. Ploog, Solid-State Electronics **37**, 609 (1994).
3.16 P. X. Zhang, D. J. Lockwood, H. J. Labbé, and J.-M. Baribeau, Phys. Rev. B **46**, 9881 (1992).

3.17 P. X. Zhang, D. J. Lockwood, and J.-M. Baribeau, Can. J. Phys. **70**, 843 (1992).
3.18 M. Giehler, T. Ruf, M. Cardona, and K. Ploog, Phys. Rev. B **55**, 7124 (1997).
3.19 D. N. Mirlin, I. A. Merkulov, V. I. Perel', I. I. Reshina, A. A. Sirenko, and R. Planel, Solid State Commun. **84**, 1093 (1992).
3.20 C. Kittel, *Quantum Theory of Solids*, (Wiley, New York, 1963).
3.21 A. A. Gogolin and E. I. Rashba, Solid State Commun. **19**, 1177 (1976).
3.22 J. Sapriel, J. Chavignon, F. Alexandre, and R. Azoulay, Phys. Rev. B **34**, 7118 (1986).
3.23 A. B. Talochkin, B. A. Markov, and M. P. Sinyukov, in *Proceedings of the 13th International Conference on Raman Spectroscopy*, edited by W. Kiefer, M. Cardona, G. Schaack, F. W. Schneider, and H. W. Schrötter, Post-Deadline Papers, p. 178, unpublished (1992).
3.24 R. Merlin, Philosophical Magazine B **70**, 761 (1994).
3.25 R. Merlin, K. Bajema, R. Clarke, F.-Y. Juang, and P. K. Bhattacharya, Phys. Rev. Lett. **55**, 1768 (1985).
3.26 K. Bajema and R. Merlin, Phys. Rev. B **36**, 4555 (1987).
3.27 R. Merlin, in *Light Scattering in Solids V*, edited by M. Cardona and G. Güntherodt, Vol. 66 of *Topics in Applied Physics*, (Springer-Verlag, Berlin, 1989), Chap. 5, p. 214.
3.28 M. A. Herman, D. Bimberg, and J. Christen, J. Appl. Phys. **70**, R1 (1991).
3.29 C. Weisbuch, R. Dingle, A. C. Gossard, and W. Wiegmann, Solid State Commun. **38**, 709 (1981).
3.30 D. Paquet, Superlattices and Microstructures **2**, 429 (1986).
3.31 M. Tanaka and H. Sakaki, J. Crystal Growth **81**, 153 (1987).
3.32 M. Maaref, F. F. Charfi, D. Scalbert, C. Benoit à la Guillaume, R. Planel, and G. Le Roux, Solid State Commun. **84**, 511 (1992).
3.33 K. Brunner, G. Abstreiter, G. Böhm, G. Tränkle, and G. Weimann, Appl. Phys. Lett. **64**, 3320 (1994).
3.34 A. Ourmazd, D. W. Taylor, J. Cunningham, and C. W. Tu, Phys. Rev. Lett. **62**, 933 (1989).
3.35 L. Däweritz and K. Ploog, Semicond. Sci. and Technol. **9**, 123 (1993).
3.36 P. Ils, J. Kraus, G. Schaack, G. Weimann, and W. Schlapp, J. Appl. Phys. **70**, 5587 (1991).
3.37 B. G. Orr, M. D. Johnson, C. Orme, J. Sudijono, and A. W. Hunt, Solid-State Electronics **37**, 1057 (1994).
3.38 R. F. Kopf, E. F. Schubert, T. D. Harris, and R. S. Becker, Appl. Phys. Lett. **58**, 631 (1991).
3.39 D. J. Mowbray, M. Cardona, and K. Ploog, Phys. Rev. B **43**, 11815 (1991).
3.40 J. Feldmann, J. Nunnenkamp, G. Peter, E. O. Göbel, J. Kuhl, K. Ploog, P. Dawson, and C. T. Foxon, Phys. Rev. B **42**, 5809 (1990).
3.41 E. Hecht and A. Zajac, *Optics*, (Addison-Wesley, Reading, 1974), p. 415.
3.42 B. P. Zakharchenya, P. S. Kop'ev, D. N. Mirlin, D. G. Polyakov, I. I. Reshina, V. F. Sapega, and A. A. Sirenko, Solid State Commun. **69**, 203 (1990).
3.43 P. S. Kop'ev, D. N. Mirlin, D. G. Polyakov, I. I. Reshina, V. F. Sapega, and A. A. Sirenko, Fiz. Tekh. Poluprovodn. **24**, 1200 (1990), [Sov. Phys. Semicond. **24**, 757 (1990)].
3.44 P. Lautenschlager, P. B. Allen, and M. Cardona, Phys. Rev. B **33**, 5501 (1986).
3.45 S. Gopalan, P. Lautenschlager, and M. Cardona, Phys. Rev. B **35**, 5577 (1987).
3.46 S. Rudin, T. L. Reinecke, and B. Segall, Phys. Rev. B **42**, 11218 (1990).
3.47 T. Takagahara, Phys. Rev. B **31**, 6552 (1985).

3.48 J. Lee, E. S. Koteles, and M. O. Vassell, Phys. Rev. B **33**, 5512 (1986).
3.49 S. Rudin and T. L. Reinecke, Phys. Rev. B **41**, 3017 (1990).
3.50 O. J. Glembocki, B. V. Shanabrook, and W. T. Beard, Surface Science **174**, 206 (1986).
3.51 S. Selci, A. Cricenti, M. Righini, C. Petrillo, F. Sacchetti, F. Alexandre, and G. Chiarotti, Solid State Commun. **79**, 561 (1991).
3.52 H. Quiang, F. H. Pollak, C. M. Sotomayor Torres, W. Leitch, A. H. Kean, M. A. Stroscio, G. J. Iafrate, and K. W. Kim, Appl. Phys. Lett. **61**, 1411 (1992).
3.53 F. H. Pollak, in *Phonons in Semiconductor Nanostructures*, Vol. 236 of NATO ASI Series E, edited by J.-P. Leburton, J. Pascual, and C. Sotomayor Torres, (Kluwer, Dordrecht, 1993), p. 341.
3.54 N. T. Pelekanos, J. Ding, M. Hagerott, V. A. Nurmikko, H. Luo, N. Samarth, and J. K. Furdyna, Phys. Rev. B **45**, 6037 (1992).
3.55 A. Tredicucci, Y. Chen, F. Bassani, J. Massies, C. Deparis, and G. Neu, Phys. Rev. B **47**, 10348 (1993).
3.56 D. Gammon, S. Rudin, T. L. Reinecke, D. S. Katzer and C. S. Kyono, Phys. Rev. B **51**, 16785 (1995).
3.57 Y. Chen, G. P. Kothiyal, J. Singh, and P. K. Battacharya, Superlattices and Microstructures **3**, 657 (1987).
3.58 V. Srinivas, J. Hryniewicz, Y. J. Chen, and C. E. C. Wood, Phys. Rev. B **46**, 10193 (1992).
3.59 J. P. Doran, J. F. Donegan, J. Hegarty, R. D. Feldman, and R. F. Austin, Solid State Commun. **81**, 801 (1992).
3.60 L. Schultheis, A. Honold, J. Kuhl, K. Köhler, and C. W. Tu, Phys. Rev. B **34**, 9027 (1986).
3.61 J. Kuhl, A. Honold, L. Schultheis, and C. W. Tu, in *Festkörperprobleme/ Advances in Solid State Physics*, Vol. 29, edited by U. Rössler, (Vieweg, Braunschweig, 1989), p. 157.
3.62 J. Feldmann, G. Peter, E. O. Göbel, P. Dawson, K. Moore, C. Foxon, and R. J. Elliott, Phys. Rev. Lett. **59**, 2337 (1987).
3.63 V. L. Alperovich, V. M. Zaletin, A. F. Kravchenko, and A. S. Terekhov, phys. stat. solidi (b) **77**, 465 (1976).
3.64 D. A. B. Miller, D. S. Chemla, D. J. Eilenberger, P. W. Smith, A. C. Gossard, and W. T. Tsang, Appl. Phys. Lett. **41**, 679 (1982).
3.65 J. Humlíček, E. Schmidt, L. Bočánek, R. Švehla, and K. Ploog, Phys. Rev. B **48**, 5241 (1993).
3.66 This value of Γ_{ep} is often quoted as being experimental. However, as has been pointed out in Ref. 11 of [3.56], it actually comes from a calculation.
3.67 The large theoretical value for Γ_{ep} in [3.46] was later attributed to a numerical error. See Ref. 22 of [3.56].
3.68 T. Hiroshima, Solid State Commun. **68**, 483 (1988).
3.69 K. T. Tsen, K. R. Wald, T. Ruf, P. Y. Yu, and H. Morkoç, Phys. Rev. Lett. **67**, 2557 (1991).
3.70 F. Ancilotto, A. Fasolino, and J. C. Maan, Phys. Rev. B **38**, 1788 (1988).
3.71 U. Rössler, Solid State Comm. **49**, 943 (1984); M. Braun, and U. Rössler, J. Phys. C **18**, 3365 (1985).
3.72 U. Ekenberg, Phys. Rev. B **40**, 7714 (1989).
3.73 R. J. Warburton, J. G. Michels, R. J. Nicholas, J. J. Harris, and C. T. Foxon, Phys. Rev. B **46**, 13394 (1992).
3.74 J. M. Luttinger, Phys. Rev. **102**, 1030 (1956).
3.75 O. Akimoto and H. Hasegawa, J. Phys. Soc. Japan **22**, 181 (1967).

3.76 C. Bosio, J. L. Staelhi, M. Guzzi, G. Burri, and R. A. Logan, Phys. Rev. B **38**, 3263 (1988).
3.77 E. T. Yu, J. O. McCaldin, and T. C. McGill, Solid State Phys. **46**, 1 (1992).
3.78 L. C. Andreani, and A. Pasquarello, Phys. Rev. B **42**, 8928 (1990).
3.79 G. C. La Rocca and M. Cardona, phys. stat. solidi (b) **167**, 115 (1991).
3.80 M. Cardona, N. E. Chistensen, and G. Fasol, Phys. Rev. B **38**, 1806 (1988).
3.81 P. V. Santos and M. Cardona, Phys. Rev. Lett. **72**, 432 (1994).
3.82 D. E. Aspnes and M. Cardona, Phys. Rev. B **17**, 726 (1978).
3.83 D. E. Aspnes and M. Cardona, Phys. Rev. B **17**, 741 (1978).
3.84 M. Hünermann, W. Richter, J. Saalmüller, and E. Anastassakis, Phys. Rev. B **34**, 5381 (1986).

Chapter 4

4.1 R. L. Aggarwal, in *Modulation Techniques*, Vol. 9 of *Semiconductors and Semimetals*, edited by R. K. Willardson and A. C. Beer, (Academic, New York, 1972), p. 151.
4.2 N. J. Pulsford, R. J. Nicholas, P. Dawson, K. J. Moore, G. Duggan, and C. T. B. Foxon, Phys. Rev. Lett. **20**, 2284 (1989).
4.3 E. O. Kane, J. Phys. Chem. Solids **1**, 249 (1957).
4.4 C. R. Pidgeon and R. N. Brown, Phys. Rev. **146**, 575 (1966).
4.5 K. Suzuki and J. C. Hensel, Phys. Rev. B **9**, 4184 (1974).
4.6 J. C. Hensel and K. Suzuki, Phys. Rev. B **9**, 4219 (1974).
4.7 H.-R. Trebin, U. Rössler, and R. Ranvaud, Phys. Rev. B **20**, 686 (1979).
4.8 R. Ranvaud, H.-R. Trebin, U. Rössler, and F. H. Pollak, Phys. Rev. B **20**, 701 (1979).
4.9 R. L. Aggarwal, Phys. Rev. B **2**, 446 (1970).
4.10 G. C. LaRocca, T. Ruf, and M. Cardona, Phys. Rev. B **41**, 12672 (1990).
4.11 C. Trallero-Giner, A. Alexandrou, and M. Cardona, Phys. Rev. B **38**, 10744 (1988).
4.12 C. Hermann and C. Weisbuch, Phys. Rev. B **15**, 823 (1977).
4.13 M. Cardona, N. E. Christensen, and G. Fasol, Phys. Rev. B **38**, 1806 (1988).
4.14 J. Schmitz, H.-R. Trebin, and U. Rössler, *TRS User's and Programmer's Guide*, Universität Stuttgart, unpublished, 1988.
4.15 S. Zollner, private communication, 1989.
4.16 A. Raymond, J. L. Robert, and C. Bernard, J. Phys. C **12**, 2289 (1979).
4.17 E. J. Johnson and D. M. Larsen, Phys. Rev. Lett. **16**, 655 (1966).
4.18 H. Fröhlich, Adv. in Physics **3**, 325 (1954).
4.19 W. Becker, B. Gerlach, T. Hornung, and R. G. Ulbrich, in *Proceedings of the 18th International Conference on the Physics of Semiconductors*, edited by O. Engström, (World Scientific, Singapur, 1987), p. 1713.
4.20 W. Becker, B. Gerlach, T. Hornung A. Nöthe, G. Spata, and R. G. Ulbrich, in *Proceedings of the 19th International Conference on the Physics of Semiconductors*, edited by W. Zawadzki, (Polish Academy of Sciences, Warsaw, 1988), p. 1505.
4.21 M. Horst, U. Merkt, and J. P. Kotthaus, Phys. Rev. Lett. **50**, 754 (1983).
4.22 R. J. Nicholas, L. C. Brunel, S. Huant, K. Karrai, J. C. Portal, M. A. Brummell, M. Razeghi, K. Y. Cheng, and A. Y. Cho, Phys. Rev. Lett. **55**, 883 (1985).

4.23 J.-P. Cheng, B. D. McCombe, G. Brozak, and W. Schaff, in *Phonons in Semiconductor Nanostructures*, Vol. 236 of NATO ASI Series E, edited by J.-P. Leburton, J. Pascual, and C. Sotomayor Torres, (Kluwer, Dordrecht, 1993), p. 163.
4.24 P. G. Harper, J. W. Hodby, and R. A. Stradling, Rep. Prog. Phys. **36**, 1 (1973).
4.25 L. Eaves, R. A. Stradling, S. Askenazy, J. Leotin, J. C. Portal, and J. P. Ulmet, J. Phys. C Solid State Phys. **4**, L42 (1971).
4.26 T. Nakashima, C. Hamaguchi, J. Komeno, and M. Ozeki, Jap. J. of Appl. Phys. **23**, Supplement 23-1, 69 (1984).
4.27 G. S. Boebinger, A. F. J. Levi, S. Schmitt-Rink, A. Passner, L. N. Pfeiffer, and K. W. West, Phys. Rev. Lett. **65**, 235 (1990).
4.28 P. G. Harper, Phys. Rev. **178**, 1229 (1969).
4.29 M. Nakayama, Physics Letters **27A**, 315 (1968).
4.30 M. Nakayama, J. of the Phys. Soc. of Jap. **27**, 636 (1969).
4.31 R. Lassnig and W. Zawadzki, Surface Science **142**, 361 (1984).
4.32 S. Das Sarma, Phys. Rev. Lett. **52**, 859 (1984).
4.33 F. M. Peeters and J. T. Devreese, Phys. Rev. B **31**, 3689 (1985).
4.34 P. Pfeffer and W. Zawadzki, Solid State Commun. **57**, 847 (1986).
4.35 T. Hornung, *Bandkantenspektren von GaAs im Magnetfeld*, Thesis, Universität Dortmund, (1984).
4.36 J. K. Furdyna, J. Appl. Phys. **64**, R29 (1988).
4.37 J. A. Gaj, R. Planel, and G. Fishman, Solid State Commun. **29**, 435 (1979).
4.38 J. A. Gaj, J. Ginter, and R. R. Galazka, phys. stat. sol. (b) **89**, 655 (1978).
4.39 J. A. Gaj in *Diluted Magnetic Semiconductors*, edited by J. K. Furdyna and J. Kossut, Vol. 25 of *Semiconductors and Semimetals*, (Academic Press, San Diego, 1988), p. 276.
4.40 S. Venugopalan, A. Petrou, R. R. Galazka, A. K. Ramdas, and S. Rodriguez, Phys. Rev. B **25**, 2681 (1982).
4.41 A. K. Ramdas and S. Rodriguez in *Diluted Magnetic Semiconductors*, edited by J. K. Furdyna and J. Kossut, Vol. 25 of *Semiconductors and Semimetals*, (Academic Press, San Diego, 1988), p. 345.
4.42 D. Gammon, R. Merlin, and H. Morkoç, Phys. Rev. B **35**, 2552 (1987).
4.43 D. Gammon, L. Shi, R. Merlin, G. Ambrazevičius, K. Ploog, and H. Morkoç, Superlattices and Microstructures **4**, 405 (1988).
4.44 D. N. Mirlin, A. A. Sirenko, and R. Planel, Solid State Commun. **91**, 545 (1994).
4.45 F. Meseguer, F. Calle, C. López, J. M. Calleja, C. Tejedor, K. Ploog, and F. Briones, Superlattices and Microstructures **10**, 217 (1991).
4.46 F. Calle, J. M. Calleja, F. Meseguer, C. Tejedor, L. Viña, C. López, and K. Ploog, Phys. Rev. B **44**, 1113 (1991).
4.47 A. Cros, A. Cantarero, C. Trallero-Giner, and M. Cardona, Phys. Rev. B **45**, 6106 (1992).
4.48 A. Cros, A. Cantarero, C. Trallero-Giner, and M. Cardona, Phys. Rev. B **46**, 12627 (1992).
4.49 A. R. Goñi, K. Syassen, Y. Zhang, K. Ploog, A. Cantarero, and A. Cros, Phys. Rev. B **45**, 6809 (1992).
4.50 N. Binggeli and A. Baldereschi, Phys. Rev. B **43** 14734 (1991).
4.51 T. Ruf, A. Cros, J. Spitzer, G. Goldoni, and M. Cardona, in *High Magnetic Fields in the Physics of Semiconductors*, edited by D. Heimann, (World Scientific, Singapore, 1995), p. 350.

Chapter 5

5.1 B. P. Zakharchenya, D. N. Mirlin, V. I. Perel', and I. I. Reshina, Usp. Fiz. Nauk **136**, 459 (1982) [Sov. Phys. - Usp. **25**, 143 (1982)].
5.2 M. A. Alekseev, I. Ya. Karlik, D. N. Mirlin, and V. F. Sapega, Fiz. Tekh. Poluprovodn. **23**, 761 (1989); [Sov. Phys. - Semicond. **23**, 479 (1989)].
5.3 G. Fasol and H. P. Hughes, Phys. Rev. B **33**, 2953 (1986).
5.4 R. G. Ulbrich, J. A. Kash, and J. C. Tsang, Phys. Rev. Lett. **62**, 949 (1989).
5.5 J. A. Kash, Phys. Rev. B **47**, 1221 (1993).
5.6 W. Hackenberg, R. T. Phillips, and H. P. Hughes, Phys. Rev. B **50**, 10598 (1994).
5.7 J. A. Kash and J. C. Tsang, in *Light Scattering in Solids VI*, Vol. 68 of *Topics in Applied Physics*, edited by M. Cardona and G. Güntherodt, (Springer, Berlin, 1991), p. 423.
5.8 D. N. Mirlin, P. S. Kop'ev, I. I. Reshina, V. F. Sapega, and A. A. Sirenko, in *Proceedings of the 20th International Conference on the Physics of Semiconductors*, edited by E. M. Anastassakis and J. D. Joannopoulos, (World Scientific, Singapur, 1990), p. 1037.
5.9 D. N. Mirlin, P. S. Kop'ev, I. I. Reshina, A. V. Rodina, V. F. Sapega, A. A. Sirenko, and V. M. Ustinov, in *Proceedings of the 22nd International Conference on the Physics of Semiconductors*, edited by D. J. Lockwood, (World Scientific, Singapur, 1995), p. 1288.
5.10 C. H. Perry, J. M. Worlock, M. C. Smith, and A. Petrou, in *High Magnetic Fields in Semiconductor Physics*, edited by G. Landwehr, Vol. 71 of *Springer Series in Solid-State Sciences*, (Springer-Verlag, Heidelberg, 1987), p. 203.
5.11 J. C. Maan, M. Potemski, K. Ploog, and G. Weimann, in *Spectroscopy of Semiconductor Microstructures*, edited by G. Fasol, A. Fasolino, and P. Lugli, Vol. 206 of *NATO Advanced Study Institute, Series B: Physics*, (Plenum, New York, 1989), p. 425.
5.12 M. Potemski, J. C. Maan, A. Fasolino, K. Ploog, and G. Weimann, Phys. Rev. Lett. **63**, 2409 (1989).
5.13 M. Potemski, R. Stepniewski, J. C. Maan, G. Martinez, P. Wyder, and B. Etienne, Phys. Rev. Lett. **66**, 2239 (1991).
5.14 K. J. Nash, M. S. Skolnick, M. K. Saker, and S. J. Bass, Phys. Rev. Lett. **70**, 3115 (1993).
5.15 R. W. J. Hollering, T. T. J. M. Berendschot, H. J. A. Bluyssen, H. A. J. M. Reinen, P. Wyder, and F. Roozeboom, Phys. Rev. B **38**, 13323 (1988).
5.16 V. F. Gantmakher, B. L. Gel'mont, V. N. Zverev, and Al. L. Efros, Zh. Eksp. Teor. Fiz. **84**, 1129 (1982), [Sov. Phys. JETP **57**, 656 (1982)].
5.17 H. Sigg, P. Wyder, and J. A. A. J. Perenboom, Phys. Rev. B **31**, 5253 (1985).
5.18 H. Sigg, J. A. A. J. Perenboom, P. Pfeffer, and W. Zawadzki, Solid State Commun. **61**, 685 (1987).
5.19 M. A. Hopkins, R. J. Nicholas, P. Pfeffer, W. Zawadzki, D. Gauthier, J. C. Portal, and M. A. DiForte-Poisson, Semicon. Sci. Technol. **2**, 568 (1987).
5.20 P. Pfeffer and W. Zawadzki, Phys. Rev. B **41**, 1561 (1990).
5.21 C. Trallero-Giner, F. Iikawa, and M. Cardona, Phys. Rev. B **44**, 12815 (1991).
5.22 Note the error in Fig. 9 of [1.43] for the calculation of $\sqrt{\epsilon_2}$ with Eq. (B7). Figure 5.14 gives the result of a new calculation as discussed in the text. This improves the agreement with the experiment.

5.23 F. Meseguer, F. Calle, C. López, J. M. Calleja, L. Viña, C. Tejedor, and K. Ploog, in *Proceedings of the 20th International Conference on the Physics of Semiconductors*, edited by E. M. Anastassakis and J. D. Joannopoulos, (World Scientific, Singapur, 1990), p. 1461.
5.24 F. Meseguer, F. Calle, C. López, J. M. Calleja, and C. Tejedor, private communication.
5.25 V. I. Belitsky, A. Cantarero, and S. T. Pavlov, Solid State Commun. **94**, 589 (1995).
5.26 R. Boroff, R. Merlin, J. Pamulapati, P. K. Bhattacharya, and C. Tejedor, Phys. Rev. B **43**, 2081 (1991).
5.27 J. B. Stark, W. H. Knox, and D. S. Chemla, Phys. Rev. Lett. **68**, 3080 (1992).
5.28 R. J. Warburton, J. G. Michels, R. J. Nicholas, J. J. Harris, and C. T. Foxon, Phys. Rev. B **40**, 7714 (1989).
5.29 D. Richards, E. T. M. Kernohan, R. T. Phillips, and B. Etienne, in *Proceedings of the 23rd International Conference on the Physics of Semiconductors*, edited by M. Scheffler and R. Zimmermann, (World Scientific, Singapore, 1996), p. 2319.
5.30 C. Schüller, R. Krahne, G. Biese, C. Steinebach, E. Ulrichs, D. Heitmann, and K. Eberl, Phys. Rev. B **56**, 1037 (1997).

Chapter 6

6.1 D. Gammon, B. V. Shanabrook, and D. S. Katzer, Phys. Rev. Lett. **67**, 1547 (1991).
6.2 M. Bernasconi, L. Colombo, and L. Miglio, Phys. Rev. B **43**, 14457 (1991).
6.3 C. Molteni, L. Colombo, L. Miglio, G. Benedek, and M. Bernasconi, Phil. Mag. **65**, 325 (1992).
6.4 M. A. Araújo Silva, E. Ribeiro, P. A. Schulz, F. Cerdeira, and J. C. Bean, J. Raman Spectroscopy **27**, 257 (1996).
6.5 R. Schorer, W. Wegscheider, and G. Abstreiter, J. Vac. Sci. Technol. B **11**, 1069 (1993);
R. Schorer, *Ramanstreuung an α-Sn, Si und Ge Übergittern*, Thesis, Technische Universität München, (1994).
6.6 J. C. Tsang, S. S. Iyer, J. A. Calise, and B. A. Ek, Phys. Rev. B **40**, 5886 (1989).
6.7 J. Spitzer, H. D. Fuchs, P. Etchegoin, M. Ilg, M. Cardona, B. Brar, and H. Kroemer, Appl. Phys. Lett. **62**, 2274 (1993).
6.8 J. Spitzer, A. Höpner, M. Kuball, M. Cardona, B. Jenichen, H. Neuroth, B. Brar, and H. Kroemer, J. Appl. Phys. **77**, 811 (1995).
6.9 W. Frank, U. Gösele, H. Mehrer, and A. Seeger, in *Diffusion in Crystalline Solids*, edited by G. E. Murch and A. S. Nowick, (Academic Press, New York, 1984), p. 63.
6.10 H. D. Fuchs, W. Walukiewicz, E. E. Haller, W. Dondl, R. Schorer, G. Abstreiter, A. I. Rudnev, A. V. Tikhomirov, and V. I. Ozhogin, Phys. Rev. B **51**, 16817 (1995).
6.11 E. Anastassakis, A. Pinczuk, E. Burstein, F. H. Pollak, and M. Cardona, Solid State Commun. **8**, 133 (1970).
6.12 E. Anastassakis, A. Cantarero, and M. Cardona, Phys. Rev. B **41**, 7529 (1990).
6.13 M. Hünermann, W. Richter, J. Saalmüller, and E. Anastassakis, Phys. Rev. B **34**, 5381 (1986).

6.14 I. De Wolf, Semicond. Sci. Technol. **11**, 139 (1996).
6.15 S. C. Jain, H. E. Maes, K. Pinardi, and I. De Wolf, J. Appl. Phys. **79**, 8145 (1996).
6.16 T. Englert, G. Abstreiter, and J. Pontcharra, Solid-State Electron. **23**, 31 (1980).
6.17 I. De Wolf, H. E. Maes, and S. K. Jones, J. Appl. Phys. **79**, 7148 (1996).
6.18 D. Wolverson, B. Schlichtherle, C. Orange, W. Heimbrodt, J. J. Davies, K. Ogata, Sg. Fujita, and T. Ruf, in *Proceedings of the 23rd International Conference on the Physics of Semiconductors*, edited by M. Scheffler and R. Zimmermann, (World Scientific, Singapore, 1996), p. 2965.
6.19 C. Orange, B. Schlichtherle, D. Wolverson, J. J. Davies, T. Ruf, K. Ogata, and Sg. Fujita, Phys. Rev. B **55**, 1607 (1997).
6.20 M. V. Klein, in *Light Scattering in Solids I*, edited by M. Cardona, Vol. 8 of *Topics in Applied Physics*, (Springer-Verlag, Berlin, 1982), Chap. 4, p. 147.
6.21 D. G. Thomas and J. J. Hopfield, Phys. Rev. **175**, 1021 (1968).
6.22 H. Siegle, L. Eckey, A. Hoffmann, C. Thomsen, B. K. Meyer, D. Schikora, M. Hankeln, and K. Lischka, Solid State Commun. **96**, 943 (1995).
6.23 W. Faschinger, J. Crystal Growth **146**, 80 (1995).
6.24 P. J. Boyce, J. J. Davies, D. Wolverson, K. Ohkawa, and T. Mitsuyu, Appl. Phys. Lett. **65**, 2063 (1994).
6.25 C. M. Townsley, B. Schlichtherle, D. Wolverson, J. J. Davies, K. Ogata, and Sg. Fujita, Solid State Commun. **96**, 437 (1995).
6.26 T. Kozawa, T. Kachi, Y. Taga, M. Hashimoto, N. Koide, and K. Manabe, J. Appl. Phys. **75**, 1098 (1994).
6.27 F. A. Ponce, J. W. Steeds, C. D. Dyer, and G. D. Pitt, Appl. Phys. Lett. **69**, 2650 (1996).
6.28 I. P. Hayward, K. J. Baldwin, D. M. Hunter, D. N. Batchelder, and G. D. Pitt, Diamond and Related Materials **4**, 617 (1995).
6.29 J. Wagner, M. Maier, R. Murray, R. C. Newman, R. B. Beall, and J. J. Harris, J. Appl. Phys. **69**, 971 (1991).
6.30 M. Ramsteiner, J. Wagner, H. Ennen, and M. Maier, Phys. Rev. B **38**, 10669 (1988).
6.31 R. Murray, R. C. Newman, M. J. L. Sangster, R. B. Beall, J. J. Harris, P. J. Wright, J. Wagner, and M. Ramsteiner, J. Appl. Phys. **66**, 2589 (1989).
6.32 H. Tsai and D. B. Bogy, J. Vac. Sci. Technol. A **5**, 3287 (1987).
6.33 P. Lespade, R. Al-Jishi, and M. S. Dresselhaus, Carbon **20**, 427 (1982).
6.34 R. O. Dillon, J. A. Woollam, and V. Katkanant, Phys. Rev. B **29**, 3482 (1984).
6.35 F. Tuinstra and J. L. Koenig, J. Chem. Phys. **53**, 1126 (1970).
6.36 N. Wada, P. J. Gaczi, and S. A. Solin, J. Non-Cryst. Solids **35** & **36**, 543 (1980).
6.37 B. Marchon, N. Heiman, M. R. Khan, A. Lautie, J. W. Ager III, and D. K. Veirs, J. Appl. Phys. **69**, 5748 (1991); and references therein.
6.38 H.-J. Reich, Dilor GmbH, company information.
6.39 K.-J. Schuchter, Renishaw GmbH, company information RTS/AN/052 (1995).
6.40 C. H. Henry, P. M. Petroff, R. A. Logan, and F. R. Merritt, J. Appl. Phys. **50**, 3721 (1979).
6.41 W. C. Tang, H. J. Rosen, P. Vettiger, and D. J. Webb, Appl. Phys. Lett. **58**, 557 (1991).
6.42 R. Puchert, A. Bärwolff, U. Menzel, A. Lau, M. Voss, and T. Elsaesser, J. Appl. Phys. **80**, 5559 (1996).

6.43 F. U. Herrmann, S. Beeck, G. Abstreiter, C. Hanke, C. Hoyler, and L. Korte, Appl. Phys. Lett. **58**, 1007 (1991).
6.44 *Proceedings of the XIII International Conference on Raman Spectroscopy*, edited by W. Kiefer, M. Cardona, G. Schaack, F. W. Schneider, and H. W. Schrötter, (J. Wiley & Sons, Chichester, 1992).
6.45 *Proceedings of the XIV International Conference on Raman Spectroscopy*, edited by N.-T. Yu and X.-Y. Li, (J. Wiley & Sons, Chichester, 1994).
6.46 *Proceedings of the XV International Conference on Raman Spectroscopy*, edited by S. A. Asher and P. Stein, (J. Wiley & Sons, Chichester, 1996).
6.47 J. Greve, G. J. Puppels, and C. Otto, in [6.44], p. 21.
6.48 D. R. T. Zahn, phys. stat. sol. (a), **179** (1995).
6.49 D. E. Aspnes, Surf. Science **307-309**, 1017 (1994).
6.50 *Optical Characterization of Epitaxial Semiconductor Layers*, edited by G. Bauer and W. Richter, (Springer-Verlag, Berlin, 1996).
6.51 M. Hünermann, J. Geurts, and W. Richter, Phys. Rev. Lett. **66**, 640 (1991).
6.52 W. Richter, N. Esser, A. Kelnberger, and M. Köpp, Solid State Commun. **84**, 165 (1992).
6.53 W. G. Schmidt and G. P. Srivastava, Solid State Commun. **89**, 345 (1994).
6.54 V. Wagner, *Characterization of Diffusion-Assisted Growth of III-V Compounds*, Vol. 17 of *Aachener Beiträge zur Physik der kondensierten Materie*, edited by B. U. Felderhof, P. Grosse, G. Güntherodt, and A. Stahl, (Verlag der Augustinus Buchhandlung, Aachen, 1995).
6.55 V. Wagner, D. Drews, N. Esser, W. Richter, D. R. T. Zahn, and J. Geurts, in *Proceedings of the 4th International Conference on the Formation of Semiconductor Interfaces*, edited by B. Lengeler, H. Lüth, W. Mönch, and J. Pollmann, (World Scientific, Singapore, 1994), p. 546.
6.56 V. Wagner, W. Richer, and J. Geurts, Appl. Surf. Science **104/105**, 580 (1996).
6.57 M. Ramsteiner, C. Wild, and J. Wagner, Applied Optics **28**, 4017 (1989).
6.58 D. Drews, M. Langer, W. Richter, and D. R. T. Zahn, phys. stat. sol. (a) **145**, 491 (1994).
6.59 V. Wagner, W. Richter, J. Geurts, D. Drews, and D. R. T. Zahn, J. Raman Spectroscopy **27**, 265 (1996).
6.60 D. W. Pohl, W. Denk, and M. Lanz, Appl. Phys. Lett. **44**, 651 (1984).
6.61 M. A. Paesler, P. J. Moyer, C. L. Jahncke, C. E. Johnson, R. C. Reddick, R. J. Warmack, and T. L. Ferrell, Phys. Rev. B **42**, 6750 (1990).
6.62 E. Betzig, J. K. Trautman, T. D. Harris, J. S. Weiner, and R. L. Kostelak, Science **251**, 1468 (1991).
6.63 E. Betzig and J. K. Trautman, Science **257**, 189 (1992).
6.64 D. Courjon and C. Bainier, Rep. Prog. Phys. **57**, 989 (1994).
6.65 M. A. Paesler and P. Moyer, *Near-Field Optics: Theory, Instrumentation, and Applications*, (Wiley, New York, 1996).
6.66 *Near-Field Optics, Selected Papers from the 2nd International Conference on Near-Field Optics*, edited by M. Isaacson, Ultramicroscopy **57** (1995).
6.67 *NFO-3, Selected Papers from the 3rd International Conference on Near-Field Optics and Related Techniques*, edited by M. A. Paesler and N. van Hulst, Ultramicroscopy **61** (1995).
6.68 *Near-Field Optics*, NATO ASI Series E: Applied Sciences – Vol. 242, edited by D. W. Pohl and D. Courjon, (Kluwer, Dordrecht, 1993).
6.69 *Photons and Local Probes*, NATO ASI Series E: Applied Sciences – Vol. 300, edited by O. Marti and R. Möller, (Kluwer, Dordrecht, 1995).
6.70 A. Kurtenbach, K. Eberl, and T. Shirata, Appl. Phys. Lett. **66**, 361 (1995).
6.71 N. N. Ledentsov et al., Phys. Rev. B **54**, 8743 (1996).

6.72 A. Ekimov, J. Luminescence **70**, 1 (1996).
6.73 *Spectroscopy of Isolated and Assembled Semiconductor Nanocrystals*, edited by L. E. Brus, Al. L. Efros, and T. Itoh, J. Luminescence **70** (1996).
6.74 A. P. Alivisatos, Science **271**, 933 (1996); A. P. Alivisatos, J. Phys. Chem. **100**, 13226 (1996).
6.75 R. D. Grober, T. D. Harris, J. K. Trautman, E. Betzig, W. Wegscheider, L. Pfeiffer, and K. West, Appl. Phys. Lett. **64**, 1421 (1994).
6.76 H. F. Hess, E. Betzig, T. D. Harris, L. N. Pfeiffer, and K. W. West, Science **264**, 1740 (1994).
6.77 H. F. Ghaemi, B. B. Goldberg, C. Cates, P. D. Wang, C. Sotomayor Torres, M. Fritze, and A. Nurmikko, Superlattices and Microstructures **17**, 15 (1995).
6.78 K. Brunner, U. Bockelmann, G. Abstreiter, M. Walther, G. Böhm, G. Tränkle, and G. Weimann, Phys. Rev. Lett. **69**, 3216 (1992).
6.79 A. Zrenner, L. V. Butov, M. Hagn, G. Abstreiter, G. Böhm, and G. Weimann, Phys. Rev. Lett. **72**, 3382 (1994).
6.80 D. Gammon, E. S. Snow, and D. S. Katzer, Appl. Phys. Lett. **67**, 2391 (1995).
6.81 S. A. Empedocles, D. J. Norris, and M. G. Bawendi, Phys. Rev. Lett. **77**, 3873 (1996).
6.82 D. P. Tsai, A. Othonos, M. Moskovits, D. Uttamchandani, Appl. Phys. Lett. **64**, 1768 (1994).
6.83 H. D. Hallen, A. H. La Rosa, and C. L. Jahncke, phys. stat. sol. (a) **152**, 257 (1995).
6.84 C. L. Jahncke, M. A. Paesler, and H. D. Hallen, Appl. Phys. Lett. **67**, 2483 (1995).
6.85 W. M. Duncan, J. Vac. Sci. Technol. A **14**, 1914 (1996).
6.86 C. L. Jahncke, H. D. Hallen, and M. A. Paesler, J. Raman Spectroscopy **27**, 579 (1996).
6.87 D. Gammon, S. W. Brown, E. S. Snow, T. A. Kennedy, D. S. Katzer, and D. Park, Science **277**, 85 (1997).
6.88 S. Nie and S. R. Emory, Science **275**, 1102 (1997).
6.89 K. Kneipp, Y. Wang, H. Kneipp, L. T. Perelman, I. Itzkan, R. R. Dasari, and M. S. Feld, Phys. Rev. Lett. **78**, 1667 (1997).
6.90 J. Grausem, B. Humbert, A. Burneau, and J. Oswalt, Appl. Phys. Lett. **70**, 1671 (1997).
6.91 E. H. Synge, Philosophical Magazine **6**, 356 (1928).
6.92 H. A. Bethe, Phys. Rev. **66**, 163 (1944).
6.93 E. A. Ash and G. Nicholls, Nature **237**, 510 (1972).
6.94 D. W. Pohl, in [6.68], p. 1.
6.95 L. Novotny and D. W. Pohl, in [6.69], p. 21.
6.96 E. L. Buckland, P. J. Moyer, and M. A. Paesler, J. Appl. Phys. **73**, 1018 (1993).
6.97 R. D. Grober, T. D. Harris, J. K. Trautman, and E. Betzig, Rev. Sci. Instrum. **65**, 626 (1994).
6.98 E. Betzig and R. J. Chichester, Science **262**, 1422 (1993).
6.99 J. K. Trautman, J. J. Macklin, L. E. Brus, and E. Betzig, Nature **369**, 40 (1994).
6.100 W. P. Ambrose, P. M. Goodwin, J. C. Martin, and R. A. Keller, Science **265**, 364 (1994).
6.101 X. S. Xie and R. C. Dunn, Science **265**, 361 (1994).
6.102 W. E. Moerner, T. Plakhotnik, T. Irngartinger, U. P. Wild, D. W. Pohl, and B. Hecht, Phys. Rev. Lett. **73**, 2764 (1994).
6.103 D. Gammon, E. S. Snow, B. V. Shanabrook, D. S. Katzer, and D. Park, Phys. Rev. Lett. **76**, 3005 (1996).

6.104 D. Gammon, E. S. Snow, B. V. Shanabrook, D. S. Katzer, and D. Park, Science **273**, 87 (1996).
6.105 W. Heller and U. Bockelmann, Phys. Rev. B **55**, 4871 (1997).
6.106 M. A. Paesler, private communication (1997).
6.107 J. F. Scott, Rep. Prog. Phys. **43**, 61 (1980).
6.108 S. Geschwind and R. Romestain, in *Light Scattering in Solids IV*, edited by M. Cardona and G. Güntherodt, Vol. 54 of *Topics in Applied Physics*, (Springer-Verlag, Berlin, 1984), Chap. 3, p. 151.
6.109 Y. Oka and M. Cardona, Phys. Rev. B **23**, 4129 (1981).
6.110 V. F. Sapega, M. Cardona, K. Ploog, E. L. Ivchenko, and D. N. Mirlin, Phys. Rev. B **45**, 4320 (1992);
V. F. Sapega, T. Ruf, M. Cardona, K. Ploog, E. L. Ivchenko, and D. N. Mirlin, Phys. Rev. B **50**, 2510 (1994).
6.111 R. Meyer, M. Dahl, G. Schaack, and A. Waag, Phys. Rev. B **55**, 16376 (1997).
6.112 A. A. Sirenko, T. Ruf, M. Cardona, D. R. Yakovlev, W. Ossau, A. Waag, and G. Landwehr, Phys. Rev. B **56**, 2114 (1997).
6.113 *Surface-Enhanced Raman Scattering*, edited by R. K. Chang and T. E. Furtak, (Plenum, New York, 1982).
6.114 A. Otto, in *Light Scattering in Solids IV*, edited by M. Cardona and G. Güntherodt, Vol. 54 of *Topics in Applied Physics*, (Springer-Verlag, Berlin, 1984), Chap. 6, p. 289.
6.115 K. Arya and R. Zeyher, in *Light Scattering in Solids IV*, edited by M. Cardona and G. Güntherodt, Vol. 54 of *Topics in Applied Physics*, (Springer-Verlag, Berlin, 1984), Chap. 7, p. 419.
6.116 A. Otto, I. Mrozek, H. Grabhorn, and W. Akemann, J. Phys.: Condens. Matter **4**, 1143 (1992).
6.117 H. Owen, in *Computer and Optically Generated Holographic Optics*, SPIE Proceedings Vol. **1555**, p. 228 (1991).
6.118 B. Yang, M. D. Morris, and H. Owen, Appl. Spectrosc. **45**, 1533 (1991).
6.119 Kaiser Optical Systems, Inc., company information.
6.120 H. Owen, D. E. Battey, M. J. Pelletier, and J. B. Slater, in *Practical Holography IX*, SPIE Proceedings Vol. **2406** (1995).
6.121 I. C. Chang, Appl. Phys. Lett. **25**, 370 (1974).
6.122 P. Katzka, in *Acousto-Optic, Electro-Optic, and Magneto-Optic Devices and Applications*, SPIE Proceedings Vol. **753**, 22 (1987).
6.123 X. Wang, Laser Focus World **28**, 173 (1992).
6.124 D. A. Glenar, J. J. Hillman, B. Saif, and J. Bergstralh, Applied Optics **33**, 7412 (1994).
6.125 P. J. Treado, I. W. Levin, and E. N. Lewis, Appl. Spectrosc. **46**, 553 (1992).
6.126 A. Yariv, *Quantum Electronics*, (Wiley, New York, 1989).
6.127 S. Marke, IFU GmbH - Privates Institut für Umweltanalysen, company information (1997).
6.128 D. M. Hueber et al., Appl. Spectrosc. **49**, 1624 (1995).
6.129 D. P. Baldwin et al., Appl. Spectrosc. **50**, 498 (1996).
6.130 F. Moreau et al., Appl. Spectrosc. **50**, 1295 (1996).
6.131 W. Fukarek and A. von Keudell, Rev. Sci. Instrum. **66**, 3545 (1995).
6.132 P. J. Treado, I. W. Levin, and N. E. Lewis, Appl. Spectrosc. **46**, 1211 (1992).
6.133 E. N. Lewis, P. J. Treado, and I. W. Levin, Appl. Spectrosc. **47**, 539 (1993).
6.134 H. R. Morris, C. C. Hoyt, and P. J. Treado, Appl. Spectrosc. **48**, 857 (1994).
6.135 H. T. Skinner et al., Appl. Spectrosc. **50**, 1007 (1996).
6.136 T. Ruf and S. Marke (1997); the spectra were obtained with IFU's AOS μchron 3 Acoustooptical Emission Spectrometer [6.127].
6.137 S. R. Goldstein et al., J. of Microscopy **184**, 35 (1996).

Index

Springer Tracts in Modern Physics

125 **Inelastic Scattering of X-Rays with Very High Energy Resolution**
By E. Burkel 1991. 70 figs. XV, 112 pages

126 **Critical Phenomena at Surfaces and Interfaces**
Evanescent X-Ray and Neutron Scattering
By H. Dosch 1992. 69 figs. X, 145 pages

127 **Critical Behavior at Surfaces and Interfaces**
Roughening and Wetting Phenomena
By R. Lipowsky 1999. 80 figs. X, Approx. 180 pages

128 **Surface Scattering Experiments with Conduction Electrons**
By D. Schumacher 1993. 55 figs. IX, 95 pages

129 **Dynamics of Topological Magnetic Solitons**
By V. G. Bar'yakhtar, M. V. Chetkin, B. A. Ivanov, and S. N. Gadetskii
1994. 78 figs. VIII, 179 pages

130 **Time-Resolved Light Scattering from Excitons**
By H. Stolz 1994. 87 figs. XI, 210 pages

131 **Ultrathin Metal Films**
Magnetic and Structural Properties
By M. Wuttig 1998. 103 figs. X, Approx. 180 pages

132 **Interaction of Hydrogen Isotopes with Transition-Metals and Intermetallic Compounds**
By B. M. Andreev, E. P. Magomedbekov, G. Sicking 1996. 72 figs. VIII, 168 pages

133 **Matter at High Densities in Astrophysics**
Compact Stars and the Equation of State
In Honor of Friedrich Hund's 100th Birthday
By H. Riffert, H. Müther, H. Herold, and H. Ruder 1996. 86 figs. XIV, 278 pages

134 **Fermi Surfaces of Low-Dimensional Organic Metals and Superconductors**
By J. Wosnitza 1996. 88 figs. VIII, 172 pages

135 **From Coherent Tunneling to Relaxation**
Dissipative Quantum Dynamics of Interacting Defects
By A. Würger 1996. 51 figs. VIII, 216 pages

136 **Optical Properties of Semiconductor Quantum Dots**
By U. Woggon 1997. 126 figs. VIII, 252 pages

137 **The Mott Metal-Insulator Transition**
Models and Methods
By F. Gebhard 1997. 38 figs. XVI, 322 pages

138 **The Partonic Structure of the Photon**
Photoproduction at the Lepton-Proton Collider HERA
By M. Erdmann 1997. 54 figs. X, 118 pages

139 **Aharonov-Bohm and other Cyclic Phenomena**
By J. Hamilton 1997. 34 figs. X, 186 pages

140 **Exclusive Production of Neutral Vector Mesons at the Electron-Proton Collider HERA**
By J. A. Crittenden 1997. 34 figs. VIII, 108 pages

141 **Disordered Alloys**
Diffuse Scattering and Monte Carlo Simulations
By W. Schweika 1998. 48 figs. X, 126 pages

142 **Phonon Raman Scattering in Semiconductors, Quantum Wells and Superlattices**
Basic Results and Applications
By T. Ruf 1998. 143 figs. VIII, 252 pages

MIX
Papier aus verantwortungsvollen Quellen
Paper from responsible sources
FSC® C105338

If you have any concerns about our products,
you can contact us on
ProductSafety@springernature.com

In case Publisher is established outside the EU,
the EU authorized representative is:
Springer Nature Customer Service Center GmbH
Europaplatz 3, 69115 Heidelberg, Germany

Printed by Libri Plureos GmbH
in Hamburg, Germany